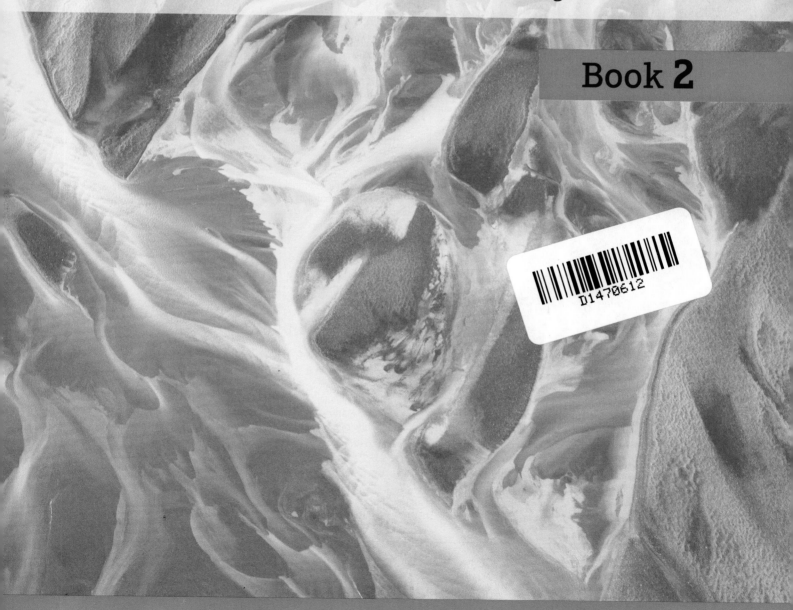

ActiveBook included

Edexcel A level
Geography

Series editor: Lindsay Frost
Lindsay Frost | Lauren Lewis | Daniel Mace | Paul Wraight

Book 2

D1470612

ALWAYS LEARNING

PEARSON

Published by Pearson Education Limited, 80 Strand, London, WC2R 0RL.

www.pearsonschoolsandfecolleges.co.uk

Copies of official specifications for all Edexcel qualifications may be found on the website: www.edexcel.com

Text © Pearson Education Limited 2016

Designed by Elizabeth Arnoux for Pearson

Typeset, illustrated and produced by Phoenix Photosetting, Chatham, Kent

Original illustrations © Pearson Education Limited 2016

Cover design by Elizabeth Arnoux for Pearson

Picture research by Rebecca Sodergren and Ned Coomes

Cover photo/illustration © Romulic-Stojcic / Lumi Images / Getty Images

The rights of Lindsay Frost, Lauren Lewis, Daniel Mace and Paul Wraight to be identified as authors of this work have been asserted by them in accordance with the Copyright, Designs and Patents Act 1988.

First published 2016

19 18 17

10 9 8 7 6 5 4 3

British Library Cataloguing in Publication Data

A catalogue record for this book is available from the British Library

ISBN 9781 292 13965 4

Websites

Pearson Education Limited is not responsible for the content of any external internet sites. It is essential for tutors to preview each website before using it in class so as to ensure that the URL is still accurate, relevant and appropriate. We suggest that tutors bookmark useful websites and consider enabling students to access them through the school/college intranet.

A note from the publisher

In order to ensure that this resource offers high-quality support for the associated Pearson qualification, it has been through a review process by the awarding body. This process confirms that this resource fully covers the teaching and learning content of the specification or part of a specification at which it is aimed. It also confirms that it demonstrates an appropriate balance between the development of subject skills, knowledge and understanding, in addition to preparation for assessment.

Endorsement does not cover any guidance on assessment activities or processes (e.g. practice questions or advice on how to answer assessment questions), included in the resource nor does it prescribe any particular approach to the teaching or delivery of a related course.

While the publishers have made every attempt to ensure that advice on the qualification and its assessment is accurate, the official specification and associated assessment guidance materials are the only authoritative source of information and should always be referred to for definitive guidance.

Pearson examiners have not contributed to any sections in this resource relevant to examination papers for which they have responsibility.

Examiners will not use endorsed resources as a source of material for any assessment set by Pearson.

Endorsement of a resource does not mean that the resource is required to achieve this Pearson qualification, nor does it mean that it is the only suitable material available to support the qualification, and any resource lists produced by the awarding body shall include this and other appropriate resources.

Contents

How to use this book

Exam questions

An important part of your learning involves preparation for examinations or other forms of assessment. For each specification enquiry question, there are medium-length A level exam-style questions in each chapter, along with guidance on how the question should be answered. These should form part of your continuous assessment preparation during your studies. In addition, at the end of each chapter there are examples of questions requiring longer written answers, together with an average student answer and detailed guidance on how that answer could have been improved. These are designed to show you how to attain a high level and so maximise your grade. Examples of possible Paper 3 synoptic questions are also provided in the separate Synopticity guidance section (pages 17–21).

Activities

There are regular questions for you to complete throughout each chapter. These are based on the text and figures or tables, and will reinforce your understanding and knowledge. They are subdivided into normal geographical questions, and skills questions linked to map work (Cartographic Skills), graphs (Graphical Skills), use of IT and GIS (IT Skills), and maths and statistical (Numerical and Statistical Skills). These questions may be used to supplement classwork, or used for homework or preparation.

Case studies

A concern that you may have is the level of knowledge required, and finding this knowledge. In this book, place detail and facts are included within the main text of each chapter, and also highlighted and expanded in case study boxes. Most of the case studies of the specification have been covered, and additional ones included in extension activities where authors believed that they would be useful. You should note that the case study examples given in the specification are suggestions, and not compulsory, and where authors could be more contemporary or knew of examples that enabled more links to be made, these have been used instead or in addition.

Extension

You will find extension boxes throughout every topic in this book. These are designed to enable A level students to expand their knowledge and understanding, or reinforce their learning of information presented in the text, figures or tables.

Literacy and maths tips

Throughout this book there are boxes highlighting advice on how you can improve your literacy and mathematical (especially statistical) abilities, helping you make the transition from AS level to A level.

Individual investigation

The individual investigation (II) is covered in a separate section in more detail (see pages 7–16). There is a requirement to complete four days of fieldwork over the two-year A level course. While the Independent Investigation is an A level requirement, it is likely that you may start this before the end of Year 12.

Synopticity

This is covered in greater detail in a separate section covering Paper 3 (see pages 17–21). There are synoptic link boxes throughout the book where authors have identified links with other geographical topics, and with players, attitudes and actions and futures and uncertainties (P, A, F).

Internet

Due to a degree of unreliability in specific website availability, the references to websites in this book have been deliberately kept broad. Where authors recommend websites, this has been done through naming the organisation or government department, for example, and the specific part of a topic to investigate. You should use the names and key words in your selected search engine to find the relevant information for an activity or extension task.

Glossary

The glossary presented at the back of the book concentrates on covering all the terms featured in the specification; these are emboldened in the text of each chapter.

A level assessment outline

The Edexcel specification topics 5, 6, 7 and 8 are covered in this book; these form part of the two-year A level course, along with topics 1, 2, 3 and 4 which were covered in Book 1 of the series (Table 1). Book 2 presents the content information for part of the full A level course. The material is a step up from Book 1, and so is suitable for second-year A level students. It consolidates the learning from Book 1 and offers the challenges necessary for students to reach an A level standard by the end of the second year of study. Learning objectives and headings in this book match the specification.

Table 1: Specification content for the two year A level course

Year 1 and Year 2 scheme of work: A
Compulsory topics: Topic 1: Tectonic Processes and Hazards Topic 3: Globalisation Topic 5: The Water Cycle and Water Insecurity Topic 6: The Carbon Cycle and Energy Security Topic 7: Superpowers
Optional topics: Topic 2A: Glaciated Landscapes and Change *or* Topic 2B: Coastal Landscapes and Change *and* Topic 4A: Regenerating Places *or* Topic 4B: Diverse Places *and* Topic 8A: Health, Human Rights and Intervention *or* Topic 8B: Migration, Identity and Sovereignty

Table 2: Specification assessment

Year 2 examinations: A
Paper 1: Dynamic Landscapes and Physical Systems and Sustainability: 2 hours and 15 minutes (105 marks): 30% of qualification Section A: Topic 1 Section B: Topic 2A *or* Topic 2B Section C: Topic 5 *and* Topic 6 Includes 20 mark extended writing questions.
Paper 2: Dynamic Places and Human Systems and Geopolitics: 2 hours and 15 minutes (105 marks): 30% of qualification Section A: Topic 3 *and* Topic 7 Section B: Topic 4A *or* Topic 4B Section C: Topic 8A *or* Topic 8B Includes 20 mark extended writing questions.
Paper 3: Synoptic Investigation: 2 hours and 15 minutes (135 minutes) (70 marks): 20% of the qualification Sections A, B, C and D are all synoptic: Based on linked geographical topics and Players, Attitudes and actions, and Futures and uncertainties, within a resource booklet about a geographical issue. There are 18- and 24-mark extended writing questions.
Paper 4: Individual Investigation (70 marks): 20% of the qualification Research-based, including fieldwork. 3,000- to 4,000-word report in appropriate sections.

Exam command words

It is important for you to know the meaning of command words in exams, so that your answer is phrased in the correct way to answer the questions. Marks can be dropped quickly if you don't answer the question, even if you write lots of correct geography.

Some AS command words are not used in the full A level examinations. Use the exam question analysis at the end of each chapter to help you.

Table A: Examination command words used on A level Papers 1, 2 and 3

Command word	Explanation
Analyse	You must use all of your academic geographical skills to investigate an issue. You must divide the issue into its main parts and make logical connections on the causes and effects of the links between these parts. You must use evidence (e.g. examples and case studies) wherever possible to support the points you make. This command word will be used with longer answer questions worth more marks.
Assess	You must use evidence to show the significance of something relative to other factors. Spend equal time on each factor in the main part of your answer, and in a conclusion identify the most important.
Complete	You must finish a diagram, map or graph using the data provided.
Draw or Plot	You must create a graphical representation of data provided.
Evaluate	You must write about the success or importance of something, and provide a balanced judgement based on evidence (e.g. data) that you have included in your answer. You should have a powerful conclusion that brings together all of the ideas, including strengths and weaknesses, and explores alternatives. This command word will be used with longer answer questions worth more marks.
Explain	You must provide reasons (e.g. how and why) for a geographical pattern or process. You must show your understanding within your answer, supported by justification and evidence – often in the form of examples.
Suggest	You must provide a reasoned explanation of how or why a geographical pattern or process may have occurred in an unfamiliar situation. You should provide justification and evidence for your suggestions based on your geographical understanding and knowledge of similar situations.

Individual investigation

Introduction

An essential part of geographical study is the collection of primary and secondary data and information to investigate a specific aspect of a geographical process or pattern. This aspect of your studies is assessed at the end of the second year of A level Geography through an Independent Investigation (II).

The Edexcel specification requires a minimum of four days of fieldwork to be completed during the two years of A level study (two in Year 1 and two in Year 2), including at least one physical geography investigation and one human geography investigation. You must use these fieldwork days to develop your skills and understanding of the enquiry and report-writing processes. You can use data from any of the fieldwork days as the basis of your II, or you can extend it by collecting more data at the same place, or a contrasting place to offer a comparison, or collect completely new data at a new place. You must choose an enquiry question linked to any of the compulsory or optional content of the specification, so it may be based on physical geography or human geography or a combination of both. You must ensure that your enquiry choice is of a manageable scale so that you can carry out primary data collection through fieldwork and that usable secondary data is available. In many cases it may be that the fieldwork for the individual investigation will be collected during the summer between the year 1 and year 2 parts of the course (i.e. June to September).

Independence

There is, as the title suggests, a very high degree of independent work expected by the examiners, so teachers and other experts have restrictions on the amount of help and guidance that they are allowed to give you. It is essential that you learn from the fieldwork days organised for you during your studies so that you are prepared to complete an independent investigation on your own. Teachers and other experts **cannot**:

- provide you with a list of titles and hypotheses
- give detailed feedback on how to improve your work to meet the assessment criteria
- provide templates or model answers for your specific work
- provisionally mark your work and tell you the standard and how to improve it
- return work to you after it has been submitted and marked
- give guidance on how you can make improvements to a draft so that you meet the assessment criteria.

To help you with the A level Independent Investigation, Book 1 provided four detailed exemplars of fieldwork enquiries completed at AS level, which could act as starters for an A level investigation. While these exemplars could not reproduce a complete II in full because of its length, they do show the stages of an investigation, both in terms of organisation and main written sections. They also show the writing style, level of detail, and quality of data presentation required. Further detailed advice is also provided in this section of Book 2.

Extension ↗

Read Table A and obtain a copy of the full Independent Investigation mark scheme from your teacher or from the Edexcel specification on their website. The descriptors are written for teachers and examiners to use, so make sure that you ask about anything in the mark scheme that you are unsure of.

ACTIVITY

1. Consider your geography fieldwork experience and which geography topics you are interested in. Compose two enquiry questions, and for each compose two aims and three hypotheses. Discuss these with your supervising teacher and then decide what your II will concentrate on.
2. Decide on the locations for your data collection and your secondary data sources. **3.** Produce a schedule for the completion of all the stages of the II using a diary or calendar.

Assessment

The Independent Investigation is worth 20% of the total A level: You will write a 3,000- to 4,000-word structured report based on fieldwork and secondary information.

Table A: Marking criteria for Geography Independent Investigation

Assessment criterion	Top level descriptors
a) Purpose of the Investigation (12 marks)	Relevant accurate knowledge and understanding of location, theory and comparative context. Applied understanding of relevant coherent links between the investigation context and wider geography. Use of a wide range of relevant geography sources for accurate information and data. Aim, question or hypothesis are based on research information and have a manageable framework and scale. Enquiry route is logically structured and comprehensive.
b) Methodology and Data Collection (10 marks)	Appropriate methods are used to collect a range of data and information. Appropriate, linked and valid sampling frameworks are used. Frequency and timings of observations and measurements are considered for accuracy. Planning shows understanding and consideration of ethical dimensions. Data and information are reliable because of accurate and precise methods.
c) Data Representation, Analysis, Interpretation, and Evaluation (24 marks)	Appropriate skills are used to present and analyse data to show evidence of connections, including accurate statistical analyses of significance. Detailed and balanced appraisal of techniques and methodology are used, including their ethical dimensions, usefulness and validity. Research findings are synthesised coherently to form logical evidence-based conclusions. Conclusions are convincing and supported by clear and technically accurate presentation of primary and secondary data.
d) Conclusion and Critical Evaluation of the Investigation (24 marks)	Relevant accurate knowledge and understanding of location, theory and comparative context is evident throughout. Understanding of relevant coherent links between conclusions and context is shown throughout. Research findings are synthesised coherently and comprehensively. Balanced appraisal of the reliability of evidence and validity of conclusions. Balanced, concise, logical, reasoned and well-developed arguments are used within a structured and comprehensive enquiry process. Accurate geographical terminology is used throughout. Conclusions are fully supported by selected relevant evidence and concepts linked to the purpose of the investigation.

Route to enquiry

Your school geography department will set the deadline for the completion of the Independent Investigation. They may also set interim deadlines for the completion of each stage. You must use these deadlines to plan a time schedule for your work. The Independent Investigation is internally assessed, but teachers have their own fixed deadline from Edexcel, and they have to mark the Independent Investigation reports and complete internal moderation, so your deadline will be well before the Edexcel deadline. You must stick to all deadlines to ensure that tasks are completed in good time and to avoid rushing and producing inferior work.

Stage 1

You must think of and research an enquiry question, aim and hypothesis, perhaps making a shortlist of two or three options. You must then choose one to concentrate on. You must read the textbooks and other academic literature, especially recent material that is relevant to your chosen geographical topic and enquiry and also the small local area(s) where you will carry out the research. This information must be used in your report and a bibliography would be useful. Write an introduction to your II outlining the *purpose of the investigation*. Help: Themes and ideas may be discussed with your teacher and others, but they cannot provide a list of titles, enquiry questions or hypotheses.

Stage 2

You must design your fieldwork methodology, perhaps using a table structure (shown in Book 1). Your methodology should include both quantitative and qualitative data collection using relevant techniques, equipment, and sampling methods (random, systematic, and stratified) to ensure the greatest accuracy and lowest bias of the data. You must consider the health and safety aspects of your fieldwork and complete a risk assessment and take steps to ensure that the fieldwork can be completed safely. As well as primary data you must plan and collect secondary data that will assist your analysis. Write your *methodology* explaining your data collection (perhaps using a comprehensive table) and amend it if necessary after the data has been collected. Help: You are allowed to discuss your planning, and group work, for the data collection, but no help is allowed with selecting secondary sources.

Maths tip

Random sampling is the least biased of all the sampling techniques as everything has equal chance of being selected; random numbers are used to select people to interview, homes to deliver postal questionnaires to, areas to record data within, or sampling distances along a transect, for example. It can be used with very large sample areas or populations, but the main difficulty with this technique is that it is usually more time consuming to complete and makes it more difficult to analyse patterns as there is not a regular database.

Systematic sampling is where people, points or areas are chosen in a systematic way and have a regular or even spacing, such as along a transect, or people passing for a questionnaire, or using a quadrat or a grid, for example. It can also be applied to timing of measurements, such as recording or observing at exact time intervals. The systematic approach helps to avoid biased choices of locations or people and it ensures regular coverage of an area or transect so that statistical analyses can be applied more easily, it is also quicker to carry out than random sampling. A main problem is when real patterns are also systematic, which can lead to important data being completely missed out, and so is not without bias.

Stratified sampling is used when there is a clear objective to study a specific geographical aspect, such as a target population group or transect through an ecosystem or a specific urban neighbourhood or transect through all urban zones, for example. Can be used in conjunction with random or systematic approaches within areas or along transects. The main advantage is that it provides specific information to match the purpose of the investigation; it also allows comparisons and correlations to be made. However, a major disadvantage can be that if the geographical understanding is incorrect at the start, then all of the data collected may become invalid. It may also take longer to complete interviews or questionnaires with this sampling technique.

Literacy tip

Primary data is original information, measurements, visual material and statistics collected by you (or your group) and *secondary data* is information, measurements, visual material and statistics that has been collected and/or analysed by someone else. There is always some debate about census information: Usually if you extract the raw data from the government's statistics website and analyse it yourself it is considered primary data, but it becomes secondary data if you use it in a form that was already analysed for you (such as from a council's planning department website).

Stage 3

Carry out the fieldwork and collect your data, remembering to add details to your methodology if necessary. Present your primary and secondary data using a range of suitable diagrams, maps, graphs, visual material, and statistical techniques (for examples see Table B). While some techniques may need ICT skills, there are still some where it is best to use your manual skills such as for choropleth or flow line maps. It is often quicker and more effective by hand, and it can be difficult to construct more complex graphs using ICT, unless more sophisticated software is available. Help: This is independent work and no help is allowed.

Stage 4

Write your analysis and explanation of the results that you have discovered; link the analysis to your data presentation and primary and secondary evidence. Make sure that you keep your enquiry question (your purpose) and hypotheses clearly in mind when writing. The specification makes it very clear that it is the quality of the thinking (reasons for links, reasoned arguments) that is important in the II: so quality is important – not quantity. Help: This is independent work and no help is allowed.

Stage 5

Write a conclusion using your knowledge and understanding of the topic(s), including relevant theories and concepts, and your interpretation of the results from the primary and secondary data. What is the significance of the results, such as do they support or reject the hypotheses? Help: This is independent work and no help is allowed.

Stage 6

Write an evaluation that critically examines the data, such as the reliability, accuracy, validity and relevance of both primary and secondary data (strengths and weaknesses). Review any ethical bias that may be present in the investigation design or responses. Examine the links between the conclusions of the investigation and wider geography, such as the extent to which the patterns found are the same or similar to elsewhere. Suggest what further research could be completed to overcome the weaknesses of the investigation or to expand the research. Help: This is independent work and no help is allowed.

Geographical and presentation skills

Between them, Book 1 and Book 2 cover the skills listed in Appendix 1 of the specification. These can be tracked through reference to the Index at the back of Books 1 and 2. There are a few that could not be included within chapters, and these are presented after Table B, including guidance on using questionnaires.

Table B: Data presentation and use

Presentation type	How to use
Cartographic	
Land use maps	Coloured according to a key, covering transect routes or whole areas.
Choropleth maps from isoline maps	Any values that can be assigned to specific points can be converted into a spatial form using lines joining up points of equal value, e.g. quality of urban environment, or micro-climate in sand dunes.
Choropleth maps for distinct areas	Census information for wards or super output areas, such as ethnicity, house tenure, socio-economic groups.
Distribution maps (dot maps)	To show spatial distribution (clustered or dispersed) in a visual format.
Flow line maps	Pedestrian flows shown with proportional arrows, wind direction, glacial movement.
Proportional symbols maps	Squares, circles, semi-circles, bars to show data in a spatial and graphic format to a scale.
Sketch maps	Outline drawings, fully annotated to specific features of small localities, such as a cliff or condition of a building.
Weather maps (synoptic charts)	Show patterns and development of weather systems, and linked to physical geography processes at the time.
Ordnance Survey maps and town centre plans	Map reading skills to extract detailed evidence of rural and urban areas, coasts and glaciated areas; maps of different ages can show change.
Graphical	
Line graphs (with or without regression or trend lines)	For showing changes over distance or time, not only trends but deviations and anomalies are shown visually, e.g. quality of urban environment, noise and air pollution, average house value).
Cross-sections	Visual representation of reality drawn to a scale, such as a sand dune profile.
Histograms or bar graphs	Show counted data but also can show trends over distance, e.g. pebble sizes or questionnaire responses.
Compound graphs	Useful for comparing data plotted on same graph with different coloured lines or a combination of line graph and bar graph, e.g. pollution readings and index of multiple deprivation, or pebble size and beach gradient.
Multiple-axes graphs	Examples are ternary graphs and radar graphs. Ternary graphs have three axes and are useful for comparing places when the data is divided into three linked parts e.g. % in employment categories primary, secondary or tertiary.
Pictographs	Bar graph with photograph or artwork superimposed to visually reflect the data being shown, e.g. height of buildings with house pictures or traffic flows with car pictures.
Pie graphs and proportional divided circles	Several pieces of data for one point or area can be shown on one graph, for example clast sizes, or types of land use.
Kite graphs	Showing changes in a number of variables over distance, such as plant succession in sand dunes.
Radial graphs (or rose graphs)	Showing data which has a directional component using compass directions (for example winds or waves).
Logarithmic scale graphs	Showing trend where data has a very wide range of values, or where the data has lots of fluctuations that make it difficult to identify a trend.
Dispersion diagrams or scattergraphs (one axis or two axes)	Analyse the distribution of values in a set of data by one or two criteria. For example, one-axis scattergraphs show range, median and quartile deviation, while two-axis scattergraphs can show the correlation between two linked data sets (with line of best-fit).
Population pyramids	Show the population structure of a place (by age, gender, and numbers).

Presentation type	How to use
Statistical	
Central tendency (mode, mean, median)	Examining the distribution of values, and show the most common features of a distribution (pattern) which then allows comparisons to be made between sets of data or within the data set.
Correlation analyses (e.g. Spearman's) (including application of significance levels)	Examining the link between two sets of data that may be correlated negatively or positively (e.g. wind speed and wave height, socio-economic data and quality of urban environment). A correlation analysis may also show that there is no correlation at all, neither positive nor negative.
Standard deviation (including interquartile range) and Student's t-test	To compare two data sets show the range of data in categories. Also useful for choropleth mapping categories.
Chi-square	Comparing distributions over an area, such as plants in a sand dune ecosystem or questionnaire responses by population group.
Gini Co-efficient or Gini Index	Measures the gap in income distribution of a place on a scale (0-100 or 0-1).
Lorenz curve	A graphical representation of the difference between a line of equal income distribution and the real income distribution in a place.
Qualitative	
Digital photographs (normal, aerial, oblique, satellite)	Showing evidence of relevant points about any topic with specific labels or annotations.
Sketch drawings	Allow comparison from a more selective point of view with specific labels or annotations.
GIS images	Combine data, maps and satellite images to create a multi-informational (layered) resource using ICT.
Interviews and oral accounts (questionnaires)	Open and structured questionnaires (see page 15) can be used to gather views and opinions through sampling, and the results coded to give categories of response and avoid bias.
Newspapers and magazines	Provide local information on issues and views and opinions of those interviewed.
Creative and social media	Blogs and comments on social media reveal issues and concerns and views, often in a trail, which can be codified to avoid biased representation. Sometimes graphics or enhanced visual material is used to emphasise points.

ACTIVITY

Which of these data presentation techniques are you unfamiliar with? Find the time to make sure that you know how to do the ones that you need to know for your independent investigation.

Ternary graphs

These graphs have three axes and are used when data is in three parts, such as where 100 per cent is divided into three categories. Commonly used with employment categories (primary, secondary, tertiary) or soil structure (sand, silt, clay), but could also be used with categories of primary energy source, river sediment loads or greenhouse gas emissions. For fieldwork it may be useful to identify which wards may be suitable for further investigation, such as shown in Figure A.

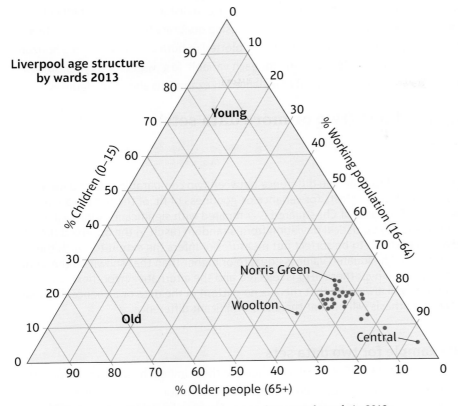

Figure A: Ternary diagram to show age structure of Liverpool wards in 2013

ACTIVITY

GRAPHICAL SKILLS

1. Study Figure A. **a)** Describe the age structure of Liverpool's census wards. **b)** Using this data, which ward(s) would you suggest would be best for an investigation into urban issues? Justify your choice(s).

Kite diagrams

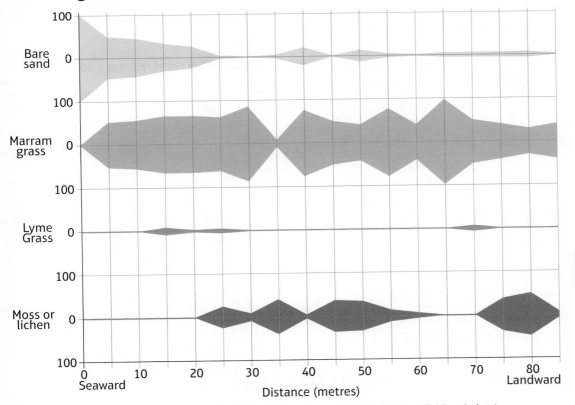

Figure B: Kite diagram for sand dune ecosystem at Holkham, North Norfolk (from fieldwork data).

ACTIVITY

GRAPHICAL SKILLS

Study Figure B. Suggest the advantages and disadvantages of this data presentation technique.

Kite diagrams are used to emphasise the pattern over a distance, most commonly for vegetation species in an ecosystem from transect records (using quadrats to estimate percentage cover). These are symmetrical line graphs with values plotted in both directions from a horizontal zero axis. There are usually parallel axes, one for each variable (e.g. vegetation type), so that the distribution pattern can be observed at a glance. An example is shown in Figure B.

Chi-squared test (two or more data sets)

> **Maths tip**
>
> *Chi-squared* is used to show the difference between two comparable sets of data, where the data is grouped into classes. It tests for variance or 'goodness of fit' of the data to a pattern predicted by hypotheses. The value of Chi-squared is compared with significance tables which confirm whether there is any statistical deviation from a random result in the observed data. It does not prove a hypothesis to be correct but measures or infers the extent to which the data fits a hypothesis. If the Chi-squared value is equal to or higher than that shown on the significance tables then the null hypothesis (H_0) is rejected and the alternative hypothesis (H_1) accepted. The null hypothesis will always be that there is no pattern in the data or a chance distribution; the alternative hypothesis will be that there is a significant difference.

Extension

Suggest the advantages and disadvantages of the data presentation techniques used in, or planned for, your Independent Investigation.

Worked example for two data sets

Between 2008 and 2013 UNESCO collected data on the numbers of school children not attending school by gender in Bolivia; they then averaged the results for every thousand households interviewed. A Chi-squared test can be used to analyse the results: The null hypothesis is that there is no difference between male and female non-attendance of school, and the alternative hypothesis is that there is a significant difference.

To calculate the expected frequencies: Column total x row total ÷ grand total. For example, for male pre-primary: 97 x 124 ÷ 264 = 45.56

Table C: Chi-squared analysis of out of school by gender in Bolivia (2008/13) per 1,000 households

	Pre-primary school		Primary school		Lower secondary school		
	Observed	Expected	Observed	Expected	Observed	Expected	Total
Male	49	45.56	49	46.97	26	31.47	124
Female	48	51.44	51	53.03	41	35.53	140
Total	97	97	100	100	67	67	264

ACTIVITY

NUMERICAL AND STATISTICAL SKILLS

There are six calculations to complete and then the 'sum of' addition: For example 'male pre-primary' =

$(49 - 45.56)^2 ÷ 45.56 = 0.2597$

Complete the other five calculations and add them together. Your sum should equal the total in the text.

Formula:

$$\chi^2 = \sum \frac{(O - E)^2}{E}$$

O = the frequencies observed
E = the frequencies expected
Σ = the 'sum of'

Degrees of freedom: number of rows – 1 x number of columns – 1 = 2-1 x 3-1 = 1x2 = 2df (= 5.991)

The result of 2.4479 is below the significance level of 5.991 and therefore the null hypothesis cannot be rejected, suggesting that there is no difference between male and female attendance at school. However, it is always important to carry out a geographical analysis of the data as well, and it is clear that male attendance is higher at the lower secondary school level.

Dot maps

This type of distribution map locates a variable through points on a map, such as showing the location of respondents to a questionnaire. The size (diameter) and colour of each dot may be varied to convey scale and categories of raw data. They create a visual impression of density and distribution of variables such as traffic or people.

Extension

Using one data set collected during your fieldwork days or a spatial planning stage, complete a dot map to show a distribution pattern.

QUESTIONNAIRE DESIGN

Questionnaires are a very useful way of obtaining both quantitative and qualitative evidence so they include questions that provide factual answers such as 'how often do you use …', and questions that ask for opinions or judgements such as 'what do you think about …'. They will have several sections of questions asking for responses. Questionnaires must be relatively quick and easy for a respondent to complete, and must be easy to understand; so they should be as short as possible and concentrate on the essential information needed for your enquiry task and hypotheses. Sections of a questionnaire are usually:

Purpose: Provide a short introduction that clearly states the purpose of the questionnaire and thanks people in advance for their assistance in completing it.

Background of the respondent: Questions asking about gender, age group, and/or ethnicity. These must be worded sensitively and may be optional. However, these factors do affect behaviours and opinions and they are useful to be able to correlate with other answers.

Precise answer questions (quantitative): These questions will provide a list of possibilities which a respondent can tick or put a number to. Often a category called 'other' is added to the end of a list just in case there is an unexpected category. These questions allow proportions to be calculated across all respondents or specific groups of respondents.

Open-ended answer questions (qualitative): These questions will leave a blank space for respondents to write their thoughts. It is important not to give any guidance or hints in a question or instructions, as these can lead the respondent and produce biased answers. These questions provide responses that can be quoted, or if there are common themes in the responses it may be possible to group them together into categories.

Organisational points

- It is usually a good idea to try your questionnaire out before using it for the data collection, perhaps on adult family members. This allows a check that it is clear and easy to follow. Make amendments as necessary.

- When deciding where to carry out your questionnaire it is important to ensure that the location chosen has the potential to provide enough respondents for a large sample size, and with characteristics suitable for the enquiry aims. The central business districts of towns and cities have the most people and so can give higher response rates, but there may be sections of a population excluded. There may need to be several locations to cover different groups of a population to create an unbiased sample.

- When deciding how to carry out your questionnaire it is important to ensure that the method will provide enough responses so that a reliable sample size is gathered and that respondents represent a true cross-section of a community. Sample size refers to the number of completed questionnaires or interviews compared to the total population of the place: In government or business, a 10 per cent sample would be considered to be statistically reliable; however, for your A level II this is likely to be unrealistic and the exact sample size will depend on how important any questionnaire data is to your enquiry. If it is a crucial part of the evidence then the sample size must be larger. Sampling technique needs to be considered in order to obtain an unbiased sample; random sampling means choosing respondents without any regular approach, systematic sampling may choose every fifth person passing, and stratified sampling would be used if data was required from a certain group of people, such as a specific ethnic background or type of residential area.

- Gathering your data will be quicker in busier areas and through a face-to-face interview approach, which involves approaching people and asking politely if they could spare the time to answer a few questions. Choosing areas where people are seated and not moving quickly from one place to another often yields higher response rates. Face-to-face interviews also allow explanation if a respondent needs help understanding a question, but the number of questions should still be kept to a minimum. Consider the effect of weather on your response rates – if it is pouring with rain then people may not want to stop and answer questions.

- If you are studying a large neighbourhood or the whole of a small settlement, then postal questionnaires can be used. This is where the questionnaire is placed through letterboxes (using a random, systematic or stratified sampling technique) with a stamped return addressed envelope. Response rates are always a concern with questionnaires; with the face-to-face technique response rates are usually over 50 per cent, whereas with a postal questionnaire response rates may be between 30 and 50 per cent. Enough numbers of questionnaires must be allocated and group data collection can be used to gather enough responses within a manageable period of time. Postal questionnaires could take longer to get responses so time needs to be allowed in the II schedule for this.

- Health and safety is important and you must never complete questionnaires on your own; always choose a public place, and never enter people's homes. Always be polite, even if someone is rude to you, and know how to seek emergency help if needed. Consider the outside weather conditions, dress sensibly and appropriately.

ACTIVITY

1. Study the photograph showing A level Geography students collecting data in a sand dune ecosystem.

a. Suggest the sampling techniques being used.
b. This is a tourist 'honeypot' (see background of photo). Suggest suitable questions for these visitors that could be used to help answer a Geography independent investigation hypothesis.

Figure D: A-level data collection along sand dune transects

Paper 3 synoptic investigation

This is a different style of examination, but one well suited to the subject of geography, as everything is interlinked in the world, and to develop a holistic understanding it is essential to see and be able to explain how everything is linked together. At GCSE level you may have experienced this approach in making geographical decisions. The specification emphasises several aspects of synopticity, and these are covered in both books in the series through Synoptic Link boxes as well as a summary at the start of every chapter. Make sure that you read and understand these as you study the topics as it may take you time to develop the ability to think synoptically, and it is not something that can be revised and learnt just before the Paper 3 examination. In this examination the examiners expect you to show your ability to combine your skills, knowledge and understanding with breadth and depth of geography. Your revision for Papers 1 and 2 will be sufficient revision of content for this examination; what is important here is your examination technique. This technique includes your ability to synthesise information from different topics, especially being able to make links and connections between them.

Synoptic themes
Players
These are individuals, groups, communities, businesses and companies, non-governmental organisations, intergovernmental organisations, and governments that have an interest in or may influence a geographical situation. In various situations the level of influence may vary, and there may be inequalities in the power of players.

Table A: Specification synoptic themes for Players (by topic)

Role of local and national government (1)	Role of governments in attracting foreign direct investment (3)	Role of planners in managing land use (5)
Role of scientists (1)	Role of TNCs (3)	Role of different players (5)
Role of emergency planners (1)	Opportunities for disadvantaged groups (3)	Role of players in reducing risk of water conflict (5)
Role of planners and engineers (1)	Role of governments (3)	Role of TNCs, OPEC, consumers and governments (6)
Role of NGOs and insurers (1)	Increasing roles of TNCs and IGOs (4A)	Role of business in developing reserves, versus environmental groups and affected communities (6)
Role of the World Trade Organization, International Monetary Fund and World Bank (3)	A government may create open or closed door policies (4A)	Role of TNCs in maintaining power and wealth (7)
Role of the European Union and the Association of South East Asian Nations (3)	Increasing roles of TNCs and IGOs (4B)	Role of powerful countries as 'global police' (7)
Role of governments in economic liberalisation (3)	The main 'gatekeeper' player affecting flows is the government (4B)	Role of emerging powers (7)
	Planners and developers may make controversial decisions (4B)	

Attitudes and actions

The players have views and opinions on the basis of which they make judgements and policies; these result in actions in support of the views and opinions. The social, political and economic culture of a place is likely to influence the views and opinions, and the relative respect for the natural environment may also influence actions.

Table B: Specification synoptic themes for Attitudes and Actions (by topic)

Attitudes range from exploitation to preservation (2A)	Attitudes on changes range from cultural erosion to enrichment (4A)	Different actions may have different impacts (4B)
Direct actions of players reduce resilience (2A)	Local communities vary in attitudes (4A)	Success depends on the attitudes of different players (4B)
Indirect actions of players alter natural systems (2A)	Players have different attitudes and approaches (4A)	Contrasting attitudes to water supply (5)
Actions range from exploitation to preservation (2A)	Government actions may prioritise national over local needs and opinions (4A)	Attitudes of global consumers to environmental issues (6)
Actions of different players may alter natural systems (2B)	Actions of local authorities will affect their success (4A)	Attitudes of different countries, TNCs and people (6)
Actions of different players may have unforeseen consequences (2B)	Differing attitudes may cause conflict (4A)	Actions and attitudes of global IGOs (7)
Attitudes of differing players may vary (2B)	Attitudes will include NIMBYism (4A)	Attitudes and actions of different countries (7)
Attitudes of pro- and anti-globalisation groups and environmental movements (3)	Actions of governments may foster or suppress diversity (4B)	Attitudes and actions in relation to resources (7)
Attitudes of pro- and anti-immigration groups (3)	Attitudes may vary (4B)	Contrasting cultural ideologies (7)
Actions of local pressure groups (3)	Urban and rural residents may differ in their attitudes towards places (4B)	
Actions of NGOs and pressure groups (3)	Intergenerational attitudes and norms may change due to global culture trends (4B)	

Futures and uncertainties

Geography is a subject that looks forward to what may happen in the future. Past situations may inform geographers of trends and likely outcomes and recent data and patterns determine what the future possible outcomes may be. Although certain things may have happened in the past, this does not guarantee that the exact same things will happen again, so there are often several scenarios for the future and uncertainty about which one may come to pass. As more evidence is gathered it is usual for the uncertainty to reduce. Ultimately players (or stakeholders) make choices, which include no change, adopting a sustainable approach or adopting a radical approach. Whatever choices are made they will affect people and the natural environment, introducing risk and depending on resilience change will take place when certain thresholds are reached.

Table C: Specification synoptic themes for Futures and Uncertainties (by topic)

Models forecasting disaster impacts with or without modification (1)	Environmental consequences of different patterns of resource consumption (3)	Projections of future drought and flood risk (5)
Climate change is creating an uncertain future and needs mitigation and adaptation (2A)	Contrasting approaches leads to uncertain futures (4A)	Projections of future water scarcity (5)
Climate change is creating an uncertain future and needs mitigation and adaptation (2B)	Future success depends on past decisions (4A)	Uncertainty of global projections (6)
Mitigation and adaptation will both be needed for future stability (2B)	Changes may create differing legacies (4B)	Uncertainty over future power structures (7)

Compulsory content areas

These P, A and F synoptic themes apply to all human and physical geography topics, but for the purposes of this examination the geographical content will only be based on the compulsory content areas of the specification, using two or more of them in the examination. These are:

Topic 1: Tectonic Processes and Hazards

Topic 3: Globalisation

Topic 5: The Water Cycle and Water Insecurity

Topic 6: The Carbon Cycle and Energy Security

Topic 7: Superpowers

There will be a synoptic assessment of geographical skills, knowledge and understanding within a place or area based on this compulsory content.

The resource booklet

An unfamiliar element of this examination, depending on your GCSE experience, is that you will be given a resource booklet with information about a selected geographical issue. This booklet will be arranged into four sections (A, B, C and D): Sections A, B and C provide information on the background geography of a place arranged in three themes, through text, data, maps, graphs and other visual materials. Section D provides information and ideas related to an issue or problem in the place and perhaps more widely, again through a range of resources.

The Paper 3 examination

There are three sections in the examination paper, with a variety of questions of different lengths, including 8-mark, 18-mark and 24-mark extended writing questions. You have to answer all of the questions, so be careful with timing because the longer questions are at the end. There will be some skills-based questions, including statistical work, so a calculator and other stationery equipment are required.

The examination is 2 hours 15 minutes long (better to think of this as 135 minutes for planning purposes) with questions totalling 70 marks; however, this includes very important reading time. There may be a temptation to start writing as quickly as possible, as you would do in Paper 1 and Paper 2, but this would be a big mistake. In this style of examination it is very important to make yourself familiar with all of the resources provided and in particular to identify where pieces of information link together. You should also be thinking about the concepts and ideas that the examiner is trying to present so that you can understand the situation presented. Edexcel

ACTIVITY

Study Tables A, B and C. Working as a group, take two or three themes from each category (P, A and F) and produce a 50-word expanded example based on the relevant topic. When completed, swap information.

Extension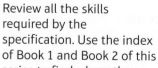

Review all the skills required by the specification. Use the index of Book 1 and Book 2 of this series to find where they are covered.

recommend 15 minutes' reading time, and this would still leave just over one minute and 43 seconds per mark, and even 25 minutes' reading would equate to almost exactly one and a half minutes per mark. It is essential that you become familiar with the material. While reading it will also be possible to plan some of your answers, so browse through the question paper and note the themes of the questions worth 8, 18 and 24 marks. These planning notes could be made on the resource booklet or any blank spaces on the question paper. In the examination there are three 4-mark questions, two 8-mark questions, one 18-mark question and one 24-mark question, so suggested timings could be: 21 minutes reading and planning, 7 minutes for each 4-mark question, 13 minutes for each 8-mark question, 29 minutes for the 18-mark question and 38 minutes for the 24-mark question.

The 4-mark question command words could include Explain, Complete, Draw or Plot, the 8-mark question command word will be Analyse, and the 18-mark and 24-mark questions command word will be Evaluate. Make sure that you understand what these command words instruct you to do (see page 6). In particular, in the evaluation questions there will not be an expected correct answer, but evidence must be used to show both sides of a situation. Your examination technique should also make full use of the resource booklet by combining pieces of information from within one section, or even from different sections to support the points you wish to make; this is where reading the booklet is so important, to spot where you can combine things. The mark scheme has phrases such as 'full and coherent interpretation supported by evidence' and 'meaningful connections to relevant geographical ideas from across the course of study'.

Key concepts

Table D: Specialist geographical concepts

Concept	Explanation
Causality	The relationship between cause and effect; the idea that a cause is responsible for something happening.
Feedback	The consequence of the action of one component of a system on another that causes a reciprocal impact; positive feedback occurs when the consequence causes more or greater actions, and negative feedback occurs when the consequence reduces the number of actions.
Globalisation	The processes through which the world has become more interconnected and interdependent, socially, economically and politically.
Identity	The sense of belonging and attachment to a place by an individual, group or community; linked to the real or perceived key characteristics of a place.
Inequality	Unequal or unjust patterns linked to lack of opportunities, treatment of people, or status of people.
Interdependence	The reliance on economic, social or political links to maintain benefits for two or more countries or groups of people.
Mitigation and adaptation	Mitigation measures are interventions by people, organisations or governments to reduce or prevent something happening, and adaptations are adjustments to reduce harmful consequences and maximise benefits.
Resilience	The ability of people, a community or a place to withstand, absorb or overcome a disturbance so that impacts are minimised.
Risk	The possibility of people, communities or places suffering damage or loss of life from a hazardous event.
Sustainability	Meeting the needs of current generations without compromising the ability of future generations to meet their needs; covering the state of the natural environment, social environment and economic environment.
System	A linkage of parts that work together through inputs and outputs to create a process; the components within the system are connected to form a working whole. Open systems are where matter and energy move into and out of the system. Closed systems are where energy may transfer into or out of the system but matter does not.
Threshold	The level at which a sudden change occurs within a system, sometimes called the critical threshold.

Sample A level exam-style questions for Paper 3

1) Explain why annual precipitation totals are important to people. (4 marks)

2) Using the data in Table 2.3, calculate the rate of change in greenhouse gas emissions between 1970 and 2010. (4 marks)

3) Study Table 3.2 and Figure 3.13. Explain why the scaling and ranking of 'soft power' may be considered unreliable. (4 marks)

4) Study Figure 1.25, and Figures 1.26, 1.32, 1.14 and 2.32. Analyse the differences in the global distribution of river flooding. (8 marks)

5) Study Figures 3.9, 3.11 and 3.32. Analyse the relationship between Kondratiev cycles and Rostow's modernisation theory. (8 marks)

6) Study Figures 2.35, 2.38 and 3.14. Evaluate the strengths and weaknesses of international agreements to combat climate change. (18 marks)

7) Evaluate the view that the superpowers have a crucial role to play in the management of the water and carbon cycles. (24 marks)

ACTIVITY

For each concept listed in Table D, identify and make a table with an expanded example of each from any compulsory topic.

Extension

After you have read Chapters 1, 2 and 3, answer the sample exam-style questions for Paper 3.

Physical systems and sustainability

The water cycle and water insecurity

Introduction

Planet Earth is often called the 'Blue Planet', because of the abundance of liquid water on its surface. Almost three-quarters of the Earth's surface is covered by water, which amounts to over a billion trillion litres in total. Water is an essential resource for life, but only a tiny fraction is available for human use because the vast majority of it – over 96 per cent – is saltwater. Of the freshwater that remains, over two-thirds is locked away in ice sheets, ice caps and glaciers and most of the rest is stored in soil, vegetation or deep underground. This leaves a relatively small proportion in streams, rivers, lakes and groundwater to provide people with the freshwater they need every day to live. This is why freshwater is regarded as a finite resource.

Water is circulated and distributed naturally via the global hydrological cycle. Water circulates between the stores on land, in the oceans and in the atmosphere, and this circulation can be studied at a variety of scales – from global to local and over short to long timescales.

As a result of natural and human processes, there are significant inequalities in the distribution of freshwater. According to a 2015 United Nations (UN) assessment, 663 million people, approximately a tenth of the world's population, lack access to safe clean water. Water insecurity has become a significant global issue for the 21st century, as human threats to water security continue to increase as agriculture expands to produce more food and because of threats from climate change (Figure 1.1) and transboundary water transfer issues. In 1995 the vice president of the World Bank warned that 'if the wars of this century were fought over oil, the wars of the next century will be fought over water, unless we change our approach to managing this precious and vital resource'.

In this topic

After studying this chapter, you will be able to discuss and explain the ideas and concepts contained within the following enquiry questions and provide information on relevant located examples:

- What are the processes operating within the hydrological cycle from the global to the local scale?

- What factors influence the hydrological system over short- and long-term timescales?

- How does water insecurity occur and why is it becoming such a global issue for the 21st century?

Figure 1.1: Human threats to water security. How are humans changing the availability and distribution of water on Planet Earth?

Synoptic links

This topic has many players involved in the use and management of water at different scales, and each group may have different attitudes towards this vital resource. Global players include international governmental organisations (IGOs) such as the UN, individual countries that share a water resource, such as the countries along the Nile, or Mexico and the USA, transnational corporations (TNCs) such as Bauer Dam Contractors, and non-governmental organisations (NGOs) such as WaterAid. National players include national governments and private water companies, such as Thames Water in the UK. At the local scale, smaller businesses and industries are involved, especially farms, as well as individuals and local pressure groups. Water managers are using a range of actions to try to increase future water security, but the many uncertainties, such as climate change and threats to sustainability, mean that water insecurity is becoming increasingly a global issue. Water resources of the future will be affected by climate change, and hydroelectric power may provide a potential solution for energy security, economic development and climate change mitigation. Traditional and emerging superpowers are involved in making decisions about resources such as water.

Useful knowledge and understanding

During your previous studies of Geography (GCSE and AS level) you may have learned about some of the ideas and concepts covered in this chapter, such as:

- The global hydrological cycle and water resources.
- Climate and weather, including climate change.
- Climate zones.
- The causes, impacts and management of flooding.
- The complex causes of drought and drought impacts.
- The variation in the global distribution of freshwater, causing areas of water surplus or deficit.
- The differences in water consumption patterns between developing and developed countries and associated water-supply problems.
- The role of agriculture in water use.
- Stakeholders' different attitudes towards the exploitation, consumption and management of water resources.
- Large-scale and small-scale management of water resources.
- The sustainable use and management of water resources at a range of spatial scales from local to international.

This chapter will reinforce this learning, but also modify and extend your knowledge and understanding of the water cycle and water insecurity.

Skills covered within this topic

- Use of diagrams showing proportional flows within systems.
- Comparative analysis of river regime annual discharges.
- Analysis and construction of water budget graphs
- Using comparative data, labelling of features of storm hydrographs.
- Use of a large database to study the patterns and trends in floods and droughts worldwide.
- Interpretation of synoptic charts and weather patterns leading to droughts and floods.
- Use of a global map to analyse world water stress and scarcity.
- Interpretation of water poverty indices using diamond diagrams for countries at different levels of development.
- Identification of seasonal variations in the regime of international rivers, such as the Nile and the Mekong, and assessment of the impact of existing and potential dams.

CHAPTER 1

What are the processes operating within the hydrological cycle from global to local scale?

Learning objectives

5.1 To understand why the global hydrological cycle is of enormous importance to life on Earth.

5.2 To understand that a drainage basin is an open system within the global hydrological cycle.

5.3 To understand how the hydrological cycle influences water budgets and river systems at a local scale.

The global hydrological cycle

The cycling of water is central to supporting life on Earth. The global hydrological cycle is typically studied and understood using a systems approach, which is a central concept for the study of physical geography. Systems theory allows us to conceptualise the main water stores and pathways at a global scale, as well as understand the role of local hydrological processes within this wider global system. This complex system adjusts and changes as a result of physical and human factors over short and long timescales.

Synoptic link

The systems approach is used to study many components of physical geography. For example, to understand the whole glacial landscape, including the glacier and its landforms, it is necessary to study the glacier as a system, with its interrelated components that are characterised by inputs (snowfall), processes (ice formation), stores (glacier) and outputs (ablation). Similarly, the shaping of a coastline depends on a system that balances inputs (wave energy), processes (erosion) and outputs (sediments for deposition). (See Book 1, page 80.)

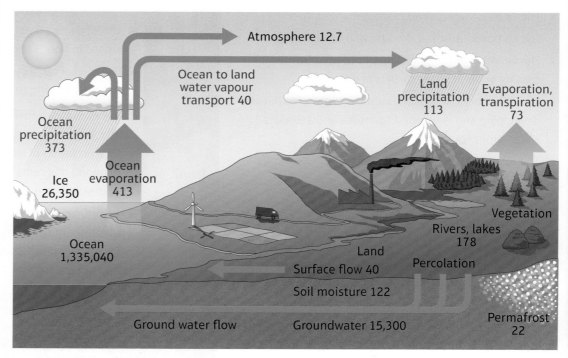

Figure 1.2: The global hydrological cycle, with estimates of the main water stores in 10^3km^3 (black type) and fluxes in 10^3km^3 per year (blue type)

A system is any set of interrelated components that are connected together to form a working whole, characterised by inputs, stores, processes (or flows) and outputs. There are two types:

- A **closed system** occurs when there is transfer of energy but not matter between the system and its surroundings (the inputs come from within the system).
- An **open system** receives inputs from and transfers outputs of energy and matter to other systems.

Figure 1.2 shows that the global hydrological cycle is a closed system because all the water is continually circulated through the stores and there is a constant amount of water in the system. The system does not change because there are no gains from or losses to other systems. This global circulation of water is driven by solar energy; heated by the Sun, the water on the Earth's surface evaporates into the atmosphere, while water is also drawn from the soil by plants and evaporated from leaves and stems by the process of evapotranspiration. When humid air rises, condensation occurs at the cooler temperatures, forming clouds, and this eventually leads to precipitation and water is returned back to the land and oceans on the Earth's surface. On land, gravitational potential energy is converted to kinetic energy as the water moves through the system by plant interception, or over land as surface runoff. Water also flows through the soil by the processes of infiltration and throughflow. Here, it may be stored as soil moisture or, if the bedrock is permeable or porous, will percolate into the rock where it is stored as groundwater. Some of this water will return to the oceans via streams and rivers, which may take some time if it is stored in lakes or glaciers en route.

Table 1.1: Estimates of the main water stores in the global hydrological cycle and typical residence times

Store	Volume 10³km³	%	Residence time
Oceans	1,335,040	96.9	3,600 years
Ice caps and glaciers (cryosphere)	26,350	1.9	15,000 years
Groundwater	15,300	1.1	Up to 10,000 years
Rivers and lakes (surface water)	178	0.01	2 weeks to 10 years
Soil moisture	122	0.01	2–50 weeks
Atmospheric moisture	13	0.001	10 days
Biological water (biosphere)	1	0.0001	1 week

The **global water budget** is the annual balance of water fluxes (flows) and the size of the water stores – oceans, atmosphere, biosphere, cryosphere, groundwater and surface water. Figure 1.2 and Table 1.1 show the size and relative importance of the water stores, and the annual fluxes between the atmosphere, ocean and land. The water stores have different residence times, but the constant circulation, albeit at variable speeds, means that water is generally considered a renewable resource, replenished naturally. However, **fossil water** is an exception; this is water that has been contained in an undisturbed space, usually groundwater in an **aquifer**, for millennia or longer. In arid regions such as the Sahara Desert, the fossil water in these aquifers may be extracted for human purposes (agriculture, industry and consumption), but there is little to no significant recharge, effectively making this type of groundwater a non-renewable resource.

ACTIVITY

GRAPHICAL SKILLS

1. **a.** Using the data in Figure 1.2, draw an appropriate graph to analyse the relative proportion of the annual fluxes between the atmosphere, ocean and land (do not include the data for the water stores).

   b. Suggest reasons for the different proportions shown in your graph.

2. Study Table 1.1.

   a. Draw an appropriate graph to display the data shown in Table 1.1.

   b. Compare and contrast the relative importance of the water stores.

   c. Consider the implications of this data for human water supply in a world with a growing population.

A level exam-style question

Explain why the global hydrological cycle is an example of a closed system. (6 marks)

Guidance

Define the term 'closed system' and use named processes and stores to show why the water cycle is closed.

A drainage basin: an open system

A **drainage basin** is an area of land that is drained by a river and its tributaries, and separated from neighbouring drainage basins by a ridge of high land called a watershed or divide (Figure 1.3). A drainage basin is an open system, so it is linked to other systems by inputs and outputs and involves a number of linked processes and stores (Figure 1.4).

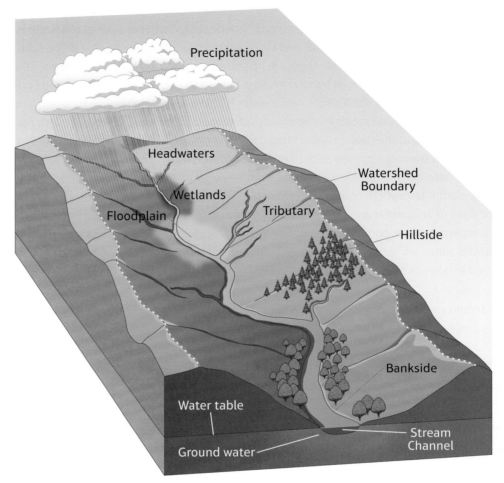

Figure 1.3: The features of a drainage basin

Physical factors affecting drainage basin inputs

Precipitation is any form of water (liquid or solid) falling from the sky. Precipitation includes rain, sleet, snow, hail and drizzle. It is the major input into any drainage basin system, and varies according to its type and intensity over time and space, linked to the climatic season and associated weather systems (Figure 1.5).

The highest precipitation inputs to drainage basins are found in the tropics, due to the Intertropical Convergence Zone (ITCZ), and in some places during the monsoon season. In the ITCZ, intense solar radiation fuels the convection of warm humid air, resulting in condensation and precipitation (convectional precipitation). For example, the highest average annual rainfall is in Mawsynram in India, with 11,873 mm of rain per year, mostly during the monsoon season between June and September. In contrast, the lowest precipitation inputs are found in stable areas of high atmospheric pressure, such as Quillagua in the Atacama Desert, which is the driest place on Earth, receiving less than 0.2 mm of rain per year.

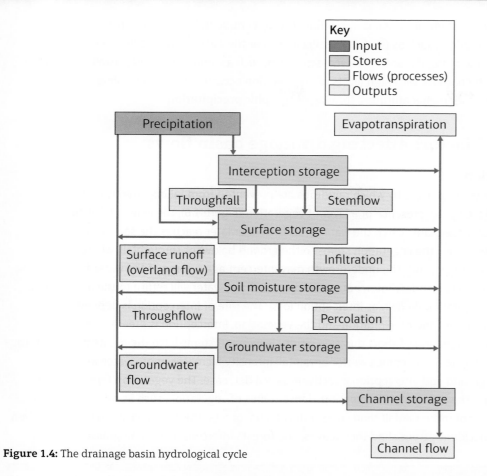

Figure 1.4: The drainage basin hydrological cycle

Extension

Research and explain, with the use of relevant diagrams, the three main types of precipitation – orographic, frontal and convectional.

ACTIVITY

CARTOGRAPHIC SKILLS

Study Figure 1.5. Explain the distribution of (a) the wettest areas of the world, and (b) the driest areas of the world.

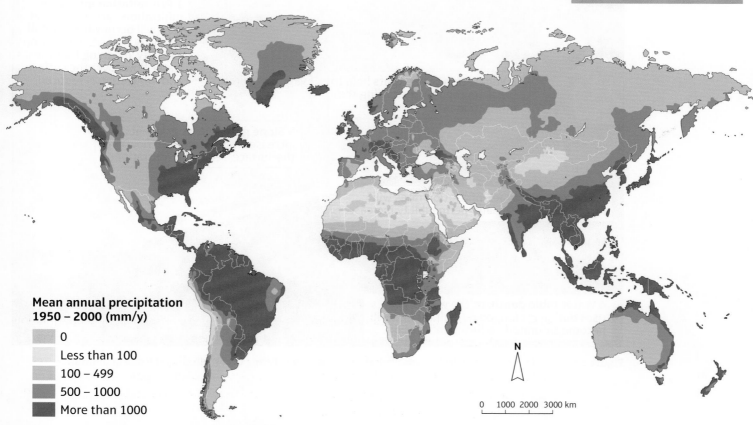

Figure 1.5: Global mean annual precipitation, 1950–2000

The distribution of precipitation is also influenced by continentality (distance from the sea), as continental interiors, such as the Gobi Desert in Asia or the Alice Springs region in Australia, are far from the moisture of maritime air masses. Relief, such as mountains, and prevailing winds complicate the pattern, with high levels of precipitation occurring where prevailing winds are forced to rise over higher altitudes, forming **orographic precipitation**.

Physical factors affecting drainage basin flows

Interception

Interception is the process by which raindrops are prevented from falling directly on to the ground surface by the presence of a layer of vegetation (Figure 1.6); the leaves, branches and stems of these plants catch the water first. The water later reaches the soil via stem flow (water flowing down the vegetation stems to the ground below) or throughfall, where the water drips to the ground. The undergrowth may intercept again some of the water falling from the canopy (secondary interception), and some of the water will return to the atmosphere via evapotranspiration (known as interception loss). The rate of interception is dependent on two physical factors: the precipitation and the vegetation. Interception is greatest when the precipitation is light and of short duration, as dry leaves and stems have the greatest water storage capacity. As vegetation becomes wetter or rainfall intensity increases, more water will drip or flow to the ground and interception effectiveness will decrease. The vegetation type and cover also influence the amount of interception: denser types of vegetation, such as many coniferous forests, intercept more rainfall than sparser deciduous forests. This is especially so in winter, when temperate deciduous trees shed their leaves; and forests intercept more than grasses or crops.

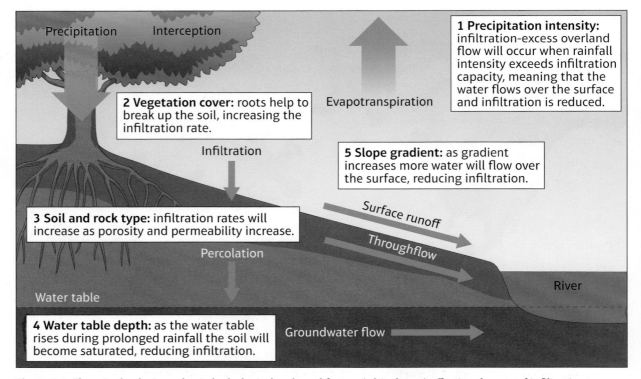

Figure 1.6: Flows in the drainage basin hydrological cycle and factors (white boxes) affecting the rate of infiltration

Infiltration and throughflow

Infiltration is the movement of water vertically downwards through pores in the soil, and the infiltration capacity is the maximum rate at which the soil can absorb precipitation. The rate of infiltration will depend on a number of factors (Figure 1.6), especially the amount of water already in the soil – the degree of saturation. Once in the soil, water moves both vertically and laterally through it by the process of throughflow – this is a downslope movement under the influence of gravity towards a stream or river.

Direct runoff (overland flow)

Water flowing over the surface of the ground is known as direct runoff or overland flow. There are two types: **saturated overland flow** and **infiltration-excess overland flow**.

Saturated overland flow occurs when water accumulates in the soil until the water table reaches or ponds on the surface, forcing further rainwater to run off the surface. This is particularly common where there are thin soils of moderate permeability. Concavities near a stream or riverbank often have high moisture levels and may produce saturated overland flow early in a rainstorm cycle.

Infiltration-excess overland flow occurs when the rainfall intensity exceeds the infiltration capacity, so the excess water flows over the ground surface. Any surface runoff will quickly deliver water into river channels, increasing the risk of flooding downstream.

Percolation and groundwater flow

When infiltrating water reaches permeable underlying bedrock, it will continue to move slowly downwards into the rock by the process of percolation. The water will fill the spaces within the permeable or porous rock, creating groundwater storage and an aquifer (a permeable rock which stores water). This will happen where the permeable layer lies above an impermeable rock layer, so that the water can percolate no further, creating a saturated zone. The upper level of this zone is known as the water table. Water may then move laterally as groundwater flow if the geological structure allows. The rate of percolation and groundwater flow will depend on the permeability of the rock, which is linked to its porosity or perviousness. Porosity relates to the total volume of pore spaces, and is greatest in coarse-grained rocks such as sandstone, while pervious rocks such as limestone have joints and bedding planes along which water can flow. Therefore percolation and groundwater flow rates will increase with porosity and perviousness; impermeable rocks such as granite, however, will prevent any percolation or water movement through the ground. The rate of groundwater flow will also increase according to the angle of the rock strata, as a steeper gradient will allow gravity to operate more effectively.

Physical factors affecting drainage basin outputs

Evaporation and transpiration

Evapotranspiration is the total amount of moisture removed from a drainage basin by processes of evaporation and transpiration; together, these processes can represent a significant output. Evaporation is the process by which liquid water is transformed into water vapour (a gas), and transpiration is the biological process by which water is drawn upwards from the soil by plants and evaporated through the minute pores, called stomata, in leaves. The factors affecting the rate of evapotranspiration are shown in Figure 1.7. It is possible to study evapotranspiration losses using the concepts of potential and actual evapotranspiration: potential evapotranspiration is the amount of evapotranspiration that could take place given unlimited supplies of water in an environment, whereas actual evapotranspiration is the amount of evapotranspiration that takes place given the actual water availability.

ACTIVITY

1. Compare the processes of infiltration and percolation.

2. Why is it important to understand these processes when analysing the balance of inputs and outputs from a drainage basin?

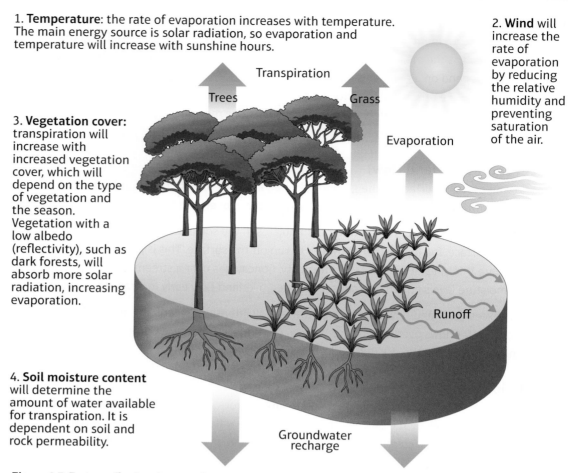

1. **Temperature**: the rate of evaporation increases with temperature. The main energy source is solar radiation, so evaporation and temperature will increase with sunshine hours.

2. **Wind** will increase the rate of evaporation by reducing the relative humidity and preventing saturation of the air.

Transpiration

Trees   Grass

3. **Vegetation cover:** transpiration will increase with increased vegetation cover, which will depend on the type of vegetation and the season. Vegetation with a low albedo (reflectivity), such as dark forests, will absorb more solar radiation, increasing evaporation.

Evaporation

Runoff

4. **Soil moisture content** will determine the amount of water available for transpiration. It is dependent on soil and rock permeability.

Groundwater recharge

Figure 1.7: Factors affecting the rate of evapotranspiration

Channel flow

Channel flow is water that has collected to flow in a rivulet, stream or river, and is another output from the drainage basin system. The discharge of a river is the volume of water passing a specific gauging station per unit of time and is measured in cubic metres per second (or cumecs). Discharge is dependent on the amount of precipitation falling directly into the channel, and on contributions from drainage basin stores via surface runoff, throughflow or groundwater flow.

Yellow = Players, Orange = Attitudes and actions, Purple = Futures and uncertainties

Human disruptions to the drainage basin cycle

The Amazonia case study shown in Figure 1.8 and Table 1.2 provides examples of how human activities can disrupt a drainage basin cycle by changing the speed of processes, creating new stores or by abstracting water. Significant disruption is also caused by hard engineering schemes, such as channelisation, to manage river flooding and, to a lesser extent, soft engineering schemes, such as overflow areas.

CASE STUDY: Human disruption to the water cycle in Amazonia

In tropical rainforests such as those found in Amazonia, South America, the dense and multiple canopies of vegetation mean that rates of interception and evapotranspiration are high. This causes high humidity and heavy local convectional rainfall and is an example of a self-sustaining cycle where water is effectively recycled within the tropical forest system. Evaporation from the forest is also important in sustaining regional rainfall in areas around the periphery of the tropical rainforest, which is important to key agricultural regions of Brazil and other countries. Research in the Amazon basin suggests that deforestation may be significantly reducing evapotranspiration and precipitation, while also increasing runoff and river discharge. For example, the Tocantins River showed a 25 per cent increase in river discharge between 1960 and 1997, coincident with increased deforestation.

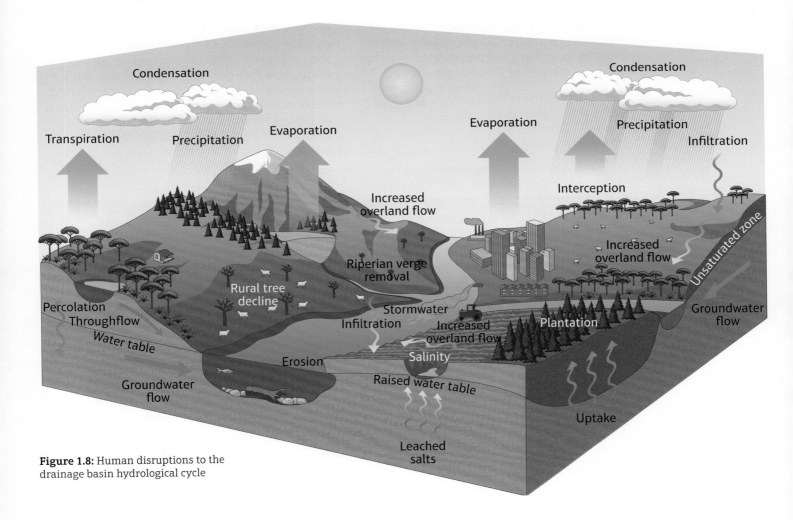

Figure 1.8: Human disruptions to the drainage basin hydrological cycle

Table 1.2: Examples of human disruptions to the drainage basin hydrological cycle

Human disruptions	Explanation	Case study examples
Cloud seeding	This is the attempt to change the amount or type of precipitation by dispersing substances into the air that serve as cloud condensation nuclei (hygroscopic nuclei). New technology and research claims to have produced reliable results that make cloud seeding a dependable and affordable water-supply practice for many regions, but its effectiveness is still debated.	China used cloud seeding in Beijing just before the 2008 Olympic Games to create rain to clear the air of pollution. It is used in the Alpine Meadows ski area in California to improve snow cover, and was used in 2015 in Texas to reduce the impact of drought.
Urbanisation	Urbanisation creates impermeable surfaces that reduce infiltration and increase surface runoff and throughflow through artificial drains; stream and river discharges often increase rapidly as a result.	Across the UK, urbanisation has increased flood risk in many towns and cities such as Winchester and Maidenhead (2014 floods) and Carlisle, York and Manchester (2015 floods).
Dam construction	Dams increase surface water stores and evaporation and reduce downstream river discharge.	Lake Nasser behind the Aswan Dam in Egypt is estimated to have evaporation losses of 10 to 16 billion cubic metres every year. This represents a loss of 20 to 30 per cent of the Egyptian water volume from the River Nile.
Groundwater abstraction	In some locations, groundwater is abstracted from aquifers faster than it is replaced, causing reduced groundwater flow and a lower water table. In other locations, reduced industrial activity or deforestation has increased groundwater storage, increasing the risk of groundwater flooding if the water table reaches the land surface.	Groundwater is used to irrigate more than 40 per cent of China's farmland and provides about 70 per cent of the drinking water in the dry northern and north-western regions. Groundwater extraction is increasing by about 2.5 billion cubic metres per year, and consequently groundwater levels in the arid North China Plain dropped by as much as a metre a year between 1974 and 2000. Groundwater rebound has occurred in some of the UK's major conurbations, including London, Birmingham, Nottingham and Liverpool, as a result of reduced abstraction for industry.

ACTIVITY

CARTOGRAPHICAL SKILLS

Research the physical and human factors affecting a drainage basin in your local area:

1. Use an Ordnance Survey map to identify a local river and its drainage basin. Using Figure 1.4, draw a similar diagram to show the inputs, flows and outputs for your chosen drainage basin.

2. Use internet research to annotate your diagram with information specific to your drainage basin, such as mean annual precipitation, vegetation types, gradients, soil and rock type, river discharge for specific gauging stations, and any other relevant information that is available.

3. In a different colour, annotate your diagram to show the human disruptions to your chosen drainage basin, such as land uses and channel alterations. Use the information in Figure 1.8, an Ordnance Survey map and internet GIS sources.

Yellow = Players, Orange = Attitudes and actions, Purple = Futures and uncertainties

Local-scale water budgets and river systems

Water budgets show the annual balance between inputs (precipitation) and outputs (evapotranspiration and channel flow). A water budget can be expressed as:

Precipitation (P) = channel discharge (Q) + evapotranspiration (E) ± change in storage (S)

This gives a direct comparison of natural water supply and demand, making it possible to identify the time periods when precipitation exceeds evapotranspiration (resulting in a positive water balance), or times when there is a negative water balance (evapotranspiration exceeds precipitation) and an increased drought risk. The soil moisture budget graph (Figure 1.9) shows annual changes in precipitation and potential evapotranspiration and this allows assessment of the impact on soil moisture availability in different climatic locations. This is useful for understanding the vulnerability of terrestrial ecosystems and the challenges for agriculture, especially in the presence of climate change.

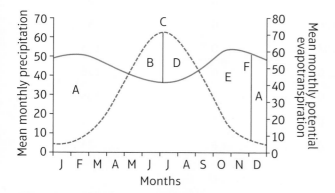

Key
— Precipitation
---- Potential evapotranspiration

Key

A = **Soil moisture surplus**: Precipitation is greater than potential evapotranspiration and the soil water store is full, so there is a surplus of soil moisture for plant use, runoff into streams and recharging groundwater supplies. The soil is said to be at field capacity.

B = **Soil moisture utilisation**: Potential evapotranspiration increases and exceeds precipitation, so there is more water evaporating from the ground surface and being transpired by plants than is falling as rain. Water is also drawn up from the soil by capillary action. The water is gradually used up.

C = **Maximum annual temperatures**: High temperatures cause maximum evapotranspiration, precipitation is at a minimum and therefore plants use up the soil moisture store. River levels will fall and crops will need irrigation.

D = **Soil moisture deficiency**: The soil water store has been used up by high rates of evapotranspiration and low precipitation. Plants can only survive if they are adapted to periods of drought or are irrigated.

E = **Soil moisture recharge**: This occurs when potential evapotranspiration decreases so that it is lower than precipitation, and the soil store starts to fill up again.

F = **Field capacity**: At this point the soil is full of water and cannot hold any more.

Figure 1.9: A typical soil moisture budget for the UK

ACTIVITY

GRAPHICAL SKILLS

1. Table 1.3 shows data to calculate the soil moisture budgets for Barrow, Alaska, USA (polar climate) and Cairo, Egypt (hot, desert climate). With reference to Figure 1.9 and Table 1.3, complete the following tasks to compare these two locations:

a. Draw a line graph to show the soil moisture budgets for Barrow and Cairo.

b. Use the following list to add relevant labels to each graph (you may not be able to use all of them): maximum precipitation, maximum evapotranspiration, soil moisture surplus, soil moisture utilisation, soil moisture deficit, soil moisture recharge, field capacity.

c. Compare the soil moisture budget for the UK (Figure 1.9) with those for Alaska and Egypt, and suggest reasons for the similarities and differences.

d. Soil moisture budgets reveal challenges for human activity. Compare the challenges that exist in the UK, Alaska and Egypt.

Table 1.3: Precipitation and potential evapotranspiration data for Barrow, Alaska, USA and Cairo, Egypt

Month	Barrow, Alaska		Cairo, Egypt	
	Mean monthly precipitation (mm)	Mean monthly potential evapotranspiration (mm)	Mean monthly precipitation (mm)	Mean monthly potential evapotranspiration (mm)
January	4	0	4	22
February	4	0	5	26
March	4	0	3	48
April	5	0	1	82
May	4	0	1	142
June	7	64	0	168
July	24	126	0	184
August	24	102	0	173
September	15	0	0	136
October	11	0	1	105
November	6	0	1	61
December	4	0	8	23

River regimes

A **river regime** describes the annual variation in the discharge of a river (Figures 1.10 and 1.11). Table 1.4 shows the physical factors affecting the regime of a river, including land use. Many rivers today do not flow under purely natural conditions, because human activities in the drainage basin affect their regimes.

The Amazon River is 6,308 km long and drains a basin of nearly 6 million km², it is in a humid tropical climate and flows over an ancient Shield area of rock: its peak discharge is in April-May and lowest discharge is in September which is linked to wet and dry seasons and snowmelt from the Andes. The Yukon River is 3,540 km long and has a drainage basin area of about 850,000 km², it is in a tundra climatic area and flows through a mountain range: its peak discharge is in May-June with a dramatic increase due to melting of snow and ice, it is low from December to May due to low precipitation and frozen conditions.

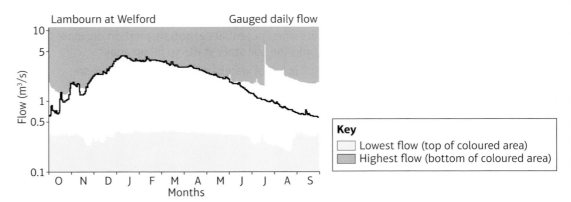

Figure 1.10: The river regime of the River Lambourn in Berkshire, October 2012 to September 2013

The river regime of the River Nile was changed significantly by the construction of the Aswan Dam in 1970. The flow of the River Nile below the dam was reduced by about 65 per cent and became regulated between the seasons so that the flood peaks in September were greatly reduced. Overall the dam is regarded to have had a severe impact on the river regime.

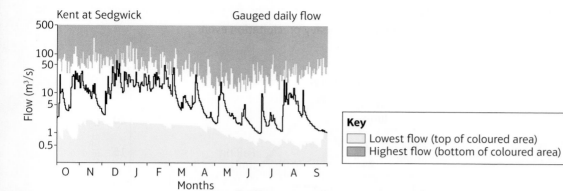

Key
☐ Lowest flow (top of coloured area)
▨ Highest flow (bottom of coloured area)

Figure 1.11: The river regime of the River Kent in Cumbria, October 2013 to September 2014

Table 1.4: Physical factors influencing the regimes of the River Kent and River Lambourn

Factor	River Kent	River Lambourn
Drainage basin area (km²)	209	234
Maximum altitude (m)	817	261
Variation in altitude (m)	817	141
Geology	Impermeable igneous rock in the north and permeable limestone in the south	Porous chalk
Mean annual precipitation (mm)	1,732	736
Mean discharge (cumecs)	8.89	1.69
Main land use	Heather moorland and peat with most land grazed	Largely arable and grassland

ACTIVITY

GRAPHICAL SKILLS

1. Use Figures 1.10 and 1.11 and Table 1.4 to answer the following questions.
   a. Compare the regimes of the River Lambourn and the River Kent.
   b. Explain the regime of the River Lambourn.
   c. Explain the regime of the River Kent.
2. Use the National River Flow Archive website to find the regime of a river in your local area. Describe and suggest reasons for this river regime.

Extension ↪

Use the internet to research:

- the regime of a river in a different climatic location, such as the Ganges or Indus, describing and suggesting reasons for the river regime discovered

- the impact of dam construction on a river regime, such as the Nile, Volta or Mekong

- the possible effects of climate change on any of the river regimes researched.

A level exam-style question

Explain why river regimes are likely to vary between drainage basins. (8 marks)

Guidance
Consider a range of physical and human factors that affect the discharge of a river over an annual cycle. Make sure that you make links and mention named located examples of river basins.

Storm hydrographs

A **storm hydrograph** (Figure 1.12) shows variations in a river's discharge at a specific point over a short period of time (usually before, during and after a storm). The shape of the storm hydrograph changes as a result of physical and human factors, and can be described as either flashy or subdued (Table 1.5). Flashy hydrographs indicate that there is a rapid increase in discharge, and perhaps a high risk of a sudden flood. Storm hydrographs are useful for predicting flood risk and comparing drainage basin responses to a heavy precipitation event.

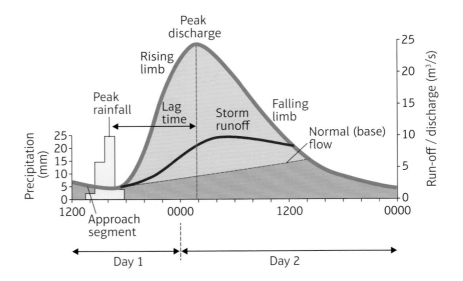

Key

Normal (base) flow is the contribution to river flow from long-term storage sources such as permeable rocks, areas of bog and marsh and peat-covered moorlands.

Rising limb is the increase in discharge in response to surface runoff and through-flow from a rainfall event, until peak flow is reached. When the storm begins, the river's initial response is negligible as the precipitation takes time to reach the channel.

Peak discharge is the maximum discharge by a stream or river in response to a rainfall event.

Bankfull discharge occurs when a river's water level reaches the top of its channel. Any further increase will result in flooding of the surrounding land.

Falling limb is the decline in discharge that occurs after peak flow. This segment is usually less steep than the rising limb because throughflow is being released relatively slowly into the channel.

Lag time is the difference in hours and minutes between the time of maximum precipitation and the time of peak discharge. The lag time varies according to drainage basin conditions. Rivers with a short lag time tend to experience a higher peak discharge and they are more prone to flooding than rivers with a long lag time.

Storm runoff is the part of the river flow derived from the immediate rainfall event. The most rapid transfer of water occurs overland and via throughflow.

Approach segment shows the discharge of the river before the storm (the antecedent flow rate).

Figure 1.12: The components of a storm hydrograph

Table 1.5: Factors affecting the discharge of a river and the shape of a storm hydrograph

Factor	Conditions likely to produce a flashy hydrograph (short lag time and high peak discharge)	Conditions likely to produce a subdued hydrograph (long lag time and low peak discharge)
Drainage basin size	Small basins: water will reach the channel rapidly, as it has a shorter distance to travel.	Large basins: water will take longer to reach the channel as it has a greater distance to travel.
Drainage basin shape	Circular basins: it will take less time for the water to reach the channel, as all the extremities are equidistant from the channel.	Elongated basins: water will take a long time to reach the channel from the extremities of the drainage basin.
Drainage basin relief	Steep slopes: water flows rapidly downhill and reaches the channel quickly.	Gentle slopes: water can infiltrate into the ground and travel slowly to the channel through the soil and rock.
Soil type	Clay soils and thin soils: clay soils have a low porosity and the grains swell when they absorb water, so water infiltrates slowly. Thin soil becomes saturated quickly.	Sandy soils and thick soils: sandy soils have a high porosity, so the water can infiltrate. Deep soils allow more infiltration.
Rock type	Impermeable rocks: water cannot percolate into the rock, increasing surface runoff to rivers.	Permeable rocks: water percolates through pore spaces and fissures into the groundwater store.
Drainage density	High drainage density: a large number of surface streams per km² means the storm water will reach the main channel rapidly.	Low drainage density: a small number of surface streams per km² means the water travels slowly through the soil and rocks to the river.
Natural vegetation	Thin grass: intercepts little water and there is little loss by evapotranspiration, so more water reaches the channel rapidly.	Forest and woodland: intercepts water and has high rates of evapotranspiration, so less water reaches the channel, and more slowly.
Land use	Urban: urban surfaces have more hard surfaces such as roads, and drains that carry the water rapidly and directly to the river.	Rural: vegetated surfaces intercept water and allow infiltration so water travels slowly to the river channel.
Precipitation intensity	High intensity: when rain falls faster than the infiltration capacity, surface runoff occurs and transports the water rapidly to the channel.	Low intensity: water can infiltrate into the soil and then travel slowly through the soil to the river channel.
Precipitation duration	Prolonged: the water table rises and the soil becomes saturated, causing surface runoff, which travels rapidly to the river channel.	Short duration: most of the water infiltrates into the soil and travels slowly through the soil into the rocks before reaching the channel.
Snowfall	Fast snowmelt: meltwater cannot infiltrate into the frozen ground, so it flows rapidly over the surface into the river channel.	Slow snowmelt: the ground thaws with the snow, so the meltwater can infiltrate into the soil and rocks before reaching the channel.
Evapotranspiration	Low rates: fewer losses from the drainage basin system will increase discharge into the river channel.	High rates: high evapotranspiration losses will reduce discharge into the river channel.

ACTIVITY

Study the factors affecting river discharge, shown in Table 1.5. Evaluate these factors, in particular explaining the one that you believe to be most important, and the one that is least important.

ACTIVITY

GRAPHICAL SKILLS

1. The River Ock and River Lambourn are adjacent catchments in Berkshire (Figure 1.13). Complete the following tasks using the data in Table 1.6 and Figure 1.13.

a. Draw a storm hydrograph with a bar chart to show rainfall totals for the storm event, and plot the discharges for the River Lambourn and River Ock as two lines on the same graph.

b. Add the following labels to your graph for both the Lambourn and the Ock rivers: peak rainfall, peak discharge, rising limb, falling limb, lag time.

c. Complete the following calculations: (i) peak discharge and lag time for the River Lambourn; (ii) peak discharge and lag time for the River Ock.

d. Describe the differences between the storm hydrographs of the River Ock and River Lambourn.

e. Using Figures 1.12 and 1.13 to help you, suggest reasons for the differences between the two storm hydrographs.

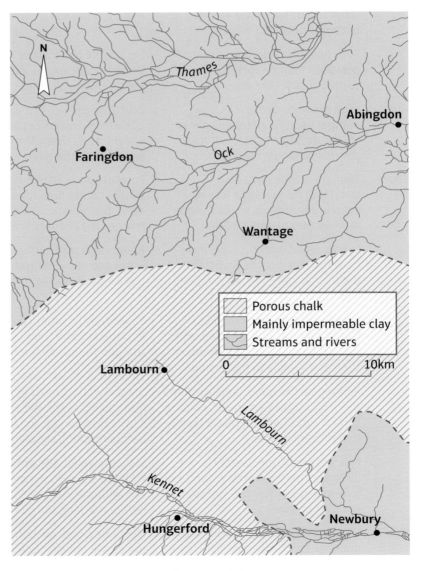

Figure 1.13: The Ock and Lambourn drainage basins, Berkshire

Table 1.6: Rainfall totals and river discharge for the River Ock and River Lambourn before, during and after a storm event

Day	Time (hours)	Rainfall (mm)	Lambourn discharge (cumecs)	Ock discharge (cumecs)
1	0	0	3.3	2.2
	4	0	3.3	2.2
	8	0	3.4	2.2
	12	0	3.5	2.3
	16	0.4	3.5	2.4
	20	2.4	3.5	2.7
2	0	1.2	3.4	3.0
	4	1.2	3.3	3.5
	8	0.2	3.4	4.0
	12	0	3.5	4.5
	16	2.4	3.5	5.2
	20	4.8	3.5	5.9
3	0	0.6	3.5	6.0
	4	0	3.4	6.1
	8	0	3.4	6.2
	12	0	3.4	6.0
	16	0	3.4	6.0
	20	0.2	3.4	6.0
4	0	0.2	3.4	6.1
	4	0	3.4	5.7
	8	0	3.3	5.4
	12	0	3.3	5.0
	16	0	3.3	4.7
	20	0	3.3	4.5
5	0	0	3.3	4.3
	4	0	3.3	4.2
	8	0	3.3	4.2
	12	0	3.3	4.0
	16	0	3.3	3.9
	20	0.2	3.4	3.8

Extension

Use the online resources available at the British Geological Survey 'Geology of Britain Viewer' to investigate the land use and geology of the River Ock and River Lambourn drainage basins. Note any additional explanations for the shape of the two storm hydrographs.

Synoptic link

The shape of the storm hydrograph and the associated river discharge and level of flood risk are affected by changes in land use within a drainage basin. Planning departments in both urban and rural areas, and landowners such as farmers, are important players in the management of flood risk. Because some rivers cross boundaries – for example the Colorado in the USA, which crosses state boundaries and then flows through Mexico, and the River Nile, which flows through several countries in north-eastern Africa – there may be sovereignty issues and disputes over control and ownership of rivers and their water.

Hydrographs are also likely to be affected by climate change, with a general prediction of more intense rainfall in many northern hemisphere areas which would create more flashy hydrographs and more flooding, while some tropical areas may become drier and hydrographs more subdued, with consequent navigation and water supply issues. (See page 66.)

What factors influence the hydrological system over short- and long-term timescales?

Learning objectives

5.4 To understand that deficits within the hydrological cycle (drought) result from physical processes but can have significant impacts.

5.5 To understand how surpluses within the hydrological cycle can lead to flooding, with significant impacts for people.

5.6 To understand why climate change many have significant impacts on the hydrological cycle globally and locally.

Deficits within the hydrological cycle (drought)

The National Drought Mitigation Centre in the USA describes **drought** as an 'insidious hazard of nature', implying that it develops gradually (it has a slow onset), with harmful impacts that vary geographically. The United Nations states that international agreement on an objective definition of drought has not yet been achieved, and so different definitions are used around the world; some quantify the rainfall deficit over a period of time, while others measure impacts such as reservoir levels or crop losses. The UN provides a general definition: 'drought is defined as an extended period – a season, a year, or several years – of deficient rainfall relative to the statistical multi-year average for a region'. Table 1.7 identifies four different types of drought: meteorological, hydrological, agricultural and socio-economic. The first three types measure drought as a physical phenomenon, whereas the last type measures drought in terms of supply and demand for human use (domestic, farming, industry).

The physical causes of drought

Figure 1.14 shows that drought has a dispersed pattern, with approximately 38 per cent of the world's land area having some level of drought exposure. This land area covers about 70 per cent of both the total global population and agricultural land. Furthermore, the areas of most severe drought exposure include 10 per cent of the land surface and 18 per cent of the global population. The physical causes of drought in regions of the world can be largely explained by the global atmospheric circulation system (Figure 1.15).

Yellow = Players, Orange = Attitudes and actions, Purple = Futures and uncertainties

Table 1.7: Types of drought

Drought type	Definition
Meteorological drought	Occurs when long-term precipitation is much lower than normal, but there is no consensus regarding the threshold of the deficit or the minimum duration of the lack of precipitation that turns a dry spell into an official drought. It is region-specific since the atmospheric conditions that result in deficiencies of precipitation are highly variable between climate types.
Agricultural drought	Occurs when there is insufficient soil moisture to meet the needs of a particular crop at a particular time. It is caused by a number of factors such as precipitation shortages, differences between actual and potential evapotranspiration, soil water deficits and reduced groundwater or reservoir levels. A deficit of rainfall over cropped areas during critical periods of the growth cycle can result in crop failures or underdeveloped crops with greatly depleted yields. Agricultural drought is typically evident after a meteorological drought but before a hydrological drought.
Hydrological drought	Occurs when there are deficiencies in surface and subsurface water supplies as measured in rivers, reservoirs, lakes and groundwater. It originates with a deficiency of precipitation but is usually out of phase with or after the occurrence of meteorological and agricultural droughts, as it takes longer for precipitation deficiencies to reach some of the components of the hydrological system such as soil moisture, stream flow and groundwater or reservoir levels.
Socio-economic drought	Occurs when the water demand for social and economic purposes (such as crop irrigation or hydro-electric power) exceeds water availability. This could be the result of a weather-related shortfall in water supply or overuse of the available water supplies. It differs from the other types of drought because its occurrence depends on temporal and spatial variations in supply and demand.

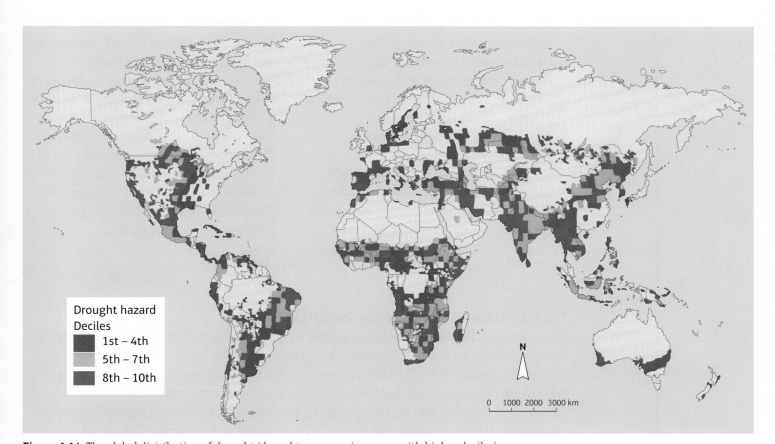

Figure 1.14: The global distribution of drought (drought exposure increases with higher deciles)

ACTIVITY

Study Figures 1.14 and 1.15. Describe the pattern of drought in the world and identify the meteorological reasons for this pattern.

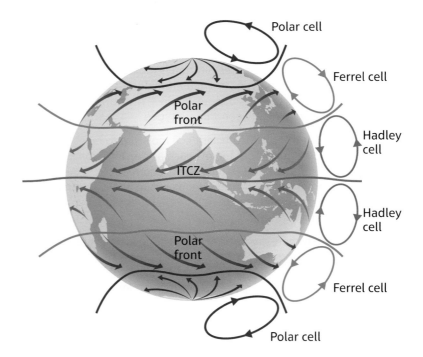

General processes taking place

1. Intense solar radiation at the Equator warms the air, which rises and starts convection. The air cools as it rises and water vapour condenses to form clouds and rain.
2. The subtropical high-pressure zone is created where air that had risen at the Equator has cooled and so sinks to form a belt of high air pressure and hot, dry conditions.
3. The air returns to ground level at the Equator, creating trade winds.
4. The trade winds meet at the Intertropical Convergence Zone (ITCZ) where the warmed air rises. The position of the ITCZ moves with the seasons – in the northern hemisphere summer (June-August), the ITCZ is north of the Equator. In December to February, the southern hemisphere is tilted towards the Sun and therefore the ITCZ is south of the Equator. This movement causes alternating wet and dry seasons in the tropics.
5. The warm air moving from the subtropics to the mid-latitudes meets cold polar air at the polar front, where the warm, less dense air rises, causing condensation and rainfall.
6. The warmer air rises into the polar front jet stream and is transferred at high altitude towards the poles, where it cools and sinks. This creates a movement of air at ground level back towards the equator.

Figure 1.15: A simplified diagram of the global atmospheric circulation system

The Intertropical Convergence Zone (ITCZ)

The ITCZ (Figure 1.15) is a belt of low atmospheric pressure located around the Equator. It moves north or south of the Equator seasonally; for example it is further north during the northern hemisphere summer. At the ITCZ the air rises as a result of intense heating by the Sun's energy; there is also high evaporation, especially from oceans. It therefore causes an alternating wet season (when it arrives) and dry season (when it moves away) in some world regions. Sometimes the subtropical high-pressure zones (Figure 1.15), associated with the descending part of the convection cell (Hadley cell) block the high humidity, rain-bearing air masses associated with the ITCZ, so that the pattern is modified. Over continental areas such as Africa, there may be lower humidity levels because less water evaporates, and if high pressure blocks the arrival of the wet season, a severe drought can occur in the Sahel (see page 46).

Yellow = Players, Orange = Attitudes and actions, Purple = Futures and uncertainties

Mid-latitude blocking anticyclones

In the mid-latitudes, frontal precipitation is created in low-pressure systems that form along the polar front, where warm tropical air rises over cold polar air (Figure 1.15). Depressions move from west to east in the mid-latitudes as a result of the Coriolis force (caused by the rotation of the Earth), and their track is directed by the polar front jet stream, which is a very fast-moving, meandering belt of air in the upper troposphere (Figure 1.15). The loops of the jet stream occasionally stabilise, or even break up, and this allows high-pressure areas (anticyclones) from the subtropics to move northwards. These anticyclones bring stable weather conditions with very little precipitation (heat waves), while the rain-bearing depressions are forced around them, usually to the north but occasionally to the south, causing drought in mid-latitude countries such as the UK. The stability of anticyclones, with their sinking air and calm conditions, means that they can persist and block weather systems from the west for up to two weeks. If this situation is repeated over the space of a few months, normal precipitation levels are greatly reduced and this may cause a drought. For example, the Met Office reported that from 2010 to early 2012 much of central, eastern and southern England and Wales experienced a prolonged period of below-average rainfall due to blocking anticyclones.

El Niño–Southern Oscillation (ENSO) cycles

The **El Niño–Southern Oscillation (ENSO)** is a naturally occurring large mass of very warm seawater in the equatorial Pacific Ocean. This warm water is normally located in the western Pacific, where it is pushed by ocean currents, trade winds and the Walker circulation cell in the atmosphere. However, on average every seven years these pushing forces weaken, and this allows the mass of warm water to move eastwards towards the west coasts of Central and South America. Wherever this mass of warm water is located, evaporation rates are higher and precipitation greater, while areas of cooler water, such as the cold current (the Humboldt Current) that flows along the Peru–Chile coastline, bring drier weather. An El Niño event reduces precipitation in the western Pacific, and the affected countries experience drought, for example the 1997 to 2009 Millennium Drought in Australia.

There are two phases of ENSO: El Niño and La Niña (Figure 1.16). La Niña occurs when the warm mass of water is pushed even further west than normal, which causes drought in Peru and California, for example. While it originates in the Pacific, ENSO is now known to cause global variations in rainfall patterns, by changing the global atmospheric circulation, creating both drought and floods in different areas of the world (Figures 1.17 and 1.18). For example, the severe East African drought of 2011 was attributed to a strong La Niña (see page 45).

Extension

a) Investigate the UK Met Office website for an example of a recent drought event in the UK. Make detailed notes on the cause of this drought.
b) The south-east of England is an area that may experience increased drought frequency in the future; explain why this would have important consequences for the country.

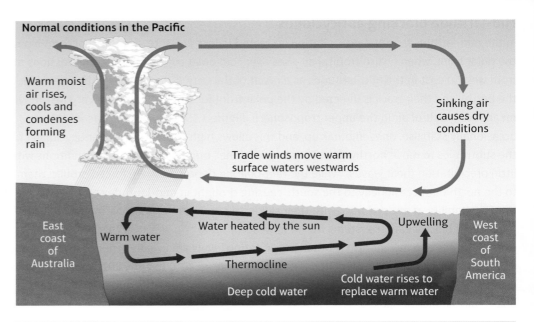

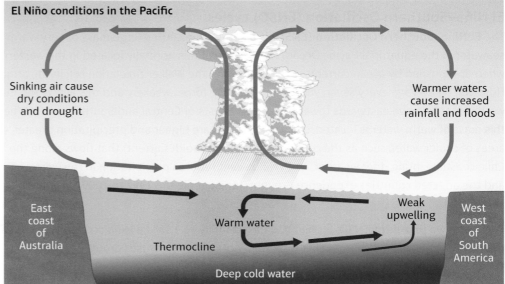

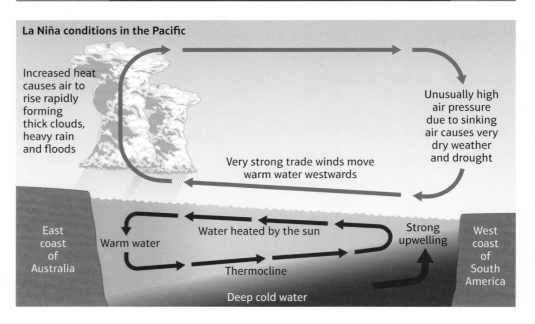

ACTIVITY

1. Referring to Figure 1.17, describe and explain the geographical distribution of drought caused by El Niño.

2. Referring to Figure 1.18, describe and explain the geographical distribution of drought caused by La Niña.

3. Use the internet to research the impacts of droughts caused by (a) the very strong 2015–16 El Niño event and (b) the moderate 2010–11 La Niña event. (Alternatively, focus on more recent events, if there have been any.)

Figure 1.16: Simplified diagrams showing normal ocean and atmospheric conditions in the Pacific, compared with El Niño and La Niña (ENSO) conditions

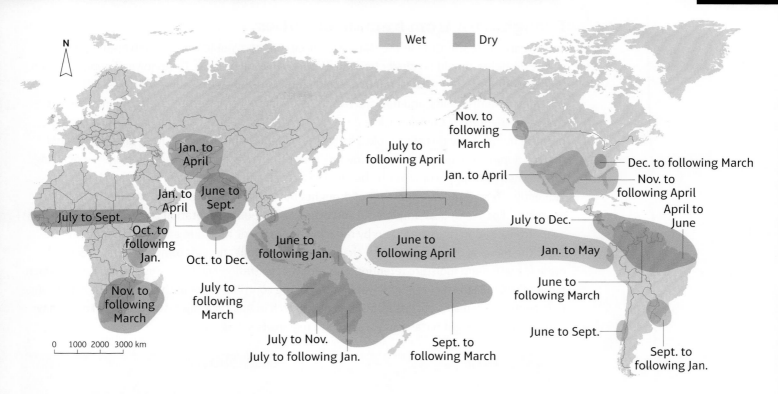

Figure 1.17: Typical global variations in rainfall due to El Niño

Sources: 1. Ropelewski, C. F., and M. S. Halpert, 1987: Global and regional scale precipitation patterns associated with the El Niño Southern Oscillation. Nom. Wea. Rev., 115, 1606–1626.
2. Mason, S. J., and L. Goddard, 2001: Probabilistic precipitation anomalies associated with ENSO. Bull. Am. Meteorol. Soc. 82, 619–638.

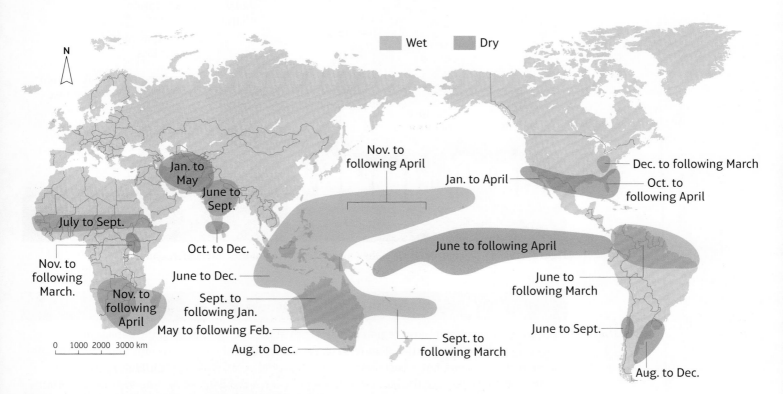

Figure 1.18: Typical global variations in rainfall due to La Niña

Sources: 1. Ropelewski, C. F., and M. S. Halpert, 1987: Global and regional scale precipitation patterns associated with the El Niño Southern Oscillation. Nom. Wea. Rev., 115, 1606–1626.
2. Mason, S. J., and L. Goddard, 2001: Probabilistic precipitation anomalies associated with ENSO. Bull. Am. Meteorol. Soc. 82, 619–638.

Drought risk from human activities

In February 2016, scientists at the University of Birmingham published research arguing that severe droughts should no longer be seen as purely natural hazards. Dr Anne Van Loon concluded that 'society is not a passive victim of drought'; human responses to water shortages influence water levels in reservoirs, aquifers and rivers. Severe droughts in settled environments, such as those experienced by China, Brazil and the USA in the 21st century, are not just natural hazards, because human activities play a role. The research identified that people have directly affected the development of droughts by abstracting water from rivers and groundwater, and by reducing the downstream supply of water by building reservoirs and water transfers. Humans have indirectly affected the development of droughts by changing land uses and altering hydrological processes. For example, deforestation and overgrazing reduce vegetation cover, so reducing evapotranspiration rates, and thereby reducing atmospheric moisture and precipitation (Figure 1.19). The removal of vegetation also changes soil conditions through compaction and reduced organic matter and moisture retention; this reduces infiltration and increases surface runoff, which reduces soil moisture content and water storage. The contribution of anthropogenic climate change to drought risk remains uncertain, but recent research in Australia suggests that it is likely to have enhanced the drought hazard in certain world regions (see page 48).

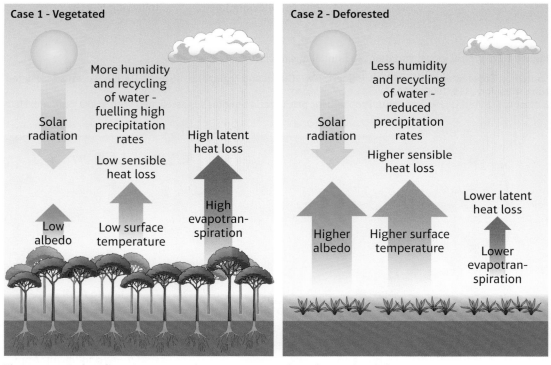

Figure 1.19: Reduced precipitation in the Amazon tropical rainforest after deforestation

CASE STUDY: Drought in the Sahel, Africa

The Sahel is a vast semi-arid region across Africa at the southern edge of the Sahara desert, covering parts of 11 countries (Figure 1.20). People living in the region usually cope with the semi-arid conditions, but occasionally severe droughts occur, making it difficult for them to adapt and survive. Since the late 1960s there has been a pronounced decline in annual rainfall, with the largest negative anomalies in the early 1980s (Figure 1.21). The most recent severe drought occurred in 2011–12. There have been multiple studies into the causes of Sahelian drought, which have identified a combination of physical and human factors.

Yellow = Players, Orange = Attitudes and actions, Purple = Futures and uncertainties

- In 2002 a study by the Commonwealth Scientific and Industrial Research Organisation (CSIRO) suggested that the Sahel droughts of the late 20th century were caused by air pollution (sulphur-based aerosols) generated in Europe and North America. These pollutants are thought to have caused atmospheric cooling, changing the global heat budget and atmospheric circulation so that tropical rains associated with the ITCZ did not arrive, therefore causing drought in the Sahel.

- In 2005 a climate modelling study by NOAA indicated that the late 20th-century Sahel drought could be the result of higher sea-surface temperatures caused by anthropogenic climate warming. The rain-bearing winds that move over the Sahel appear to fail when the sea-surface temperatures in the tropical Atlantic Ocean are warmer than average.

Professor Adam Scaife of the Met Office Hadley Centre forecast that the 2015/16 El Niño event would increase water stress in the western Sahel, where temperatures were 2°C higher than average in December 2015 and January 2016, with lower rainfall. These are conditions similar to the 1972–3 Sahel drought and famine. A pulse of easterly winds in the Pacific caused effects in other world regions, and cooler waters in the North Atlantic added to the drying conditions with reduced evaporation.

The Sahel region has one of the world's highest poverty rates and lowest development levels, with countries experiencing some of the highest population growth rates in the world (annual growth rates range from 2.5 per cent to nearly 4 per cent). The demand for food and fuel wood is therefore accelerating, and the natural dryland ecosystems are progressively being converted into farmland, which is being overcultivated and overgrazed, causing desertification. Where effects of reduced rainfall and human pressures are combined, there is reduced vegetation cover and soil moisture, which may create self-perpetuating severe droughts. While the exact relationship between drought and desertification remains unclear, desertification certainly increases human vulnerability to the drought hazard.

Extension

In 2011 an unusually strong La Niña interrupted seasonal rains in East Africa. Research this and make notes on (a) the rainfall differences from normal in 2011, (b) the causes of the rainfall differences, and (c) the scale of the impacts on people in East Africa.

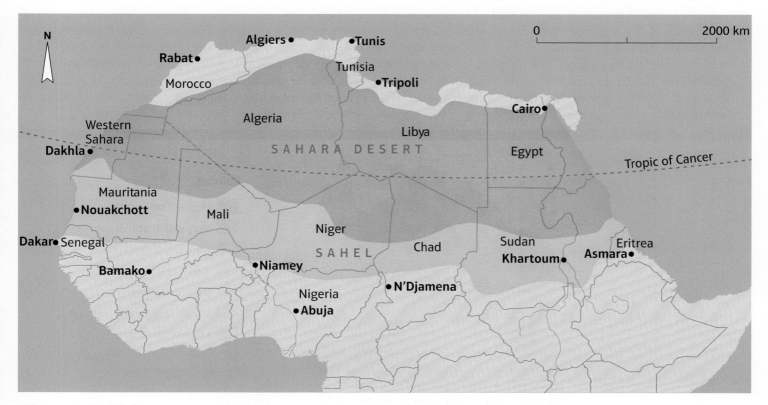

Figure 1.20: The Sahel

ACTIVITY

1. On a copy of Figure 1.21 add a trend line to show how the rainfall pattern has changed between 1900 and 2013.

2. To what extent have the Sahel drought years since 1970 been due to human activities?

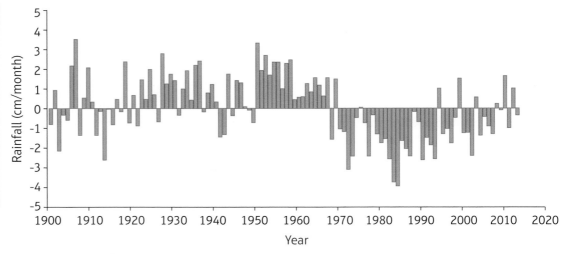

Figure 1.21: Rainfall patterns in the Sahel, 1900–2013

CASE STUDY: The Millennium Drought in south-eastern Australia, 1997–2009

The Millennium Drought was the longest uninterrupted series of years with below median rainfall in south-eastern Australia since at least 1900, with annual rainfall between 1997 and 2009 12.4 per cent below the 20th-century mean (Figure 1.22). The drought was the result of multiple physical and human causes.

- El Niño events in 2002–3 and 2006–7 were partly responsible, with research indicating that the prevailing El Niño conditions explained about two-thirds of the rainfall deficit in eastern Australia. Strengthening of the high-pressure belt, known as the subtropical ridge (STR), is estimated to have accounted for around 80 per cent of the rainfall decline in south-eastern Australia. This ridge of high pressure blocked storm tracks (depressions), forcing them towards higher latitudes and thereby reducing frontal rainfall.

- Research indicates that there may be changes to the Hadley Cell and the STR associated with anthropogenic global warming; in particular, the STR appears to have intensified as global surface temperatures increase. Scientists in Australia have conducted simulations of global climate over recent decades, using a global climate model that includes different external climate drivers (natural and human). In these simulations, the climate model was only able to reproduce STR strengthening when human emissions of greenhouse gases were included. It is thought that anthropogenic warming is reducing the temperature gradient between the Equator and the Pole, which would reduce the energy available for mid-latitude storm systems and the polar front jet stream. However, as with many climate-change scenarios, there is insufficient evidence to prove this relationship, which is an area of ongoing investigation.

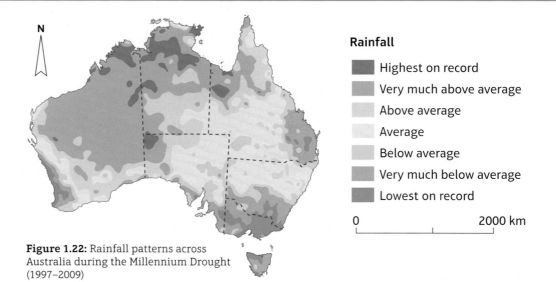

Figure 1.22: Rainfall patterns across Australia during the Millennium Drought (1997–2009)

Yellow = Players, Orange = Attitudes and actions, Purple = Futures and uncertainties

The impacts of drought on ecosystems

Ecosystems are vital for human wellbeing, providing us with innumerable and invaluable services, from tangible amenities such as clean air and water, food and fuel, to intangible amenities such as aesthetic and psychological benefits. Ecosystems are defined as interconnected communities of living organisms within a defined physical environment. The term **ecosystem functioning** refers to the biological, chemical and physical processes that take place within the ecosystem. All ecosystem components and processes are dependent on supplies of water so ecosystems are vulnerable to drought. Levels of **ecosystem resilience** to drought vary, with desert and semi-arid ecosystems being the most resilient.

Wetland ecosystems

A wetland is a land area saturated with water, either permanently or seasonally. Wetlands include marshes, swamps, bogs and fens. The Everglades in Florida are an example of a large wetland, while Minsmere in Suffolk is a small example. Figure 1.23 summarises the ecosystem functions of a wetland.

The Environment Agency has researched the impact of drought on wetland ecosystems in the UK. The Agency found that as the supply of water is reduced, areas of open water shrink or dry up altogether, resulting in progressive loss of habitat. As soil moisture is reduced, extended drying can lead to soil erosion, for example by the wind, and a reduced ability to store water in times of flood (potentially increasing downstream flood risk) or to release water in times of drought. Organic soils may oxidise, releasing carbon into the atmosphere; as water availability diminishes, concentrations of dissolved nutrients or pollutants may increase.

Birds, terrestrial wetland vegetation and invertebrates show varying responses to drought, depending on species' resistance and resilience, competitive and predatory interactions, availability of food, and timing and characteristics of a drought. Drought can alter communities by eliminating some species and creating gaps in food webs in which other species can establish themselves. For example, as open water is lost, population numbers of aquatic birds such as ducks and moorhen will decline, while some non-aquatic birds such as swallows and martins are also likely to be reduced as nesting sites are lost. Species such as snipe and thrush will also be affected because the dry soil surface will be less penetrable, limiting their ability to feed on soil invertebrates.

ACTIVITY

1. List five specific ways in which drought may affect a wetland ecosystem.

2. List five specific ways in which drought may affect a forest ecosystem.

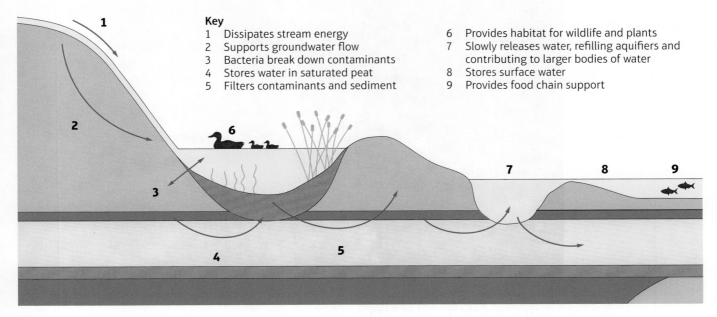

Key
1 Dissipates stream energy
2 Supports groundwater flow
3 Bacteria break down contaminants
4 Stores water in saturated peat
5 Filters contaminants and sediment
6 Provides habitat for wildlife and plants
7 Slowly releases water, refilling aquifers and contributing to larger bodies of water
8 Stores surface water
9 Provides food chain support

Figure 1.23: Wetland ecosystem functioning

In general, river-fed wetlands experience a wider range of water levels naturally and therefore support communities that may be more adapted to such fluctuations. By contrast, groundwater-fed wetlands experience a smaller range of water levels, supporting communities that prefer this type of environment but that are therefore more vulnerable to prolonged drought over several seasons.

Forest ecosystems

Forests are highly valued for their ecological functions and services, which include water storage and regulation of hydrological processes, timber production, wildlife habitat, carbon sequestration and recreational opportunities. Research published in 2015, from over 1,300 forest sites worldwide, revealed that living trees take an average of two to four years to recover and resume normal growth rates following a period of drought. Resilience to drought was lower for some species such as pine, which tend to keep using water at a high rate, even during a drought. The researchers suggest that the long-term harm to drought-stressed trees includes foliage loss, impairing growth, increased accumulation of pests and diseases, and lasting damage to vascular tissues, impairing water transport.

Between 2000 and 2003 a combination of severe drought and unusually high temperatures led to a significant die-off of Piñon pines in the Four Corners region of the south-west USA (Figure 1.24). The hot dry conditions made the Piñons more susceptible to pine bark beetle attacks, and in some areas more than 90 per cent of the Piñons died, resulting in major ecosystem changes over a large area. Warmer winter temperatures have also contributed to bark beetle outbreaks, because more beetles are able to survive through the winter and then reproduce.

Forests are important for carbon sequestration, with the Global Footprint Network (GFN) calculating that forests store an average of 0.73 tonnes of carbon per hectare per year, which is important for regulating climate change.

A level exam-style question

Using named examples, suggest two impacts of drought on ecosystem functioning. (6 marks)

Guidance

Use case studies to explain two different impacts to specified ecosystem functions, such as loss of pine trees in New Mexico reducing carbon sequestration. Consider making a list of ecosystem functions and choosing the two that you know best.

Figure 1.24: Piñon pines die off due to drought. Photos taken from the same vantage point near Los Alamos, New Mexico, in 2002 (left) and 2004 (right)

Yellow = Players, Orange = Attitudes and actions, Purple = Futures and uncertainties

Hydrological cycle surpluses and flooding

Freshwater flooding affects over one-third of the world's land area (Figure 1.25), especially on the floodplains of rivers. In addition to a number of meteorological causes of flooding, human actions can also increase the flood risk by altering physical factors.

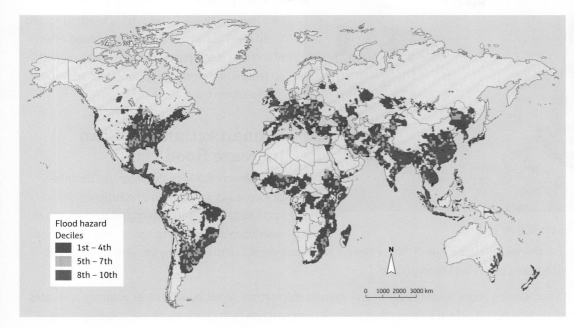

Figure 1.25: The global distribution of floods (hazard magnitude increases with higher deciles)

ACTIVITY

CARTOGRAPHICAL SKILLS

1. Using Figure 1.25, describe and explain the global distribution of floods.

2. Use the International Disaster Database (EM-DAT) to research the pattern and trends in river floods and droughts worldwide (including the changes in hazard frequency, deaths and damage over time).

The meteorological causes of flooding

Flash floods happen very quickly, often without warning. They may be caused by intense heavy rainfall associated with severe thunderstorms or tropical storms. In November 2013 thunderstorms on the island of Sardinia caused flash floods resulting in at least 18 deaths and more than US$1.14 billion in damage. In contrast, prolonged rainfall, perhaps caused by a series of mid-latitude depressions passing over the same place, can cause a slower build-up to river flooding.

Much of the river flooding in the UK (see pages 53–54) is the result of mid-latitude depressions. Each depression usually brings two bands of rain – showers and rain with the warm front and then heavier rain with the cold front. Initially the ground may be able to absorb some of the rainfall, but when throughflow and groundwater flow cannot transfer the water away quickly enough, it becomes saturated. Once rain falls on saturated ground the only other transfer is runoff, which quickly moves water to river channels and increases their discharge. Once the capacity of the river channel is exceeded, water will spill over the banks and spread over the floodplain. This meteorological situation may arise when the polar front jet stream is meandering steadily in the mid-latitudes.

A **monsoon** is a seasonal change in the direction of the prevailing winds of a world region; this involves wet and dry seasons throughout subtropical areas of the world, especially those in the vicinity of oceans. In India and South East Asia, the summer monsoon is associated with very heavy rainfall (Figure 1.26). It usually happens between April and September when warm, moist air from the south-west Indian Ocean blows towards India, Sri Lanka, Bangladesh and Myanmar, bringing a humid climate and torrential rain. This happens because the ITCZ moves northwards and the warm moist air follows behind it. In July 2015 unusually heavy monsoon rain in Myanmar caused 103 deaths and critically affected over a million people.

Figure 1.26: Monsoon flooding in Bangladesh (July 2015)

Flooding can also arise after winter snowfall, when a sudden rise in air temperatures in spring causes rapid **snowmelt**, especially in mountain environments, resulting in increased surface runoff and flash floods. In 2009, the Red River in North Dakota, USA, reached record flood levels as a result of the combined effects of increased temperatures causing snowmelt on frozen impermeable ground and additional precipitation from a rainstorm.

Human actions that can increase flood risk

The human causes of flooding can be linked to increasing population pressure, which encourages vegetation removal for agricultural development and living space, and these land-use changes increase runoff. There can be mismanagement of drainage basins and floodplains, which also increases vulnerability.

Deforestation reduces interception and evapotranspiration, resulting in greater volumes and rates of surface runoff, which ensures that precipitation reaches river channels faster, creating more flashy hydrographs. Furthermore, deforestation and intensive crop-growing expose soil to greater rates of erosion, increasing river sediment load and deposition within channels; this reduces the capacity of the river to carry water and increases the likelihood of flooding. For example, deforestation in Nepal and Tibet is known to be increasing the frequency and magnitude of floods in Bangladesh from the Ganges and Brahmaputra rivers.

Urbanisation causes a number of changes that also increase flood risk (Figure 1.27). For example, the expansion of impermeable surfaces such as roads, roofs and patios increases the rate of

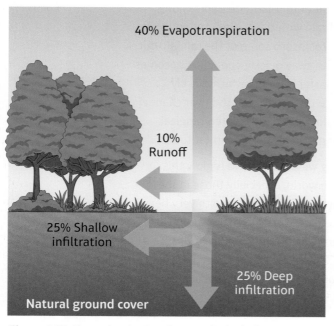

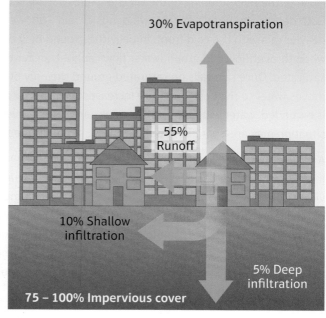

Figure 1.27: How urbanisation changes the hydrological cycle, increasing flood risk

Yellow = Players, Orange = Attitudes and actions, Purple = Futures and uncertainties

surface runoff into rivers via the urban drainage system. River lag times are shortened by urban drainage systems, which aim to transfer water efficiently into watercourses so that streets do not flood or have standing water. In addition, bridges and culverts – underground channels designed to divert river water under infrastructure such as roads – often reduce river capacity.

Floodplain drainage is common in developed countries to provide land for agriculture and to expand urban areas. The drainage process reduces the natural storage capacity of the floodplain, especially where natural wetlands are lost. The land may shrink as it dries out, getting lower and thus even more susceptible to flooding. Flood mismanagement may also occur, as alterations at one point in a drainage basin cause negative consequences further downstream. In most cases, flood management by hard engineering is designed to reduce the frequency of floods. Hard engineering includes the building of embankments (artificial levees) to increase channel capacity, but it may simply operate by transferring the discharge to unprotected areas or choke (narrow) points downstream.

For example, there has been much criticism of the hard engineering of the River Mississippi. Research suggests that channelisation (widening or deepening the river), straightening the river by cutting through meanders and the construction of artificial levees to increase the river's capacity and velocity have actually increased the flood risk. While river engineers who worked on the Mississippi schemes claim that natural forces overwhelmed the defences in the major 1993 flooding, conservation groups and many living on the floodplain claim that the levees and artificial channels constricted the river, greatly magnifying the effects of the 1993 flood. In 2011 there was another major flood on the Mississippi on the same scale as in 1993, and this may become a regular occurrence because of the changing weather patterns in the USA associated with climate change.

> ### A level exam-style question
>
> Explain the meteorological causes of river flooding. (8 marks)
>
> **Exam guidance**
> Your answer should include the weather systems and the characteristics of the precipitation that cause flooding in different locations, for example frontal, monsoonal and snowmelt.

CASE STUDY: The impact of the 2015–16 floods in the UK

In 2015 the Met Office decided to start naming severe depressions, in alphabetical order, starting with Storm Abigail in November 2015.

The 2015–16 UK winter floods were caused by a series of storm events (depressions) that created intense heavy rainfall and widespread flooding.

- **Storm Desmond** brought severe gales and heavy rain to the UK in early December 2015. This led to localised flooding in north-west England, southern Scotland, north Wales and parts of Northern Ireland. Cumbria was the worst-hit county, with more than a month's rainfall in one day; the main rivers across Cumbria exceeded the highest discharge levels ever recorded. Honister in Cumbria received 341.1 mm of rain in 24 hours between 4 and 5 December. Storm Desmond involved very low air pressure – 946 millibars (Figure 1.28) – and its fronts brought exceptionally prolonged and heavy rainfall as the air was forced to rise across the high ground of the Lake District. Convectional and orographic precipitation was combined and, together with already saturated ground, this created the conditions for extreme flooding. The existing flood defences were unable to deal with the water levels.

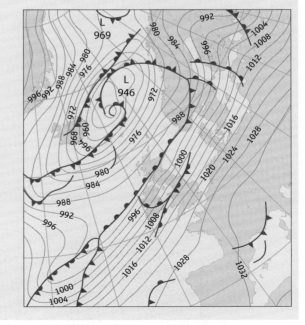

Figure 1.28: Storm Desmond: a deep Atlantic low-pressure system east of Iceland, with associated fronts stretching across northern Britain

- **Storm Eva** brought heavy rainfall and severe flooding over Christmas 2015. On Boxing Day residents in West Yorkshire and Lancashire were evacuated from their homes, and flooding hit Leeds, Greater Manchester and York.

- **Storm Frank** followed at the end of 2015 (29–30 December), bringing storms and severe gales to western parts of the UK, particularly north-west Scotland. Further flooding occurred, leading to many homes being evacuated.

A number of physical and human factors increased the flood risk from these storm events. The soils and rock were already saturated by previous heavy rainfall in November from the passage of Storm Abigail and the remains of Hurricane Kate. New hard flood defences (completed in June 2010) failed in cities in northern England when they could not cope with the river flows, and even held the water up, prolonging the flooding. Academics from the universities of Liverpool, Southampton and King's College London highlighted the role of deforestation and intensive agriculture in increasing sedimentation of channels and landslide risk. In addition to more carefully engineered flood defences, these academics called for action to manage landscapes, rivers and streams to release water more slowly, through afforestation schemes and floodplain restoration.

The impacts of Storm Desmond

Storm Desmond caused heavy rain and flooding of homes and infrastructure, as well as extensive disruption to travel and evacuation of properties. According to Environment Agency estimates, over 5,200 homes were affected by flooding in Cumbria, and the BBC reported two deaths related to the flooding. Around 40 schools were closed in Cumbria, and routine appointments across NHS hospitals were cancelled. Lancaster University closed early for Christmas because of power cuts, and many homes and businesses in Cumbria were also left without electricity. The West Coast mainline rail route to Scotland was closed because of flooding. The government provided £50 million to Cumbria and Lancashire county councils to provide money to affected households, so that they could make their homes more flood resilient.

The impacts of Storm Eva

The Secretary of State confirmed that about 9,000 properties were flooded as a result of Storm Eva and residents were evacuated. Some of the flooded properties suffered structural damage and partial collapse. A total of 7,574 homes across the north of England were affected by power cuts, and there were explosions in places where gas mains were ruptured. There was considerable disruption to travel: for example, Irish Sea ferries were cancelled because of the stormy weather, main roads such as the M62 were closed because of flooding, and flooding and landslides led to the suspension of many rail services across northern England. A number of bridges collapsed, isolating communities and businesses. On Boxing Day, dozens of football fixtures and a horse race were cancelled. Emergency measures to reduce floodwaters included the transportation of 20 water pumps and the installation of 2 km of temporary flood barriers in northern England.

The impacts of Storm Frank

Storm Frank caused severe flooding in Scotland, resulting in transport disruption to road, rail and air travel. More than 300 properties were evacuated and there were power cuts and school closures. In early January, flooding, combined with freezing temperatures and snow, reduced the ability of services to rescue people. The BBC reported one death as a result of the Storm Frank floods.

Overall impacts

In total, around 16,000 properties in England were flooded. The Association of British Insurers published data for the period 3 December 2015–3 January 2016 showing that £24 million had been spent on emergency payments to households and businesses, 3,000 families were helped into alternative accommodation, and the average cost of a domestic property flood claim was £50,000. Some of the victims of the floods were uninsured, but could seek government support from a number of schemes, and the government announced £200 million in additional funding for flood recovery. The government also committed £2.3 billion for flood management projects over the next six years. The accountancy firm KPMG estimated that the total economic cost of the floods would be approximately £5 billion.

Yellow = Players, Orange = Attitudes and actions, Purple = Futures and uncertainties

How climate change may affect the hydrological cycle

The Intergovernmental Panel on Climate Change (IPCC) predicts considerable changes to the global hydrological cycle (inputs, outputs and stores) as a result of increasing greenhouse gas concentrations (Figure 1.29). In the UK, the Living With Environmental Change Network (LWEC) has studied the latest scientific evidence to produce a report predicting the impact of climate change on the water cycle in the UK by 2050 (Figure 1.30).

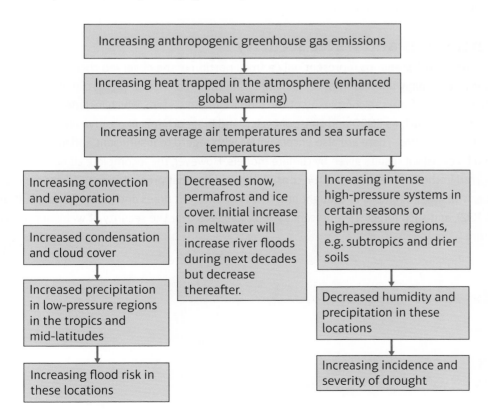

Figure 1.29: Flow diagram showing possible changes to the global hydrological system as a result of climate change

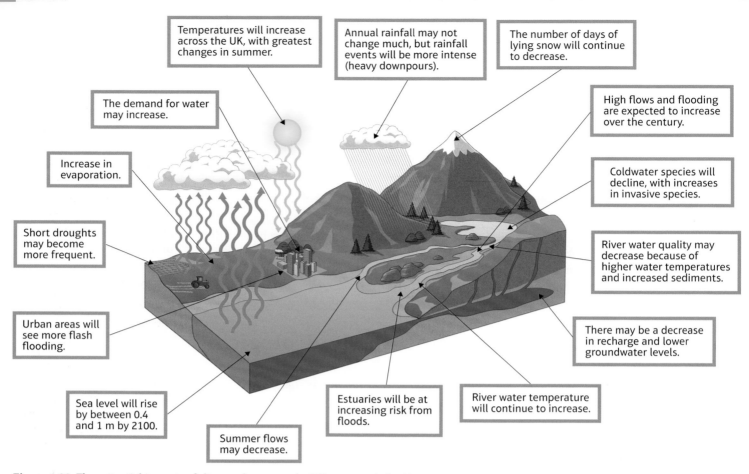

Temperatures will increase across the UK, with greatest changes in summer.

Annual rainfall may not change much, but rainfall events will be more intense (heavy downpours).

The number of days of lying snow will continue to decrease.

The demand for water may increase.

High flows and flooding are expected to increase over the century.

Increase in evaporation.

Coldwater species will decline, with increases in invasive species.

Short droughts may become more frequent.

River water quality may decrease because of higher water temperatures and increased sediments.

Urban areas will see more flash flooding.

There may be a decrease in recharge and lower groundwater levels.

Sea level will rise by between 0.4 and 1 m by 2100.

Estuaries will be at increasing risk from floods.

River water temperature will continue to increase.

Summer flows may decrease.

Figure 1.30: The potential impacts of climate change on the UK water cycle by 2050

Uncertainties and water security

The IPCC and LWEC scenarios represent only a small portion of the observed changes in the water cycle, and many uncertainties remain regarding the prediction of future climate. These uncertainties arise from the complexity of the global climate system, which includes natural short-term oscillations such as ENSO and sunspot cycles, and biogeochemical positive and negative feedback mechanisms. There are also uncertainties with regard to insufficient and incomplete data sets and inconsistent results given by climate models. However, the incomplete climate models, as reported in the IPCC reports of 2014, consistently predict that precipitation will become more variable, with increased risks of drought and floods at different times and places (Figure 1.31 and Figure 1.32), and such forecasts represent serious concerns over the security of water supplies in the future.

Yellow = Players, Orange = Attitudes and actions, Purple = Futures and uncertainties

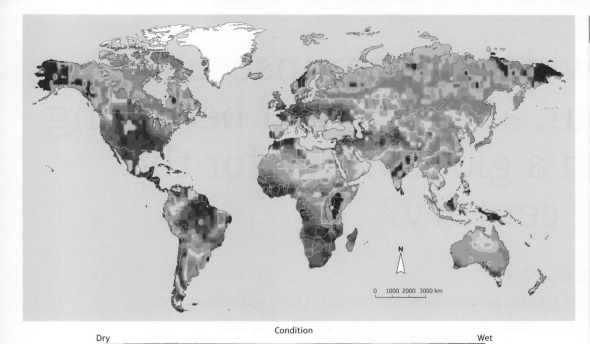

Dry | Condition | Wet

-10 -8 -7 -6 -5 -4 -3 -2 -1 -0.5 0 0.5 1 2 3 4 5 6 7 8 10

Positive numbers show when conditions are unusually wet for a particular region, and negative numbers when conditions are unusually dry. A reading of −4 or below is considered extreme drought.

Figure 1.31: Projections of future drought risk by 2099

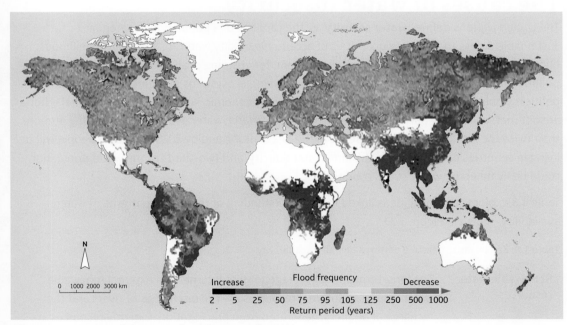

Increase | Flood frequency | Decrease

2 5 25 50 75 95 105 125 250 500 1000

Return period (years)

Figure 1.32: Projected change in flood frequency by 2100

Synoptic link

A multitude of future issues are associated with climate change. Water security will be affected by variable precipitation, bringing increased risks of drought and floods in world regions. Food supplies will also be affected, and so will the survival of ecosystems that provide essential benefits for people. The comprehensive IPCC reports of 2014, assembling data from a huge number of scientists and experts, state that the cryosphere may be affected the most, as temperatures are increasing at a faster rate in polar environments than elsewhere. People need to adapt to these changes and mitigate the worst effects of climate change by switching energy sources and adopting consumer habits that use less energy. (See Table 2.7, page 97 and pages 105–110.)

57

How does water insecurity occur, and why is it becoming such a global issue for the 21st century?

Learning objectives

5.7 To understand the physical causes and human causes of water insecurity.

5.8 To understand the consequences and risks associated with water insecurity.

5.9 To understand the different approaches to managing water supply, some more sustainable than others.

The causes of water insecurity

The United Nations defines **water security** as 'the capacity of a population to safeguard sustainable access to adequate quantities of acceptable quality water for sustaining livelihoods, human wellbeing and socio-economic development, for ensuring protection against water-borne pollution and water-related disasters, and for preserving ecosystems in a climate of peace and political stability'. **Water insecurity** occurs when these economic, social and environmental criteria are not met, or are only partially met. According to UN Water, water use has been growing at more than twice the rate of population increase in the last century, and by 2025 1.8 billion people will be living in countries or regions with absolute water scarcity, and two-thirds of the world population could be living under water stress conditions.

Table 1.8 explores the main ideas linked to an understanding of the global patterns of water availability, shown in Figure 1.33.

Table 1.8: Key terms to describe patterns of water availability

Renewable water resources	The long-term annual average total of internal and external renewable water resources. Internal resources are the discharges of rivers and the recharge of aquifers, generated from precipitation. External resources are those generated outside a country and they include inflows from upstream countries and parts of a water body (lake or river) divided by a border.
Water stress	Countries experience water stress if renewable water resources are between 1,000 and 1,700 m^3 per capita. Symptoms are widespread and include frequent and serious restrictions on water use, growing tension and conflict between users and competition for water, declining standards of reliability and service, and harvest failures and food insecurity.

Yellow = Players, Orange = Attitudes and actions, Purple = Futures and uncertainties

Water scarcity	Countries experience water scarcity if their renewable water resources are low, between 500 and 1,000 m³ per capita. Symptoms include unsatisfied demand, open tension and conflict between users, competition for water, over-extraction of groundwater and insufficient flows to the natural environment.
Absolute water scarcity	Countries experience absolute water scarcity if their renewable water resources are very low, less than 500 m³ per capita. This situation leads to widespread restrictions on water use and rationing.

ACTIVITY

GRAPHICAL SKILLS

Describe and explain the global variations in the patterns of water stress and scarcity shown in Figure 1.33.

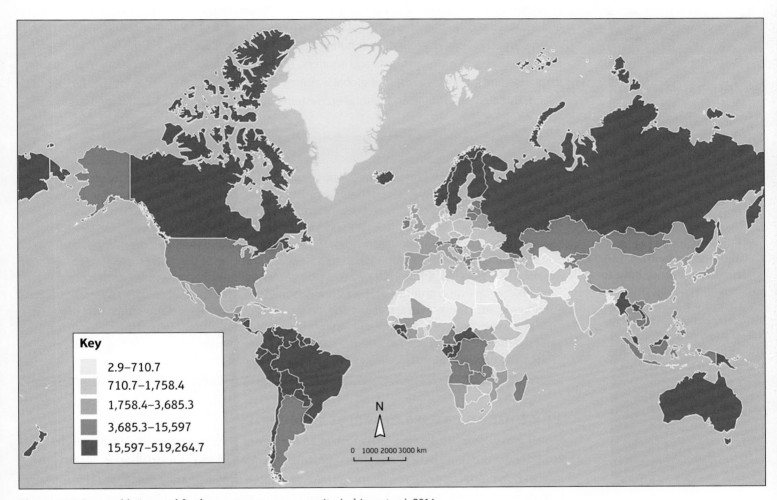

Figure 1.33: Renewable internal freshwater resources per capita (cubic metres), 2014

The physical causes of water insecurity

There are significant global variations in the distribution and availability of freshwater resources, due to natural **climate variability** between arid and humid climates and over wet and dry seasons. Climate change may increase these variations and the risks associated with scarce supplies. Increasing variability in precipitation patterns, which many countries have already begun to experience, is leading to direct and indirect effects on the whole of the hydrological cycle, with changes in runoff and aquifer recharge and in water quality.

The warmer climate at some locations will increase rates of evaporation and transpiration, leading to less effective precipitation (the amount of precipitation that is added and stored in the soil after losses). In addition, the higher water temperatures of a warmer climate and localised industrial discharges of warm waste water increase many forms of pollution. Warmer waters encourage the growth of bacteria and other organisms that are harmful to human health. The quality of water may be affected by sedimentation, nutrient enrichment, dissolved organic carbon, pathogens, pesticides and salt, as well as thermal pollution, with possible negative impacts on ecosystems, human health and water system reliability and operating costs.

Under natural conditions, the seaward movement of freshwater reduces **saltwater encroachment** in coastal zones, and soil moisture and groundwater remain fresh. However, global sea-level rise and localised abstraction of groundwater are increasing the risk of saltwater intrusion into many coastal areas. Extensive groundwater pumping from freshwater wells lowers the water table, and allows saltwater to move into soils and aquifers. Thermal expansion of the sea, along with melting ice sheets and glaciers – a result of global warming – are enabling saltwater to intrude further inland, threatening farming and natural ecosystems, such as the Sundarbans in Bangladesh.

Synoptic link

Saltwater encroachment will occur in coastal areas due to sea-level rise associated with global climate change, but further than this, all coastal processes will change, with stronger storm events and associated storm surges, increased coastal erosion, and flooding of low-lying land. This will put pressure on people and whole countries, and superpowers may have to help developing countries to relocate their people (environmental refugees) or provide funds for mitigation and adaption measures. (See Book 1, page 155.)

CASE STUDY: Saltwater encroachment in the Pacific Islands

Many of the small, low-lying Pacific Islands depend on small aquifers for their freshwater supply; for example, 35 per cent of Samoa's water supply is drawn from aquifers. The aquifers are increasingly threatened by saltwater intrusion as a result of over-abstraction, climate variability and sea-level rise (the IPCC estimates that average global sea levels rose by 0.19 m between 1901 and 2010). Population growth has increased water demand, resulting in an increase in groundwater pumping and a consequent lowering of the water table. Climate variability associated with ENSO cycles is causing recurrent droughts and floods; during the 1997/8 El Niño event, groundwater withdrawals in the Marshall Islands nearly tripled as a result of drought, and there were serious concerns over saltwater intrusion. Crop production depends on freshwater irrigation, making saltwater intrusion a serious threat to health, food security and livelihoods. This is one reason why some island communities are considering completely relocating to other countries, such as Kiribati to New Zealand.

The human causes of water insecurity

Global gross domestic product (GDP) rose at an average of 3.5 per cent per year from 1960 to 2012. During the same period, population growth, urbanisation and industrialisation, along with increases in production and consumption, have generated ever-increasing demands for freshwater resources. Strong income growth and rising living standards (a growing middle class) have led to sharp and perhaps unsustainable increases in water use, especially where supplies are vulnerable or scarce and where the distribution, price, consumption and management of water are poorly managed or regulated. Changing consumption patterns that typically involve increased water use include the rising demand for meat (water is needed for fodder crops as well as for the animals), for larger homes (water is used in concrete manufacture) and for motor vehicles, appliances and other energy-consuming devices (needing HEP and cooling water for power stations).

Agriculture is the human activity with the largest use of water (70 per cent globally and 90 per cent in developing countries), and excessive water withdrawals for agriculture have created problems, from California to India. Freshwater withdrawals for energy production account for 15 per cent of the world's total because nearly all forms of energy require some input of water as part of their production process. Thermal power generation and hydropower account for 80 per cent and 15 per

Yellow = Players, Orange = Attitudes and actions, Purple = Futures and uncertainties

cent of global electricity production respectively, and require large quantities of water. Inefficient use of water for crop production depletes aquifers, reduces river flows, degrades wildlife habitats, and has caused salinisation of 20 per cent of the global irrigated land area.

Groundwater plays a substantial role in the water supply. Worldwide, 2.5 billion people depend solely on groundwater resources to satisfy their basic daily water needs, and hundreds of millions of farmers rely on groundwater to sustain their livelihoods and ensure food security. It is estimated that groundwater provides drinking water for at least 50 per cent of the global population, and 43 per cent of the irrigation water. **Over-abstraction** occurs when too much water is removed from groundwater so that supplies diminish; an estimated 20 per cent of the world's aquifers are over-exploited. Groundwater levels are declining in several of the world's intensive farming areas, such as the North China Plain, and around numerous megacities such as Beijing.

Water availability is also affected by pollution and **contamination**. Most problems related to water quality are caused by intensive agriculture (chemical fertilisers and pesticides), industrial production (wastes and chemicals), mining (dangerous metals), untreated sewage (harmful bacteria), and urban runoff and waste water. Many cities in developing countries have inadequate sewerage and water infrastructure to collect and treat sewage and separate it from rivers; it is estimated that up to 90 per cent of all waste water in developing countries is discharged untreated directly into rivers, lakes or oceans. Expansion of commercial agriculture (agribusinesses) has led to increases in nitrate and phosphate fertiliser applications, causing eutrophication of freshwater ecosystems and significant environmental and health risks.

In 2015 1.35 million m³ of contaminated water was released by the US Environmental Protection Agency into the Animas River in Colorado. This water was from mining operations and contained arsenic at 300 times normal levels, and lead at 3,500 times normal levels. People and livestock downstream were at risk and access to the river was closed; people with wells were told not to use the water without testing. A local state of emergency was declared, for example by the Navajo Nation Commission on Emergency Management. Following widespread criticism of their error, the EPA suspended their clean-up operations at ten abandoned mines in four states where the situation was similar to the Animas River mine in Colorado. These suspensions were to allow the EPA time to develop emergency plans in case similar accidental pollution occurred. The EPA also opened a temporary water treatment plant at the Animas River mine, at a cost of US$1.8 million, to treat the polluted waste water.

Rising demand and future water scarcity

The United Nations World Water Development Report 2015 has projected an increase in global water demand of 55 per cent by 2050 (Figure 1.34). This is mainly due to a growing demand from secondary industries, thermal electricity generation and domestic use, all of which are linked to increasing urbanisation in developing countries. The UN projects a 40 per cent global water deficit by 2030 under the 'business-as-usual' scenario, which would pose serious challenges in some locations (Figure 1.35).

> **Synoptic link**
>
> Water contamination is a particular risk in the informal settlements (shanty towns) of megacities, where there are high population densities with inadequate infrastructure for effective sewage treatment. Water supplies may be from occasional standpipes, rather than a complete network to every home. This creates poor living conditions, with health risks. Poor sanitation encourages the spread of waterborne diseases such as cholera and typhoid. These are major challenges for the countries that consider themselves to be emerging superpowers. (See Book 1, page 199.)

A level exam-style question

Explain why there is an increasing global demand for water. (8 marks)

Guidance

Figure 1.34 provides hints for this question. Think about what is happening in different types of country, and make sure that you explore each selected reason fully, with some located examples and facts to support your points.

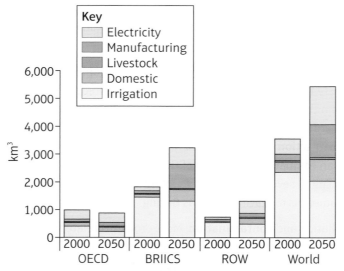

Figure 1.34: Global water demand in 2000 and 2050

Note: BRIICS (Brazil, Russia, India, Indonesia, China, South Africa); OECD (Organisation for Economic Co-operation and Development); ROW (rest of the world). This graph only measures 'blue water' demand and does not consider rainfed agriculture.

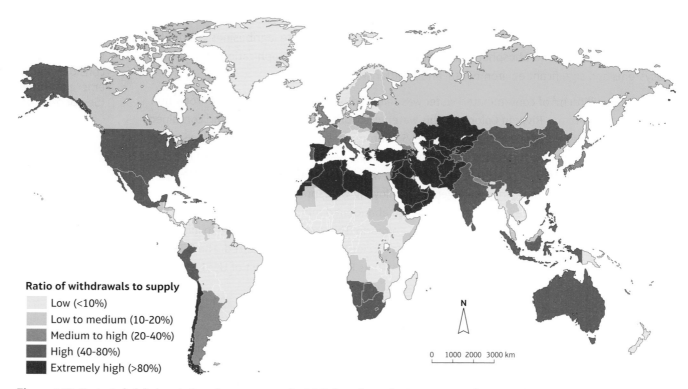

Figure 1.35: Projected global variations in water stress by 2040 (based on a business-as-usual scenario)

The consequences and risks of water insecurity
The global pattern of water scarcity

The Food and Agriculture Organisation (FAO) identifies three dimensions of water scarcity:

- Availability – physical scarcity of clean freshwater resources to meet demand.
- Access – scarcity due to the failure of institutions to ensure a reliable water supply through water management.
- Utilisation – scarcity arising from inadequate infrastructure to use water resources due to financial constraints.

Yellow = Players, Orange = Attitudes and actions, Purple = Futures and uncertainties

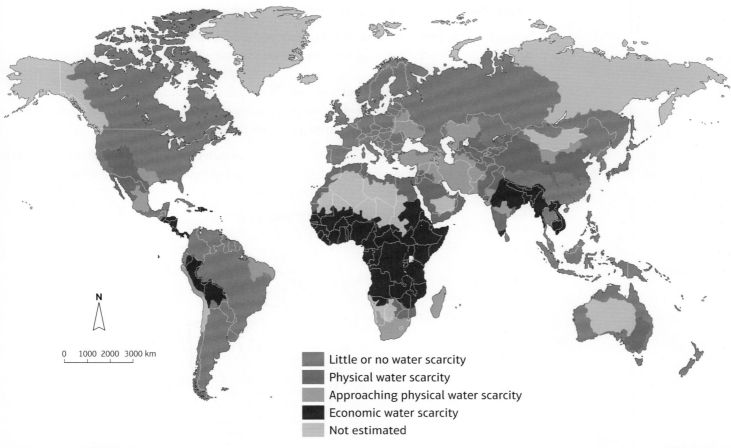

Little or no water scarcity
Physical water scarcity
Approaching physical water scarcity
Economic water scarcity
Not estimated

Definitions and indicators
• Little or no water scarcity. Abundant water resources relative to use, with less than 25% of water from rivers withdrawn for human purposes.
• Physical water scarcity (water resources development is approaching or has exceeded sustainable limits). More than 75% of river flows are withdrawn for agriculture, industry, and domestic purposes (accounting for recycling of return flows). This definition – relating water availability to water demand – implies that dry areas are not necessarily water scarce.
• Approaching physical water scarcity. More than 60% of river flows are withdrawn. These basins will experience physical water scarcity in the near future.
• Economic water scarcity (human, institutional, and financial capital limit access to water even though water in nature is available locally to meet human demands). Water resources are abundant relative to water use, with less than 25% of water from rivers withdrawn for human purposes, but malnutrition exists.

Figure 1.36: Global patterns of physical and economic water scarcity

Water scarcity therefore consists of **physical water scarcity**, due to the lack of availability, and **economic water scarcity**, due to a lack of access and poor resource management (Figure 1.36).

The price of water

The cost of supplying water varies significantly between developed and developing nations, but prices are rising all around the world. Providing access to clean drinking water requires the construction and maintenance of robust infrastructure systems that also dispose of and treat used dirty water. Such services are expensive and therefore difficult for many developing nations to provide, especially in areas of poverty and high population density. In developing countries, many people are forced to rely on street vendors for their water, and the cost can be up to 100 times more expensive than if the water were supplied to the home. During the 1980s the World Bank and International Monetary Fund (IMF) gave loans that required many developing countries to privatise their water supply system, in the hope that competition would reduce costs. However, many of these projects have since been cancelled due to public pressure, because the price of water became unaffordable for a large portion of the population.

The Water Poverty Index (WPI) uses five measures to indicate levels of water insecurity:

1. Resources – the physical availability of surface and groundwater and its quality.
2. Access – the accessibility of safe water for human use, including domestic, industrial and agricultural use.
3. Capacity – the effectiveness of water management to ensure affordability.
4. Use – the use of water for different purposes, including domestic, industrial and agricultural use.
5. Environment – water management strategies to ensure ecological sustainability.

Each of these measures is scored out of 20 to give a maximum possible score of 100; water poverty decreases as the score increases. Figure 1.37 shows the Water Poverty Index for the UK, and Table 1.9 provides water quality data for a range of countries.

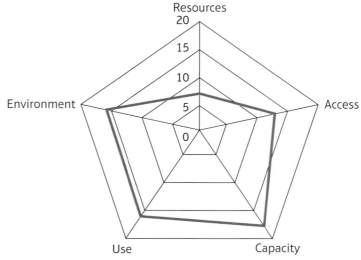

Figure 1.37: The Water Poverty Index for the UK

GRAPHICAL SKILLS

1. Make a copy of the UK's Water Poverty Index (WPI) shown in Figure 1.37 and add the data for the countries shown in Table 1.9.

2. Describe the variations in the WPI for these countries at different levels of development.

3. Suggest physical and human reasons for the WPI variations shown.

Table 1.9: Water Poverty Index data for selected countries

Country	Resources	Access	Capacity	Use	Environment	WPI
Haiti	6.1	4.8	10.5	4.3	7.0	32.7
Ethiopia	6.6	3.1	8.0	7.7	8.6	34.0
China	7.1	8.6	13.2	18.0	9.7	56.6
India	6.8	9.6	12.1	20.0	9.7	58.2
USA	10.3	14.1	16.7	1.3	16.2	58.5
Brazil	13.5	10.1	12.5	11.6	11.1	58.8
UK	7.3	13.5	17.8	16.4	16.0	71.0

The importance of water supply

Water is essential for sustainable development because it is needed for almost everything that humans do. The UN World Water Development Report 2015 states that 'progress in each of the three dimensions of sustainable development – economic, social and environmental – is bound by the limits imposed by finite and often vulnerable water resources and the way these resources are managed to provide services and benefits'.

Water is an essential resource in the production of most types of goods and services, including food, energy and manufactured products, and so is needed to ensure **economic sustainability**.

Yellow = Players, Orange = Attitudes and actions, Purple = Futures and uncertainties

Industrial water supply (quantity and quality) must be reliable and predictable to support financially sustainable investments in economic activities. Global water demand for these economic activities is increasing, particularly for manufacturing industry, where demand is expected to increase by 400 per cent between 2000 and 2050 (Figure 1.34), with the majority of this increase occurring in emerging economies and developing countries. Energy production is water-intensive, and meeting ever-growing demands for energy will generate increasing stress on freshwater resources, with repercussions for other uses. Increasing industrial production will also have potential impacts on water quality: in localised areas where water use for industrial production is not well regulated or enforced, pollution could increase dramatically, closely linking increasing economic activity with the degradation of environmental services.

Access to water supplies for domestic purposes is critical for a family's health (**human wellbeing**) and social dignity (**social sustainability**), and access to water for economic productive uses such as agriculture or family-run businesses is critical for generating income and contributing to overall economic productivity. Therefore investment in improved water management and sanitation services can help reduce poverty and sustain economic growth through better health, reduced health costs, increased productivity and time savings. Access to safe drinking water and sanitation is a human right, but its limited realisation throughout the world often has disproportionate impacts on the poor and on women and children and other disadvantaged groups.

Water for environmental sustainability

Freshwater ecosystems provide essential services for economic, social and environmental wellbeing. Wetlands, for instance, reduce the incidence of floods, store water and provide other direct economic benefits such as fisheries and recreation, yet half of them have been lost. Another example is forested highlands, which serve a key role in recharging aquifers and ensuring clean water flows for agriculture, hydropower and other uses. Most economic models do not value the essential services provided by freshwater ecosystems, partly because they are difficult to quantify, which may cause natural environments to be overlooked and lead to unsustainable use of water resources and then ecosystem degradation. The disruption of ecosystems through urban sprawl and urbanisation, over-abstraction for agriculture, deforestation and pollution may undermine an environment's capacity to sustain itself.

Degraded ecosystems may have difficulty regulating and restoring themselves, losing resilience, which then accelerates water-quality degradation and reduces water availability for nature and people. An example is the Salton Sea in the dry south-eastern part of California. The Sea has a productive ecosystem, with fish in its waters and invertebrates in its muds, providing a food source for over 400 species of birds, including pelicans and egrets. Millions of birds use the Sea as a stopping point when migrating along the Pacific Flyway. Since 2005 water levels have been falling as evaporation from its shallow waters (15 metres deep) exceeds inflow, causing salinity levels to increase (over 50 g/l). Water is in great demand in southern and central California, so water that should reach the Salton Sea is diverted to agriculture, and some inflow consists of contaminated runoff from fields. The Salton Sea is affected by the regional water agreements, and from 2017 to 2018 there could be an 18 per cent decline of inflow, as water will no longer be diverted to it from the Imperial Irrigation District, causing the Sea to shrink by up to 260 km².

Conflicts between water users

Competition for freshwater between water 'uses' and water 'users' increases the risk of localised or international conflict over inequalities in water access or allocation. During these conflicts or tension situations, the need to maintain water and ecosystem sustainability, for life and economic development, is often overlooked. Often, the natural environment and marginalised and vulnerable people are the biggest losers in the competition for water.

Conflict may occur over local water sources or over **transboundary water** – where a river, lake or aquifer crosses one or more political borders (either a border within a nation, such as a state or province, or an international border). According to UN Water, approximately 40 per cent of the world's population lives within river or lake drainage basins that extend over two or more countries, and the 263 transboundary lake or river basins account for about 60 per cent of global freshwater flow. In addition, about 2 billion people worldwide depend on groundwater, which includes

ACTIVITY

Using the River Nile and research on the Mekong River, answer the following questions:

1. Explain the general social, economic and environmental advantages and disadvantages of large scale schemes using rivers.

2. Explain how conflicts between countries over water management can arise.

CASE STUDY: River Nile transboundary water conflicts

The 6,700-km long River Nile is the longest river in the world, so it is not surprising that it flows through several countries. It is an international transboundary river, whose water resources are shared by 11 countries – Tanzania, Uganda, Rwanda, Burundi, Congo-Kinshasa, Kenya, Ethiopia, Eritrea, South Sudan, Sudan and Egypt. The Nile has two major tributaries, the White Nile and the Blue Nile, which meet near Khartoum in the Sudan. Much of the river and its tributaries flow through semi-arid (Sahel) and arid (Sahara) areas, which increases their importance as other water resources are scarce. The river is an increasingly valuable resource for these Nile countries, providing water for domestic, industrial and agricultural purposes and it is under increasing pressure from rapid population growth, economic development and climate change. The river is particularly important to Egypt and Sudan, where it is the primary water source; Egypt depends on the River Nile for 95 per cent of its water needs.

There is a long history of disagreement over the use of the Nile, arising originally from historic water allocation agreements that favoured the downstream states of Egypt and Sudan, and more recently as a result of ambitious dam-building programmes that have the potential to reduce downstream flows. In 1929 a Nile agreement between Egypt and the UK (the colonial power in Sudan at that time) granted significant water allocations to Egypt and Sudan, making no allowance for the water needs of the other Nile states. It also granted Egypt veto power over construction projects on the Nile and its tributaries, in an effort to minimise interference with the flow of water in the Nile. This historic agreement has led to conflicts between the Nile river basin countries; upstream countries are increasingly harvesting the waters of the River Nile to meet the needs of their growing populations and economies, yet Egypt has maintained that the water rights it acquired through the 1929 agreement must be honoured, and that no construction project may be undertaken on the River Nile or any of its tributaries without prior approval from its government.

In 1999, the River Nile countries, except Eritrea, signed the Nile Basin Initiative (NBI) in an effort to enhance cooperation over the use of the Nile water resources. This initiated work on developing a permanent legal framework for governing the Nile River Basin, with equitable water allocation – the Cooperative Framework Agreement (CFA). In May 2010, Burundi, Ethiopia, Kenya, Rwanda, Tanzania and Uganda signed the CFA agreement on sharing Nile water, which raised strong opposition from Egypt and Sudan over fears that it would reduce their water rights and allocations.

In April 2011 the Ethiopian Prime Minister announced the construction of the Grand Ethiopia Renaissance Dam on the Blue Nile, which met with an angry response from the Egyptian president, who stated that Egyptians would not accept any projects on the Nile River that threatened their livelihood. Fortunately, further discussions led to a recent agreement that is expected to resolve some of the issues between Ethiopia and Egypt; in March 2015 Egypt, Ethiopia, and Sudan met and signed an agreement on the development of the Grand Ethiopia Renaissance Dam. However, this agreement did not resolve the broader, contentious issues of sharing the Nile waters among all Nile countries, as shown by Egypt not signing the CFA.

Yellow = Players, Orange = Attitudes and actions, Purple = Futures and uncertainties

approximately 300 transboundary aquifer systems. The UN also reports that 158 of the world's 263 transboundary water basins lack any type of cooperative management framework, a factor that significantly increases the potential for conflict.

CASE STUDY: Conflict between water users in Ethiopia

Ethiopia has an ambitious and controversial dam-building programme, designed to turn the country into the 'powerhouse of Africa' and fuel economic growth. The programme has not only caused international conflict (see the Nile case study above) but it has also caused significant internal conflict, such as controversy over the Gilgel Gibe III Dam.

The Gilgel Gibe III Dam and hydroelectric power plant is on the Omo River in Ethiopia. This US$1.8 billion project began in 2008 and started to generate electricity in October 2015, becoming the third-largest hydroelectric plant in Africa. The project is controversial because of local negative environmental and social impacts.

The environmental impact assessment was not published until two years after construction started. According to critics, the dam will be potentially devastating to the downstream indigenous population as it will prevent seasonal floods. It is estimated that more than 200,000 people rely on the Omo River below the dam for subsistence agriculture, and they are dependent on the seasonal floods to replenish the dry soils for planting. Many of the ethnic groups such as the Mursi and Nyangatom already live in chronic hunger, so the dam not only threatens their livelihoods but their very survival. Many tribespeople are armed to defend themselves against neighbouring tribes, and there are fears that water shortages could cause violent conflict.

Those supporting the dam construction argue that artificial floods can be released from the reservoir, and irrigation projects are planned for massive cotton and sugar cane plantations in the lower Omo Valley, which will improve the livelihoods of the downstream population. However, critics claim that these plantations will benefit only Ethiopian state-owned companies, and there are reports of human rights violations by the Ethiopian army against locals who oppose the sugar plantations in the lower Omo Valley, including beatings and intimidation.

In June 2011 UNESCO's World Heritage Committee called for the construction of the dam to be halted, to review its impact on Lake Turkana, a natural World Heritage Site straddling the Kenya–Ethiopia border. It is feared that the dam could reduce the level of Lake Turkana by up to 10 metres, affecting up to 300,000 people as well as the wildlife. This could increase the salinity of the water, threatening the drinking water supply, the fishing industry and the lake ecosystem.

ACTIVITY

1. Research the roles and attitudes of the players involved in international water conflicts such as those involving the River Nile, River Colorado, River Jordan, River Niger or the River Danube:

   a. The global players (such as UN Water, Water Aid, groups of countries and other IGOs, NGOs and TNCs).

   b. The national players (such as water companies and government departments).

   c. The local players (such as communities and local pressure groups).

Different approaches to managing water supply

Effective water management is essential to minimise the risk of water insecurity and reduce the potential for conflict. It can involve **technological fixes,** human innovations to increase supply, or attitudinal fixes in which people change their behaviour to reduce demand through conservation strategies.

Hard-engineering schemes

Hard-engineering schemes use artificial structures to increase water supply, including **water transfer** projects such as China's South–North Water Transfer, **mega dams** such as Ethiopia's Gilgel Gibe Dam, and **desalination plants**.

CASE STUDY: China's South–North Water Transfer

Northern China is suffering from water scarcity and the problem is most acute in the Hai basin, where Beijing is located. Half of China's people and two-thirds of the farmland are in the north of the country, while 80 per cent of its water is in the south, notably in the Yangtze River basin. Beijing, the capital, has similar levels of water scarcity to Saudi Arabia, at just 100 cubic metres per person a year. The water table under the capital has fallen by 300 metres since the 1970s, and the problem is compounded by poor water quality in up to 60 per cent of river water as a result of pollution, which further reduces the supply of clean water for drinking and domestic use.

China has designed a huge hard-engineering solution, the largest inter-basin water transfer scheme in the world: the South–North Water Transfer Project. This project has the capacity to deliver 25 billion m³ of freshwater per year from the Yangtze River in China's south to the drier north by two routes (central and eastern), each covering a distance of over 1,000 km (Figure 1.38). A proposed third western route remains in the planning stage.

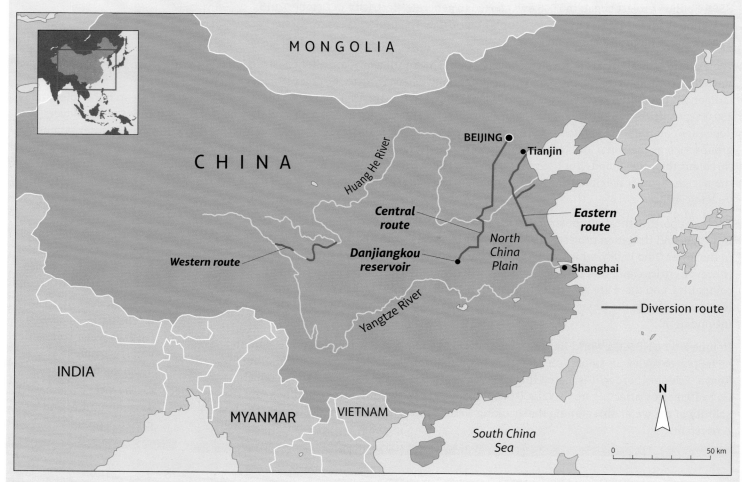

Figure 1.38: China's South–North Water Transfer Project

The huge project has raised a number of concerns:

- Economic – it is very expensive, with reservoir and canal construction costs reaching US$80 billion in 2015 and continuing high maintenance costs and water prices. Many farmers claim that the water will be too expensive and that therefore they will continue to exploit groundwater.

- Social – more than 300,000 people were displaced during the construction of the central route, as the water level of the Danjiangkou reservoir had to be raised by 13 metres.

- Environmental – the transfer of water does not address the underlying causes of water shortages in the north: pollution and inefficient agricultural, industrial and urban use. Extracting water from the Yangtze drainage basin may further reduce discharge levels and affect river ecosystems in addition to the Three Gorges Dam impacts.

Yellow = Players, Orange = Attitudes and actions, Purple = Futures and uncertainties

Furthermore, in the past decade two severe droughts in the Yangtze basin have caused water shortages in south China, so the South–North Transfer Project would increase the risk of more frequent water scarcity in the future, in addition to the effects of urbanisation and climate change. This could have significant negative economic, social and environmental consequences in the Yangtze basin and create the potential for conflict within China itself.

Critics of the project suggest that, with more effective water management, North China could be self-sufficient in water without the transfer of water from the south. They claim that rainwater harvesting and waste-water recycling could meet much of the demand in cities. In agriculture, losses could be reduced by lining irrigation canals with concrete and using smart irrigation techniques.

Sustainable water supply schemes

Sustainable water management schemes are usually less controversial than other schemes as they aim to balance economic, social and environmental needs by working with local people to develop soft-engineering projects that work with natural processes to restore water supplies, and often involving water conservation. The organisation UN Water advocates the use of 'natural infrastructure', such as the components of the hydrological cycle, as sustainable water management solutions. This means harnessing the water-related services provided by forests, wetlands and floodplains (Figure 1.39).

Water management issue (primary service to be provided)		Natural infrastructure solution	Watershed	Floodplain	Urban	Coastal	Corresponding built infrastructure solution (at the primary service level)
Water supply regulation (including drought mitigation)		Re/afforestation and forest conservation	X				Dams and groundwater pumping Water distribution systems
		Reconnecting rivers to floodplains		X			
		Wetland restoration/conservation	X				
		Constructing wetlands	X				
		Water harvesting[1]	X				
		Green spaces (bio-retention & infiltration)			X		
		Permeable pavements[1]			X		
Water quality regulation	Water purification and biological control	Re/afforestation and forest conservation	X				Water treatment plant
		Riparian buffers		X			
		Reconnecting rivers to floodplains		X			
		Wetland restoration/conservation	X				
		Constructing wetland	X				
		Green spaces (bio-retention & infiltration)			X		
	Erosion control	Re/afforestation and forest conservation	X				Reinforcement of slopes
		Riparian buffers		X			
		Reconnecting rivers to floodplains		X			
Moderation of extreme events (floods)	Riverine flood control	Re/afforestation and forest conservation	X				Dams and levees
		Riparian buffers		X			
		Reconnecting rivers to floodplains		X			
		Wetland restoration/conservation	X				
		Constructing wetland	X				
		Establishing flood bypasses		X			
	Coastal flood/ storm control	Protecting/restoring mangroves, coastal marshes and dunes				X	Sea walls
		Protecting/restoring reefs (coral/oyster)				X	

[1] Built elements that interact with natural features to enhance water-related ecosystem services.

Figure 1.39: Sustainable natural infrastructure solutions to water management issues

CASE STUDIES: water management

Sustainable water management schemes in Singapore

Singapore is a city-state with an area of only 710 km² and a population of 5.4 million inhabitants in mid-2015. It receives abundant rainfall (2,400 mm per year). However, because of the limited land for the collection and storage of rainwater, the high evaporation rate due to tropical climate and the lack of groundwater resources, Singapore is considered to have water scarcity. Singapore's national water agency (PUB) has therefore invested in research and technology to create a diversified water supply, comprising local catchment water, imported water, **recycled water** known as NEWater and desalinated water.

- Local catchment water involves collecting rainwater through a network of drains, canals, rivers, storm-water collection ponds and reservoirs before it is treated to supply drinking water.

- Singapore has an agreement until 2061 to import water from Malaysia.

- NEWater is high-grade recycled water produced from used water that is treated and further purified using advanced membrane technologies and ultraviolet disinfection, making it ultra-clean and safe to drink. Together, Singapore's four NEWater plants can meet up to 30 per cent of the nation's current water needs. By 2060 PUB aims to meet 55 per cent of water demand with NEWater.

Smart irrigation in China and Australia

Smart irrigation provides crops with a suboptimal water supply causing mild stress during crop growth stages that are less sensitive to moisture deficiency. This technique has been found to conserve water without a significant reduction in yield. A six-year study of winter wheat production on the North China Plain showed water savings of 25 per cent or more through the application of smart irrigation, combined with acceptably high yields and net profits. In Australia, regulated smart irrigation of fruit trees increased water productivity by approximately 60 per cent, with a gain in fruit quality and no loss in yield. Controlled irrigation, such as drip-feed, ensures that water goes directly into the soil next to the roots of crops, which prevents evaporation losses.

Rainwater harvesting jars in Uganda

WaterAid is an international non-governmental organisation that raises funds to improve access to safe water, sanitation and hygiene for some of the world's poorest people. An example of their work can be seen in Kitayita village, Uganda, where 3,000 people lack access to safe water. Local builders have been trained in the construction of rainwater harvesting jars (Figure 1.40), which are made from locally available materials and have a capacity of 1,500 litres. They are usually designed to collect rainwater from roofs and store water for drier periods. The objective has been to help the community construct on-site water supplies, close to home, removing the need for old or infirm people to travel long distances across difficult terrain to collect water. The jars have a long life, and once constructed can provide a stable water source for many years.

Yellow = Players, Orange = Attitudes and actions, Purple = Futures and uncertainties

Figure 1.40: Rainwater harvesting jar in Uganda

ACTIVITY

1. Make a table to show the advantages and disadvantages of the different approaches to managing water supply using examples such as China's South–North Water Transfer, Ethiopia's Gigel Gibe III mega dam, Singapore's desalination plants, Singapore's water recycling, China's smart irrigation and the rainwater harvesting jars in Uganda.

2. Using Figure 1.39 and the internet, research a case study of a 'natural infrastructure' solution to water management issues.

Integrated drainage basin management

According to the World Bank, **integrated river basin management** aims to establish a framework for coordination whereby all administrations and stakeholders involved in river basin planning and management come together to develop an agreed set of policies and strategies. The aim is to achieve a balanced and mutually acceptable approach to land, water, and natural resource management. This has been attempted at various scales and involving different players, including **water-sharing treaties** such as the multiple River Nile agreements and water frameworks or the Colorado River Compact, and, at a global scale, the Berlin Rules and the European Water Framework Directive.

CASE STUDIES: Colorado integrated river management

The 2,330 km long Colorado River flows through a basin of 637,000 km^2. Ninety-seven per cent of the basin is in the USA and 3 per cent in Mexico. The source is in the Rocky Mountains, it flows through the semi-arid areas of the Colorado plateau and has its mouth at the Gulf of California in Mexico. Precipitation in the basin varies, but has recently been prone to drought conditions due to climate change. Other challenges include increased urbanisation, population growth and agricultural needs for irrigation water. Even in the 1920s it was recognised that the growing western states needed water, especially southern California which has no major river to use. In 1922 seven states agreed a compact and the basin was divided into upper (source areas) and lower (demand areas) sections for management purposes and water was allocated by state. There were further agreements and compacts to define priorities and a water treaty with Mexico in 1944 to allocate water to the country. In 1956 a comprehensive development plan was developed to cover river regulation, HEP production, water rights and irrigation. There are 29 dams and numerous diversion projects now in existence. Sometimes there has not been enough water to meet all allocations and this has led to disputes that still exist today.

In 1990 the lower basin US states used their full allocation for the first time making the issue worse, but with drier conditions and reservoir levels falling there simply is not enough water to meet demands. Environmental protection laws have also complicated management, such as the Grand Canyon Protection Act 1992. Individual states have been forced to look at alternatives; for example, Nevada has tried negotiating for extra water allocation (especially for Las Vegas), California has introduced irrigation restrictions and is investigating desalination, and Arizona established a Water Banking Authority in 1996 with the aim of using its water share efficiently and storing surpluses in aquifers within the state (creating a water bank). All of the 21st century has seen discussion of solutions, but without a new agreement being reached.

CASE STUDY: The Berlin Rules on Water Resources

The Berlin Rules on Water Resources were approved by the International Law Association in 2004. The Rules outline international law relating to freshwater resources, whether within a nation or crossing international boundaries, and they replace the earlier 1966 Helsinki Rules. There are nine water management principles that apply to all countries, including national and internationally shared waters:

1. Participatory water management – the public have a right to be involved in decision-making.

2. Coordinated use – surface water and groundwater resources to be managed to maximise the availability and reliability of water supplies.

3. Integrated management – all components of the drainage basin to be considered, such as vegetation, watercourses, settlements and all stakeholders.

4. Sustainability – economic, social and environmental needs to be met, now and in the future.

5. Minimisation of environmental harm – such as pollution.

6. Cooperation over shared water resources – between regions in a country or between countries.

7. Equitable utilisation of shared water resources – all groups and communities to receive a fair share.

8. Avoidance of transboundary harm – control of actions upstream.

9. Equitable participation – all countries or players to have equal status.

The ongoing conflicts between the countries of the River Nile Basin illustrate the difficulties in implementing the Berlin Rules and the global challenge to achieve effective water-sharing treaties in several river basins.

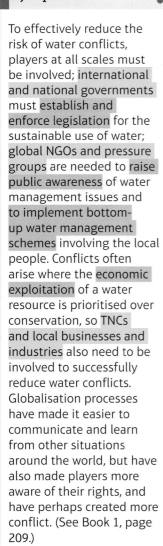

Synoptic link

To effectively reduce the risk of water conflicts, players at all scales must be involved; international and national governments must establish and enforce legislation for the sustainable use of water; global NGOs and pressure groups are needed to raise public awareness of water management issues and to implement bottom-up water management schemes involving the local people. Conflicts often arise where the economic exploitation of a water resource is prioritised over conservation, so TNCs and local businesses and industries also need to be involved to successfully reduce water conflicts. Globalisation processes have made it easier to communicate and learn from other situations around the world, but have also made players more aware of their rights, and have perhaps created more conflict. (See Book 1, page 209.)

Yellow = Players, Orange = Attitudes and actions, Purple = Futures and uncertainties

Summary: Knowledge check

Through reading this chapter and by completing the tasks and activities, as well as your wider reading, you should have learned the following and be able to demonstrate your knowledge and understanding of the water cycle and water insecurity (Topic 5).

a. Explain why the global hydrological cycle is considered to be a closed system.

b. Explain the inputs, outputs, stores and flows in the global hydrological cycle.

c. Explain the difference between the global water budget and local-scale water budgets.

d. Assess how physical factors affect drainage basin inputs, flows and outputs.

e. Explain how humans may disrupt the drainage basin cycle.

f. Suggest reasons for the variations in river regimes.

g. Describe how physical and human factors affect the shape of a storm hydrograph.

h. Explain the causes of different types of drought.

i. Explain how human activities can increase the risk of drought.

j. Explain the meteorological and human causes of flooding.

k. Explain the environmental impacts of droughts and floods.

l. Assess how climate change affects global and local hydrological cycles.

m. Explain the physical and human causes of water insecurity.

n. Suggest the possible consequences of future water insecurity.

o. Explain the different approaches to managing water supply.

As well as checking your knowledge and understanding through these questions, you also need to be able to make links and explain the following ideas:

- The hydrological cycle is studied as a system with inputs, outputs, stores and flows. At a global scale it is a closed system, whereas at a local scale it is an open system.

- Hydrological cycles are being disrupted by human activities at different levels, from local deforestation to global emissions of greenhouse gases, and these have proved difficult to control.

- The hydrological cycle influences water budgets, river regimes and storm hydrographs, which in turn influence and are affected by human activities.

- Deficits within the hydrological cycle stores have complex causes with disrupted pathways, resulting in drought, which can have significant ecological impacts.

- Surpluses within the hydrological cycle can be difficult to manage, resulting in flooding, and may be caused by meteorological and/or human factors, with significant environmental and socio-economic impacts.

- There are natural and human causes of climate change operating over long timescales, which will affect global and local hydrological cycles by changing inputs, stores and outputs.

- The risk of water insecurity is increasing in the 21st century due to physical and human causes such as population increase, economic development and changes in weather patterns.

- Water is an essential resource for human activities and the natural environment, but many water resources are transboundary and therefore, as water insecurity increases, there is the potential for conflict between users within countries and internationally.

- There are different approaches to managing water supply, including hard and soft engineering schemes and legislation at national and international levels. Management methods vary in their effectiveness and sustainability due to the aims of the players involved.

Preparing for your A level exams

Sample answer with comments

Assess the significance of environmental and economic impacts caused by river floods. (12 marks)

Flooding occurs when a river's discharge exceeds the bankfull discharge, resulting in environmental and socio-economic impacts. The significance of the environmental and economic impacts will be assessed by considering the geographical scale of their impact, with national-scale impacts more significant than local impacts; also there may be differences between countries at different stages of development.

> This partial definition of flooding makes good use of storm hydrograph terms but does not consider the broadest definition, which would include groundwater flooding. Some appropriate criteria are given to assess significance but only in general terms; others could be the significance of the impacts over different timescales. Types of environmental and economic impact could also have been identified.

In December 2015 and January 2016, a series of intense storms led to widespread flooding in Great Britain. The first storm was called Desmond and caused flooding in the north-west of England, southern Scotland, north Wales and parts of Northern Ireland. The second storm was called Eva and caused flooding in Leeds, Greater Manchester and York. The third storm was called Frank and caused flooding in north-west Scotland. The storms were caused by deep Atlantic depressions, which are low-pressure systems. However, human factors also contributed, as deforestation has increased river discharges and new flood defences failed.

> An appropriate example, but this paragraph is full of information that does not answer the question. The answer does not need a detailed description of the location and causes of the storms, but rather an assessment of their impacts.

The economic impacts of these floods were considerable. In December 2015, £24 million was spent on emergency insurance payments to households and businesses, and the average cost of a flood claim was £50,000 per home. The accountancy firm KPMG estimated that the total economic cost of the floods would be approximately £5 billion. There were also environmental impacts in northern England due to a number of landslides, which were caused by the movement of saturated soil over steep gradients. Sewers overflowed, causing water pollution, and there was considerable sediment deposition on both the natural floodplain and the built environment, e.g. roads and car parks. The environmental and economic impacts of floods are more significant in developing countries, which lack effective flood management strategies to prepare, protect and respond to flood events.

In conclusion, economic impacts of flooding are more significant than environmental impacts because the natural environment quickly recovers when the flood waters retreat. The environmental and economic impacts of floods are likely to increase due to human-induced climate change. As air temperatures increase, evaporation and humidity will increase in some locations, resulting in increased condensation and precipitation and therefore increased flood risk.

> This conclusion refers to the question and makes a simple statement about the significance of the impacts, but there is limited assessment. The student makes valid comments about the role of climate change in increasing future environmental and economic impacts.

> Strong use of relevant case study data and detail here, explaining the economic and environmental impacts of the case study but no assessment of their significance. The economic impacts appear to be more significant than the environmental impacts as they have a national impact over a longer timescale. The statement about developing countries is relevant but needs case study evidence.

Verdict

This is an average answer that shows some relevant knowledge and understanding, but it could have been improved with an example from a developing country. This would have allowed more connections to be made and enabled assessment of a wider range of environmental and economic impacts. It is a partial answer to the question, but it lacks the interpretation required at A level. The candidate should have identified a range of criteria to assess the significance of the impacts, including geographical scale, duration, numbers of people affected, deaths and injuries, disruption to transport, and water and food supply issues. Environmental impacts could also have been linked to ecosystem services. This range would have provided balance and allowed a judgement to be made on which impacts are the most significant, and why.

Preparing for your A level exams

Sample answer with comments

Evaluate the contribution of large-scale schemes to increasing water security. (20 marks)

Large-scale schemes use technology to increase water security in areas where demand is high or supplies are limited. Sometimes these schemes work against natural processes to restore water supplies. The benefits and costs of schemes will be assessed to determine their contribution to increasing water security.

> This is a simplistic introduction. The criteria for the evaluation need to be identified, such as possible strengths and weaknesses linked to contrasting places. The candidate refers to benefits and costs but does not expand on these. Links need to be established between the use of large-scale technologies and the need for water security.

In China, the South-North Water Transfer Project is being built to increase water security in the north, by transferring water from the Yangtze River in the south. The project has the potential to make a significant long-term, positive contribution to the water security in Beijing and the surrounding industrial and farming areas, which have been experiencing similar levels of water scarcity to Saudi Arabia, with a 300 m fall in the water table since the 1970s. However, the project creates the risk of more frequent water scarcity in the Yangtze basin, which has experienced water shortages in the past decade due to two severe droughts, perhaps linked to climate change. Critics claim that soft engineering schemes such as rainwater harvesting could make a more sustainable contribution to increasing water security in the north, without negative impacts in the south.

> A few details of the actual engineering would have helped strengthen this paragraph, but the example is valid. There is some evaluation, but an idea of how much water will be transferred from south to north would have been useful.

Across the world, mega dams such as the Hoover Dam on the Colorado are being used to increase water security by damming large rivers to create big reservoirs providing a water supply for domestic, industrial and agricultural purposes. They can make a very positive contribution to increasing water security locally, but they often reduce water availability downstream, threatening water security in downstream communities.

> A useful continuation from the previous paragraph, but key data is missing, which makes an evaluation more difficult – a lost opportunity.

Desalination is a controversial process, which offers the benefit of producing large amounts of freshwater. By 2015, there were over 18,000 desalination plants worldwide in 150 countries, producing more than 86 million cubic metres of water per day and with more than 300 million people dependent on desalinated water for some or all of their daily needs. However, the plants are expensive to build and operate and they produce harmful waste by-products that can kill plants and animals if they are released without correct treatment. They often use lots of fossil fuels for energy and therefore produce high carbon emissions, which contribute to climate change. Sustainable water management schemes are often less controversial than hard engineering schemes as they aim to balance economic, social and environmental needs by working with local people.

> Desalination is another relevant scheme supported by data, but the focus has shifted to assessing benefits and costs, rather than evaluating the scheme in terms of water security.

In conclusion, large-scale technology can be used to increase water security, their success has been long established in developed countries such as the USA and UK, but the costs are high for developing countries and so they are not always appropriate.

> This is an evaluation statement but it does not arise from the points made in the rest of the answer. Evaluation should have been continuous.

Verdict

This is an average answer because there is insufficient evaluation. Measures of success in increasing water security were needed so that an evaluation of success could be made. The answer contains some relevant geographical knowledge and understanding but does not always apply this to the question. There should have been case study evidence in factual form to support an evaluation of each large scheme, and the candidate needed to allow time to develop a full conclusion that clearly answers the question, which is important in these 20-mark questions.

THINKING SYNOPTICALLY

Read the following extract carefully and study Figure A. Think about how all the geographical ideas link together or overlap. Answer the questions posed at the end of the article. This extract is from an article in the *Observer/Guardian* online source dated 9 February 2014 by Suzanne Goldenberg. Figure A is from an Observer/Guardian article dated 8 and 12 March 2015 by Robin McKie.

WHY GLOBAL WATER SHORTAGES POSE THREAT OF TERROR AND WAR

On 17 January [2014], scientists downloaded fresh data from a pair of NASA satellites and distributed the findings among the small group of researchers who track the world's water reserves. … That same day, the state governor, Jerry Brown, declared a drought emergency and appealed to Californians to cut their water use by 20%. 'Every day this drought goes on we are going to have to tighten the screws on what people are doing', he said. … There are other shock moments ahead – and not just for California – in a world where water is increasingly in short supply because of growing demands from agriculture, an expanding population, energy production and climate change. Already a billion people, or one in seven people on the planet, lack access to safe drinking water. … 'The middle latitudes … are already the arid and semi-arid parts of the world and they are getting drier' [said University of California, Irvine, hydrologist James Famiglietti]. On the satellite images the biggest losses were denoted by red hotspots, he said. And those red spots largely matched the locations of groundwater reserves. … The Middle East, North Africa and South Asia are all projected to experience water shortages over the coming years because of decades of bad management and overuse.

Watering crops, slaking thirst in expanding cities, cooling power plants, fracking oil and gas wells – all take water from the same diminishing supply. Add to that climate change – which is projected to intensify dry spells in the coming years – and the world is going to be forced to think a lot more about water than it ever did before. The losses of water reserves are staggering. In seven years, beginning in 2003, parts of Turkey, Syria, Iraq and Iran along the Tigris and Euphrates rivers lost 144 km^3 of stored freshwater – or about the same amount of water in the Dead Sea, according to data compiled by the Grace [satellite] mission and released last year [2013]. … But the majority of the water lost, 90 km^3, or about 60%, was due to reductions in groundwater. Farmers, facing drought, resorted to pumping out groundwater – at times on a massive scale. The Iraqi government drilled about 1000 wells to weather the 2007 drought, all drawing from the same stressed supply. In South Asia, the losses of groundwater over the last decade were even higher. About 600 million people live on the 2000-km swath that extends from eastern Pakistan, across the hot dry plains of northern India and into Bangladesh, and the land is the most intensely irrigated in the world. Up to 75% of farmers rely on pumped groundwater to water their crops, and water use is intensifying. Over the last decade, groundwater was pumped out 70% faster than in the 1990s. Satellite measurements showed a staggering loss of 54 km^3 of groundwater a year. Indian farmers were pumping their way into a water crisis.

The US security establishment is already warning of potential conflicts – including terror attacks – over water. In a 2012 report the US director of national intelligence warned that overuse of water – as in India and other countries – was a source of conflict that could potentially compromise US national security. The report focused on water basins critical to the US security regime – the Nile, Tigris-Euphrates, Mekong, Jordan, Indus, Brahmaputra and Amu Darya. It concluded: 'During the next 10 years, many countries important to the United States will experience water problems – shortages, poor water quality, or floods – that will risk instability and state failure, increase regional tensions, and distract them from working with the United States.' Water, on its own, was unlikely to bring down governments. But the report warned that shortages could threaten food production and energy supply and put additional stress on governments struggling with poverty and social tensions. Some of those tensions are already apparent on the ground. The Pacific Institute, which studies issues of water and global security, found a fourfold increase in violent confrontations over water over the last decade. …

There are dozens of potential flashpoints, spanning the globe. … Egypt has demanded Ethiopia stop construction of a mega-dam on the Nile, vowing to protect its historical rights to the river at 'any cost'. The Egyptian authorities have called for a study into whether the project would reduce the river's flow. Jordan, which has the third lowest reserves in the region, is struggling with an influx of Syrian refugees. … Prince Hassan, the uncle of King Abdullah, warned that a war over water and energy could be even bloodier than the Arab spring. The United Arab Emirates, faced with a growing population, has invested in desalination projects and is harvesting rainwater. At an international water conference in Abu Dhabi last year, Crown Prince General Sheikh Mohammed bin Zayed al-Nahyan said: 'For us, water is [now] more important than oil. … As water shortages become more acute beyond the next 10 years, water in shared basins will increasingly be used as leverage; the use of water as a weapon or to further terrorist objectives will become more likely beyond 10 years.'

… Water tensions would erupt on a more local scale. 'I think the biggest worry today is sub-national conflicts – conflicts between farmers and cities, between ethnic groups, between pastoralists and farmers in Africa, between upstream users and downstream users on the same river,' said Gleick. 'We have more tools at the international level to resolve disputes between nations. We have diplomats. We have treaties. We have international organisations that reduce the risk that India and Pakistan will go to war over water but we have far fewer tools at the sub-national level.' And new fault lines are emerging with energy production. America's oil and gas rush is putting growing demands on a water supply already under pressure from drought and growing populations. More than half the nearly 40,000 wells drilled since 2011 were in drought-stricken areas, a report from the Ceres green investment network found last week. About 36% of those wells were in areas already experiencing groundwater depletion.

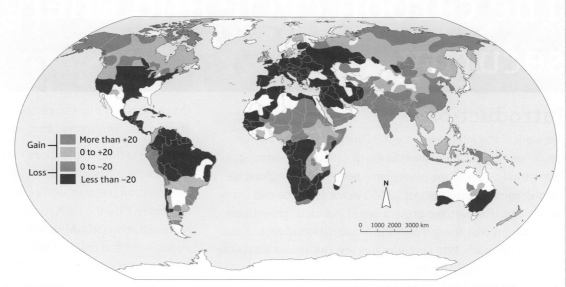

Figure A: Water stress and climate change (% change in water availability (1961–1990 average to 2050) (IPCC climate change scenario A1))

ACADEMIC SKILLS

1. Assess the effectiveness of the use of data and factual material to support the points made by the author.

2. Study Figure A. Evaluate the strengths and weaknesses of this map and suggest how you would improve it.

3. Compare this online newspaper article by Suzanne Goldenberg, US Environment Correspondent, with a later one on the same issue entitled "Why fresh water shortages will cause the next great global crisis" by Robin McKie, Science Editor. Find this article of 8th March 2015 on the Guardian website. Which is the more useful article from a geographical perspective? Explain your answer.

4. Investigate NASA's GRACE (Gravity Recovery and Climate Experiment) mission on NASA's website. Why may this form of technology be regarded as essential for evaluating global issues?

ACADEMIC QUESTIONS

5. Explain the range of geographical factors outlined in the article that are causing water stress around the world.

6. Why may the countries of the Middle East consider their water reserves to be more important than their oil reserves?

7. Assess the extent to which places in the mid-latitudes should be concerned about links between climate cycles and water shortages.

8. Suggest how a superpower, such as the USA, could be affected by global water shortages.

9. Explain why international interventions by superpowers or emerging powers may result from water conflicts.

Physical systems and sustainability

The carbon cycle and energy security

Introduction

The work of the Intergovernmental Panel for Climate Change (IPCC) has shown clearly that the Earth's climate is warming up. The natural cycles of the planet have been mostly responsible for this warming over the last 10,000 years, since the end of the last ice age. However, the present era of industrial growth and development that started with the Industrial Revolution in the late 18th and early 19th centuries has seen an almost insatiable appetite for energy. The age of coal and steam was quickly followed by the oil age, and the human reliance on fossil fuels has changed the carbon cycle balance more quickly than ever before. Coal, oil and natural gas are found in deep-level sinks where the Earth's long-term geological processes have stored carbon. But humans have 'short-circuited' the carbon cycle by extracting and burning the fossil fuels to get energy. During the combustion process, carbon dioxide is released into the atmosphere, which traps heat and enhances natural warming.

There are serious implications for humans due to this enhanced global warming, such as sea level rise, stronger storms, changing precipitation patterns, and of course higher temperatures. In January 2016 the US National Oceanic and Atmospheric Administration (NOAA) stated that the level of CO_2 in the atmosphere had reached 402.59 parts per million (ppm), the highest for 800,000 years. The 2014 IPCC reports confirmed that each decade is getting warmer: from 1880 to 2015 the world average temperature has increased by 1°C, with precipitation also increasing over northern hemisphere land areas since 1901. The World Economic Forum (WEF) *Global Risks Report* (2016) identified the 'failure of climate change mitigation and adaptation' as the greatest risk to the world.

In this topic

After studying this chapter, you will be able to discuss and explain the ideas and concepts contained within the following enquiry questions, and provide information on relevant located examples:

- How does the carbon cycle operate to maintain planetary health?

- What are the consequences for people and the environment of our increasing demand for energy?

- How are the carbon and water cycles linked to the global climate system?

Figure 2.1: Oil drilling in Prudhoe Bay, Alaska, USA. Why do you think oil companies are drilling in such challenging natural environments? What are the implications of this activity?

Synoptic links

There are many players at different levels, from the IPCC at a world scale, to national governments responsible for energy and environmental policies, industries that produce or use energy, and consumers who have energy-intensive lifestyles. The combined actions of these players will determine climate and energy futures. This topic links with coastal change through sea level rise, the submergence of coastlines and the creation of weather extremes, all of which will cause environmental refugee movements. Changes to climates will cause changes to ecosystems, such as forests, and to the cryosphere, such as the melting of ice sheets and glaciers. There are links to water supplies and the hydrological cycle due to the changing patterns of precipitation and increased evaporation. Environmental changes need global action with sovereign states working together; this has implications for sovereignty and superpower control of energy resources and their role in 21st-century international decision-making.

Useful knowledge and understanding

During your previous studies of Geography (GCSE and AS level) you may have learned about some of the ideas and concepts covered in this chapter, such as:

- Climate change and global warming.
- Rock types and the rock cycle.
- Fossil fuels and types of energy.
- Ecosystems and photosynthesis.
- Tropical rainforests.
- Coral reefs.
- Polar regions.
- Weathering and erosion.
- Physical geography processes (glaciation, coasts, rivers).
- Plate tectonics and volcanic activity.
- Weather and climate patterns.
- Energy production and consumption patterns.

This chapter will reinforce this learning and also modify and extend your knowledge and understanding of the carbon cycle, energy use and climate change.

Skills covered within this topic

- Interpretation and use of proportional flow diagrams showing carbon fluxes.
- Interpretation and use of global temperature and precipitation maps.
- Interpretation of graphs showing energy mixes of different countries and their change over time.
- Analysis of maps showing global energy trade and flows.
- Interpretation of emissions data for different energy sources.
- Interpretation and use of GIS images to analyse land-use changes over time.
- Analysis of maps produced by climate models showing future risk of water shortages and floods.
- Plotting of data on graphs to show CO_2-level change over time, including statistical analysis and rates of change.

CHAPTER 2

How does the carbon cycle operate to maintain planetary health?

Learning objectives

6.1 To understand long-term biogeochemical carbon cycle and geological processes.

6.2 To understand short-term biological carbon storage processes.

6.3 To understand why a balanced carbon cycle is important in sustaining Earth systems and why it is under threat from human activities.

Long-term biogeochemical cycles
Carbon stores

Carbon is a common element in the composition of the planet Earth. It exists in gas, liquid and solid forms, and in biotic (living) and abiotic (non-living) forms. Carbon moves between these forms (the **carbon pathway**) through natural **biogeochemical** processes over a very long geological timescale. The balance of atmospheric gases has changed over geological time because of changes in the Earth's systems and processes. During the Precambrian geological period, volcanic activity added carbon dioxide (CO_2), water (H_2O) and sulphur dioxide (SO_2) to the atmosphere at an exponential rate, forming the basic composition of the atmosphere today (Table 2.1).

Table 2.1: Major gas emissions from an active shield volcano (Hawaii) and air composition today

Gas	Percentage of total volcanic emissions (Hawaii)	Percentage of unpolluted natural air
Water vapour (H_2O)	73.5	0 to 3
Carbon dioxide (CO_2)	**11.8**	**0.03**
Nitrogen (N_2)	4.7	78.08
Sulphur dioxide (SO_2)	6.6	Trace
Oxygen (O_2)	0	20.94

When primitive bacteria such as cyanobacteria started photosynthesising 3 billion years ago, they added oxygen to the atmosphere and absorbed CO_2 from it. The higher oxygen levels that resulted allowed more complex organisms to develop about 2 billion years ago. CO_2 was dissolved in the early oceans and then stored in sedimentary rocks, a process that accelerated when land-based (terrestrial) ecosystems developed about 400 million years ago (mya). The Earth established its present carbon cycle balance about 290 mya, at the time of the Carboniferous tropical rainforests; however, this balance has been altered since about 1800 by human activities such as deforestation and the burning of fossil fuels, which release the stored carbon.

ACTIVITY

1. Which of the following increases or decreases the amount of carbon dioxide in the atmosphere?

- The evolution of land-based ecosystems.
- Emissions from shield volcanoes during eruptions.
- Evolution of large land-based animals.
- Photosynthesis by phytoplankton.
- Precipitated calcium carbonate ($CaCO_3$) in oceans.

2. Explain the contributions of volcanic activity to the composition of the atmosphere in the past and the present.

Yellow = Players, Orange = Attitudes and actions, Purple = Futures and uncertainties

Overall, crustal rocks have a small amount of carbon (320 ppm) compared to other elements, especially oxygen (466,000 ppm) and silicon (277,000 ppm). Sedimentary rocks, however, have much higher concentrations: limestone, for example, is about 42 per cent calcium carbonate by weight, while sandstone is 5 per cent and shale (mudstone) is 3 per cent. During the Carboniferous period the formation of coal stored carbon underground and reduced concentrations in the atmosphere for 300 million years. In oceans, carbon is only 0.003 per cent of the mass, while chlorine and sodium are 1.9 and 1.06 per cent respectively, with water making up 96.7 per cent.

ACTIVITY

1. Study Figure 2.2.
   a. Which is the largest store of carbon?
   b. Which is the largest flux (transfer) of carbon?
   c. Based on the data (petagrams of carbon (PgC)), explain the relative importance of terrestrial, atmospheric and oceanic stores.
   d. Based on the data (PgC/yr), explain the relative importance of terrestrial, atmospheric and oceanic fluxes.
   e. Explain the importance of geological carbon stores in balancing the carbon cycle.
2. Calculate the net ocean flux per year and the net terrestrial flux per year (in PgC/yr).
3. Using the carbon flux data from Figure 2.2, create a simple proportional flow diagram to show them. The width of the arrows should be drawn to a scale to match the data (the data range is 0.1 to 123 PgC/yr).

Literacy tip

For this section of the course it is useful to know some geological terminology. For example, the Palaeozoic was a geological era that lasted 340 million years after the Precambrian era, which was the first era up to 590 mya. The Carboniferous *period* (360 to 285 mya) was part of the Palaeozoic *era*. The present geological period is the Holocene *epoch* (from about 11,700 years ago), but because humans have had such a large influence during this period, the term Anthropocene age has also been used (*anthro* meaning human).

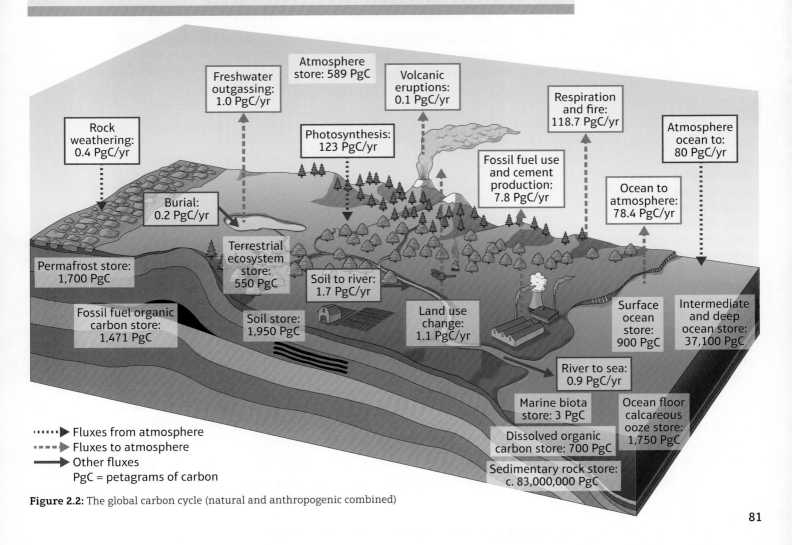

Figure 2.2: The global carbon cycle (natural and anthropogenic combined)

Literacy tip

Abbreviations may be used in academic writing. The accepted protocol is to write the term in full the first time you use it, followed by the abbreviation in brackets. Thereafter the abbreviation on its own can be used, for example mya or ppm.

Extension

Investigate the Palaeocene-Eocene Thermal Maximum (PETM), which took place about 55.8 mya when huge amounts of carbon were released into the atmosphere, causing the surface layer of the oceans to increase in temperature by between 4°C and 9°C. What caused this change, and could it happen again today?

Synoptic link

Weathering and erosion processes, such as those found at a coast or in a glaciated area, provide sediments and other elements that are deposited on the seabed to form sedimentary rocks, some of which are important for storing carbon in a geological sink. This is part of the rock cycle. (See Book 1 Chapter 2 pages 92 and 96, and Chapter 3 pages 121–123 and page 130.)

There are small **carbon stores**, such as the organic part of marine ecosystems, and large carbon stores such as in the ocean (Figure 2.2). The stores are sometimes referred to as 'sinks' or 'reservoirs'. There is an exchange of carbon between these stores over a yearly timescale, called annual fluxes, but exchanges also take place over a longer timescale.

Sedimentary carbonate rocks

Limestone rocks contain a high concentration of calcium carbonate, which is formed partly from the shells and skeletons of marine creatures, such as corals, that extracted the mineral from the seawater, and also from marine **phytoplankton** that absorb carbon through photosynthesis. Their remains accumulate on the seabed where over long periods of time they are cemented and compacted into organic limestone rock (Figure 2.3). Limestone may also form from direct precipitation of calcium carbonate from salt or freshwater (chemical limestone rock) or from the evaporation of seawater, which leaves behind calcium carbonate deposits (evaporite limestones). Limestone rocks are vulnerable to chemical weathering: rain becomes a weak carbonic acid when it falls through the air and it dissolves the calcium carbonate (a weathering process), allowing erosion processes to transfer dissolved carbon for deposition on the seabed (Figure 2.4).

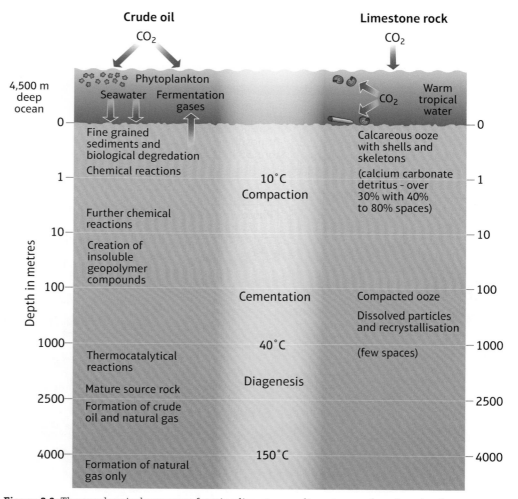

Figure 2.3: The geochemical processes forming limestone sedimentary rock and crude oil

Geological concentrations of carbon compounds (carbon-hydrogen chains, or hydrocarbons) represent the remains of living organisms deposited during the formation of sedimentary rocks. The remains of land-based plants may form peat and then coal; such conditions existed 300 mya in tropical coastal swamps that eventually formed major coalfields. The formation of coal takes tens to hundreds of millions of years, depending on temperatures and pressures. The highest temperatures and pressures concentrate carbon to produce anthracite, which has a high energy potential. The source material must contain about 2 per cent organic carbon. Anaerobic reactions convert over 90 per cent of this organic carbon into a liquid – crude oil – that moves or migrates, due to its lower density, into permeable or porous rock layers where it may be trapped within anticlines (upfolds) of impermeable rock. Crude oil contains about 85 per cent carbon and 13 per cent hydrogen. Natural gas, such as methane (CH_4), is created as a by-product during coal and oil formation, and may be trapped within the same sedimentary rock layers. Black shales also contain organic material that may become shale oil and natural gas.

ACTIVITY

1. Describe and explain how carbon is stored during the **diagenesis** of sedimentary rocks.

2. Human activity has released carbon through the extraction of coal and crude oil. Identify and describe one other human activity that releases carbon from sedimentary rocks.

Literacy tip

Always use the correct geographical terms in your written answers, when there is one to use. For example, diagenesis is the long-term process where sediments are changed into sedimentary rocks. Make sure you know the meaning of anaerobic, anticline, phytoplankton, degassed, diffusion, autotrophs and Anthropocene.

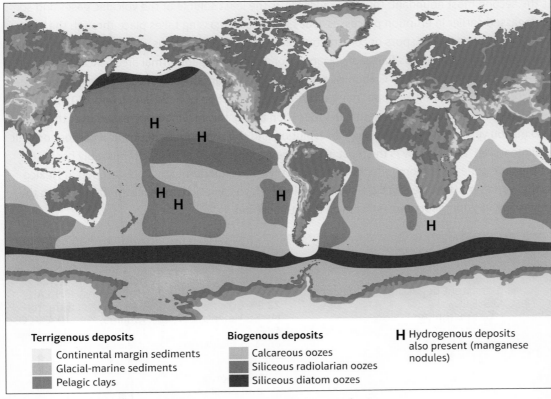

Terrigenous deposits	Biogenous deposits	
Continental margin sediments	Calcareous oozes	**H** Hydrogenous deposits also present (manganese nodules)
Glacial-marine sediments	Siliceous radiolarian oozes	
Pelagic clays	Siliceous diatom oozes	

Figure 2.4: Calcareous oozes and other oozes and sediments in ocean basins

ACTIVITY

1. Study Figure 2.4. Compare the distribution of calcareous oozes in the different ocean basins.

2. Explain this pattern.

3. What does this information suggest about the future long-term storage of carbon?

Geological carbon release

Several geological processes release carbon into the atmosphere or into the rock cycle over hundreds of millions of years (Figure 2.2). Weathering is one of these, especially chemical weathering of carbon-rich rocks such as limestone. Rainwater is a carbonic acid, absorbing CO_2 from the air and then dissolving rock minerals to form new minerals such as calcium carbonate. Rivers carry these minerals to the sea, where they are deposited and buried, eventually forming new rock (Figure 2.3). Tectonic forces may bring carbon-rich sedimentary rocks into contact with extreme heat, which causes chemical changes and the release of CO_2 back into the atmosphere. Volcanic activity at subduction zones, constructive plate boundaries or intra-plate locations causes gases, including CO_2, to be released into the atmosphere (see Table 2.1). This **out-gassing** is also common in geothermal areas, such as the CO_2-rich hot springs in Yellowstone, Iceland and New Zealand.

It is estimated that volcanic activity releases about 300 million tonnes of carbon dioxide every year, with a tiny amount of carbon monoxide. The most common volcanic gas is water vapour, but CO_2 is the second most abundant; this is because CO_2 is the least soluble of the volcanic gases and so is degassed earlier in eruption processes. Out-gassing or degassing takes place through the main vent of a volcano, at hot spots or constructive plate boundaries, perhaps with gases direct from the mantle, through porous flanks of volcanoes, or diffusion in geothermal areas, including from lakes oversaturated with CO_2. Carbon is recycled at subduction zones with carbonate rocks dragged into the mantle, creating an upper mantle carbon concentration of between 50 and 250 ppm. CO_2 is released at a shallow crustal depth at subduction zones; this may have increased over the last 180 million years as a result of biological evolution and the consequent deposition of more organic sediments. Mt. Etna in Italy is the most actively degassing volcano in Europe, probably because of the limestone and dolomite rocks from the Tethys Ocean underneath; Popocatepetl in Mexico is the world's most active degassing volcano. However, volcanic activity is intermittent and so the degassing has only short-term effects on the atmosphere.

Synoptic link

The theory of plate tectonics explains the locations of volcanic activity and plate boundary movements, such as convection and slab-pull causing subduction. These determine whether carbon is released in gaseous form, or carried into the rock cycle. Emissions of CO_2 gas may also form a localised hazard; for example high densities of CO_2 emissions in hollows can kill vegetation and animal life. (See Book 1, page 27.)

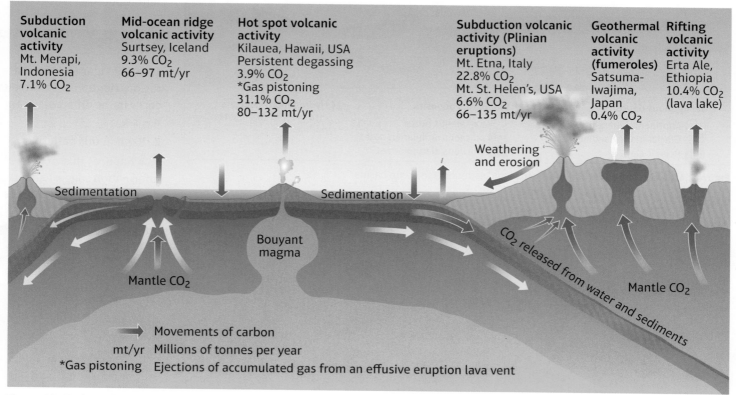

Figure 2.5: Geological processes in the carbon cycle

Short-term biological processes
Surface and deep ocean waters

The biological carbon cycle has a shorter timescale than the geological carbon cycle, measured from days to thousands of years. One of the main stores (sinks) is the ocean, and there are significant fluxes due to photosynthesis and respiration. Phytoplankton in surface waters use sunlight to turn carbon into organic matter through photosynthesis (see page 88), and carbon enters the food web via other organisms that use carbon to make their shells and skeletons (calcium carbonate) – such as corals, oysters, crabs and zooplankton. This process is sometimes called the biological carbon pump. By using carbon from the seawater, more carbon dioxide can then enter the sea from the atmosphere, although some is returned through respiration (see page 88). In this way, living organisms move carbon from the atmosphere to the shallow ocean, and then to the deep ocean when they die and sink. This carbonate material accumulates on the seabed and eventually turns into sedimentary rocks, in a process sometimes called the marine carbonate pump. Oceans are also a small natural source of methane gas (CH_4), especially shallow coastal offshore areas, where gas seeps from a nutrient-rich seabed through the ocean to the atmosphere.

Since colder water, and also deep water under pressure, can hold more gas than warm or shallow water, the Southern Ocean around Antarctica is an important carbon sink. It has been estimated that the Southern Ocean accounts for 25 per cent of the diffusion of CO_2 from the atmosphere to the oceans: the deep cold water rises very gradually to the surface, where it absorbs CO_2 from the atmosphere. Some research suggests that this process may be slowing, which would mean that the Southern Ocean may not be such an effective carbon sink in the future. The Southern Ocean shows that there is a solubility or physical carbon pump involving upwelling and downwelling currents, which move dissolved CO_2. Cold, denser seawater sinks into the deep ocean, where slow-moving deep ocean currents hold the CO_2. These deep currents eventually return to the surface, where the seawater is warmed and CO_2 is diffused back into the atmosphere. Carbon compounds are transported between the world's oceans in this way along the deep ocean conveyor known as the **thermohaline circulation** (Figure 2.6).

see page 88

see page 88

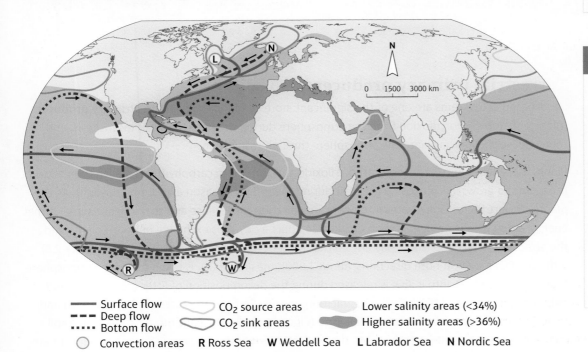

—— Surface flow	▽ CO_2 source areas
- - - Deep flow	△ CO_2 sink areas
···· Bottom flow	
○ Convection areas	**R** Ross Sea **W** Weddell Sea **L** Labrador Sea **N** Nordic Sea

Lower salinity areas (<34%)

Higher salinity areas (>36%)

Figure 2.6: The thermohaline circulation

ACTIVITY

1. Study Figure 2.5.

   a. Which tectonic situations create the largest releases of carbon dioxide into the atmosphere?

   b. Why do these situations release larger quantities of CO_2?

Extension

Investigate the following carbon fluxes (flows or exchanges) recognised in the 2014 IPCC report. Make a table to show which of these are fast fluxes (a few years to thousands of years), and which are slow fluxes (10,000 years to millions of years): erosion of land; volcanic activity; ocean surfaces; soils; lakes; sedimentary rock formation; terrestrial vegetation; atmosphere; chemical weathering.

Literacy tip

When faced with a new long word, it is sometimes possible to work out its meaning by looking for shorter, more common words within it. For example, thermohaline consists of *thermo* – meaning temperature – and *haline* – meaning amount of salt content. Breaking down a word in this way may also help you remember correct geographical terminology.

Ocean currents transfer heat energy, with large gyres moving warm tropical water towards the poles, and colder water towards the equator. These flows are linked to a world-scale thermohaline circulation, which starts in deep ocean areas and is driven by differences in density, with cold dense salty water sinking, and warmer water upwelling from intermediate depths to replace it. In this way not only are nutrients, oxygen and heat brought to the surface, but also fresh CO_2 to diffuse into the atmosphere. The thermohaline circulation is clearest in the Atlantic Ocean (Figure 2.7). Radioactive carbon-isotope dating of deep ocean water shows that it takes hundreds, even thousands, of years, for deep ocean water to return to the surface. This is because the flow rate is very slow, between 1 and 3 km per day due to its high density; however, the volume of water moving in the deep ocean is huge – about 400,000 km³.

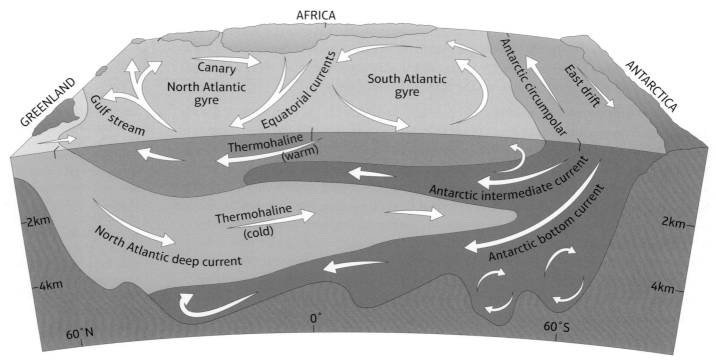

Figure 2.7: Ocean currents in the Atlantic Ocean

Terrestrial primary producers and soils

Land-based ecosystems are one of the significant stores and fluxes of carbon. **Primary producers** (plants) absorb carbon dioxide from the atmosphere during the process of photosynthesis, and then release the gas back into the atmosphere through respiration.

Plants use solar energy to change carbon dioxide and water into carbohydrates, which then allow the plants to grow; this process is greatest during the growing season and during daylight. As the plants convert the carbohydrates into energy, respiration takes place and they release some of that energy, along with water and carbon dioxide, 24 hours a day. This exchange between plants and the atmosphere has a seasonal pattern, so that winter CO_2 concentrations are higher (Figure 2.8). Although CO_2 moves in both directions, about 1,000 times more CO_2 is taken out of the atmosphere than is released back into it. CO_2 is released during the decomposition of litter (dead leaves and twigs or humus) by soil biota, including microorganisms. Carbon structures are broken down and stored in the biomass of the biota or released (Figure 2.9). High organic matter content in a soil increases its ability to store carbon, but then soil biota may release more CO_2 into the atmosphere.

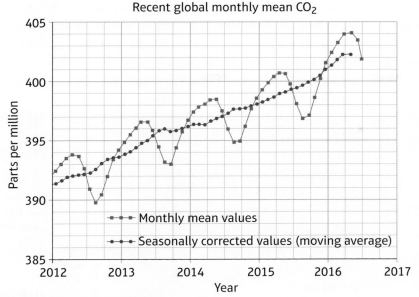

Figure 2.8: Global monthly mean CO_2

Synoptic link

Chemical and biological weathering at the coast may be affected by the changing acidity of the ocean, with the increased solution of limestone rocks. The bases of coral reefs may be dissolved, which may affect not only their ability to protect coastlines from destructive waves, but also this marine ecosystem. (See page 114.)

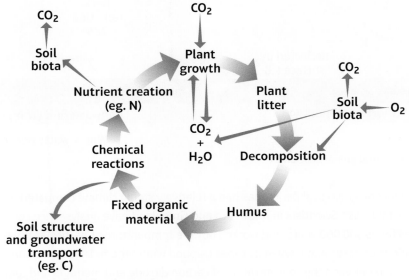

Figure 2.9: The soil carbon cycle

Maths tip

- Photosynthesis by plants: energy (sunlight) + $6CO_2$ + $H_2O \rightarrow C_6H_{12}O_6 + 6O_2$

- Respiration by plants: $C_6H_{12}O_6$ (organic matter) + $6O_2 \rightarrow 6CO_2 + 6H_2O$ + energy

- Methanogenesis in wetland soils: $C_6H_{12}O_6 \rightarrow 3CO_2 + 3CH_4$

A level exam-style question

Explain how geological processes store carbon for a long time period. (8 marks)

Guidance

Make sure that you do not just describe what happens, but offer a genuine explanation of how. Be careful also not to move away from geological processes to biological ones.

Methane (CH_4) is produced in wetland environments by anaerobic decomposition of organic matter and by termite digestive processes. The quantities are extremely small compared with carbon dioxide and vary spatially and over time, although methane is a more potent greenhouse gas.

Earth systems and human activities
The natural greenhouse effect

The Earth has sometimes been described as being in the 'Goldilocks position' in the solar system in terms of temperature: it is just right. The heat energy from the Sun has warmed the planet to a point that can support life, a natural process modified by the orbit, tilt and wobble of the Earth as it moves around the Sun over long timescales (Milankovitch cycles). As the Sun's energy enters the atmosphere, clouds reflect some of it back, so that only about half of it reaches the Earth's surface and lower atmosphere. This energy is able to pass through the denser gases of the lower atmosphere because of its short wavelength. Heat energy is then reflected back towards space from the Earth's surface, but at a longer wavelength, which means that it has difficulty travelling through the denser gases such as carbon dioxide and methane, and so the atmosphere absorbs the heat (Figure 2.10). This warms the lower atmosphere, the land and the sea, albeit with a range of temperatures from tropical to polar. Without the greenhouse gases such as water vapour, carbon dioxide and methane, the average temperature of the Earth would be −6°C rather than +15°C.

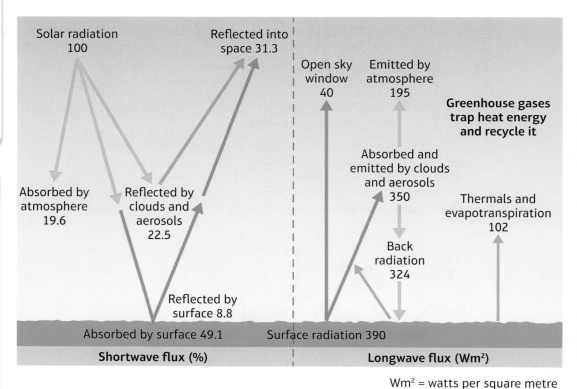

Figure 2.10: The natural greenhouse effect

The present Holocene geological period has had a relatively stable climate compared with the Earth's more distant past. Scientists in Antarctica and Greenland have analysed ice cores that extend back as far as 800,000 years, and worked out the temperature at particular times from the concentrations of gases and oxygen isotopes trapped within them. They have found a strong positive correlation between the concentration of carbon dioxide and methane and temperature, which corresponds to ice ages and inter-glacial periods in the Earth's geological history. Up until about 1750, natural geological and biological processes controlled the levels of greenhouse gases in the atmosphere, but after this date anthropogenic emissions have altered natural stores and fluxes.

Oceanic and terrestrial photosynthesis

Carbon is necessary to biosynthesis on land and in the sea, and photosynthesis is the process on which almost all life on Earth depends. Carbohydrates are created in the cells of autotrophs by combining carbon dioxide, water and light energy from the Sun. Carbon dioxide is consumed during this process and oxygen is produced:

$$\text{Carbon dioxide + water} \xrightarrow[\text{Chlorophyll}]{\text{Light energy}} \text{Carbohydrate + oxygen}$$

The process of respiration balances photosynthesis by producing energy when combining oxygen with the carbohydrate and releasing carbon dioxide and water. Photosynthesis and respiration both

Yellow = Players, Orange = Attitudes and actions, Purple = Futures and uncertainties

cycle CO_2 and O_2 between oceans, the atmosphere and the biological world. CO_2 is stored in organic matter that is buried in the soil.

Over geological time, the amount of oxygen in the atmosphere increased until it reached today's balance between oxygen, carbon dioxide released by the weathering of sedimentary rocks and water vapour.

The implications of fossil fuel consumption

Before the Industrial Revolution, the fast fluxes or exchanges of carbon were fairly constant and the carbon cycle was balanced. However, human activity has transferred considerable amounts of carbon from fossil stores, where exchanges are very slow, into the fast category, significantly disturbing the climate cycle. Whereas after an ice age it usually took the planet about 5,000 years to warm up again (by 4 to 7°C), the present warming rate is estimated to be eight times faster than previous changes.

The IPCC has been collecting evidence of natural forcing of heat, such as Milankovitch cycles, Sun cycles, El Niño, ocean oscillations and volcanic activity – and none of these account for the recent rapid warming of the Earth. What has been increasing at the same time is the concentration of CO_2 in the atmosphere, with levels now higher than ever before (when compared to bubbles of air trapped in Antarctic ice going back 800,000 years). CO_2 enters the atmosphere when fossil fuels, extracted from the geological store are combusted to provide energy for human activities such as transport and electricity production. This process has continued since the Industrial Revolution and accelerated through the oil age, changing the chemistry of the atmosphere.

Before the Industrial Revolution, the CO_2 concentration in the atmosphere had varied between 180 ppm and 290 ppm over the previous 2 million years, the highest level being after the last ice age. However, in 2016 the concentration passed 400 ppm. The IPCC estimates that between 1750 and 2011 the concentration of CO_2 in the atmosphere increased by 40 per cent, with combustion of fossil fuels adding 375 PgC to the atmosphere and deforestation adding 180 PgC. The carbon cycle has absorbed this quickly in different ways, with 240 PgC going into the atmosphere, 155 PgC into the oceans and 160 PgC into terrestrial ecosystems. Between 15 and 40 per cent of the CO_2 remains in the atmosphere for up to 2,000 years, so climate change will continue for some time to come. CO_2 emissions from fossil fuels have increased by over 3 per cent a year since 2000 – with a temporary reduction resulting from the world financial crisis of 2008 – while emissions related to land-use change have been slowly decreasing over recent decades.

ACTIVITY

NUMERICAL AND STATISTICAL SKILLS

1. Using the data (PgC) in the paragraph on this page, calculate the proportions of anthropogenic carbon emissions from fossil fuels and deforestation that have been absorbed between 1750 to 2011 by (a) the atmosphere, (b) the oceans and (c) terrestrial ecosystems. Give answers in percentages to two decimal places.

Extension

Investigate World Meteorological Organisation (WMO) press releases for the current and previous years. Make notes on any new understanding of climate and weather changes that have arisen from research.

Maths tip

Combustion formulae:

- Coal: $C + O_2 = CO_2 +$ energy (Note: 1g of C creates 3.67g of CO_2)

- Natural gas: $CH_4 + 2O_2 = CO_2 + 2H_2O +$ energy

- Petrol: $2C_8H_{18} + 25O_8 = 16CO_2 + 18H_2O +$ energy

Literacy tip

Expand your vocabulary with wider reading. For example, learn the meaning of anthropogenic, meteorological, thermohaline, Holocene, gyres, extratropical and ablation.

Table 2.2: The possible implications for the climate, ecosystems and the hydrological cycle of a 2°C temperature increase

Climate	Ecosystems	Hydrological cycle
• Atlantic and Southern Ocean thermohaline circulation may weaken, altering the transfer of heat by oceans. • Antarctic ice shelves will melt, adding more freshwater to the Southern Ocean, changing density and convection. • Extratropical low-pressure-system (depressions) tracks will move northwards with climate pattern shift. • Temperate and tropical zones may experience stronger storm activity as a result of more heat energy and moisture in the atmosphere, including more intense tropical cyclones and stronger mid-latitude westerly winds. • Precipitation will increase in higher latitudes and decrease in lower latitudes. Worldwide patterns will change, with wetter eastern parts of North and South America, northern Europe and northern and central Asia. • The Sahel, the Mediterranean, South Africa and South Asia will become drier, with drought more common in the tropics and subtropics – but some uncertainties remain, as there are multiple factors (such as El Niño). • The hottest year on record was 2015, with the average world temperature 1°C above that of the pre-industrial era; 2011 to 2015 were the hottest five years on record. The number of cold days and nights will decrease, and warm ones will increase. There have been fewer extreme cold events over the last 50 years, but more extreme heat events. • The average Arctic temperature has already increased at twice the global average over the last 200 years. Snow and ice cover will contract with the ablation of glaciers.	• Habitat changes will mean 10 per cent of land species with limited adaptability will face extinction as the climate gets warmer and either wetter or drier. Rates of extinction could rise to 15 and 40 per cent of all species, especially in high-risk polar regions. Arctic and Antarctic fauna will be affected (e.g. polar bears, caribou, emperor penguins, gyrfalcons, lemmings, snowy owls, rainbow trout). • Biodiversity will be affected as habitats shift poleward or into deeper ocean waters or higher altitudes. In north Brazil and central-southern Africa, lower rainfall and soil moisture, which cause changes to soil and oxygen, will reduce biodiversity. Tundra biome will be affected by thawing permafrost. • By 2080 shifting temperatures may reduce bird habitats in North America, affecting 314 species, with ocean and coastal habitats affected the most (e.g. Pacific golden plover), also partly because of coastal flooding and salt encroachment. • Butterflies are a good example of the shift northwards of climate zones (6.1 km per decade). Marine diversity may be lost as fish move away from warming sea temperatures and about 80 per cent of coral reefs could be bleached (e.g. the Great Barrier Reef). Acidification of seawater (carbonic acid) will threaten corals and the shells of marine creatures will get smaller and thinner. • Plant changes will lag behind animal changes, as they cannot move, and they will face pests and diseases where there is less cold weather to kill them.	• Rivers will dry up in regions where precipitation is reduced or less effective because of higher evaporation rates. • A shift of subtropical high-pressure areas northwards will cause a 20 to 30 per cent decrease in water availability in Mediterranean climate zones. • Small glaciers will disappear (e.g. in the Andes and Himalayas), decreasing river discharges once they have gone. • Humidity levels in the atmosphere will increase, consistent with what warmer air can hold. • Extreme heavy precipitation events will become common, with precipitation increases over northern-hemisphere land areas. • Uncertainty remains over increased river flooding, since multiple factors are involved, but more flash flooding is likely as a result of more intense precipitation. • Permafrost areas will thaw and add more water to Arctic rivers.

ACTIVITY

GRAPHICAL SKILLS

1. Study Table 2.3. Present the data on anthropogenic greenhouse gas emissions in a graph. Justify your choice of graph.

    a. Using your graph, analyse the changes over time (including the rate of change).

Table 2.3: Anthropogenic greenhouse gas emissions

Type of greenhouse gas	1970 (%)	1990 (%)	2010 (%)
Carbon dioxide (fossil fuels and industry)	55	59	65
Carbon dioxide (land uses)	17	16	11
Methane (CH_4)	19	18	16
Nitrous oxide (N_2O)	7.9	7.4	6.2
Fluorinated gases	0.4	0.8	2.0
Total in gigatonnes of CO_2 equivalent per year	27 Gt	38 Gt	49 Gt

What are the consequences for people and the environment of our increasing demand for energy?

Learning objectives

6.4 To understand why energy security is a key goal for countries, with most relying on fossil fuels.

6.5 To understand why reliance on fossil fuels for economic development is still the global norm.

6.6 To understand the costs and benefits of the alternatives to fossil fuels.

Energy security
Consumption patterns and energy mix

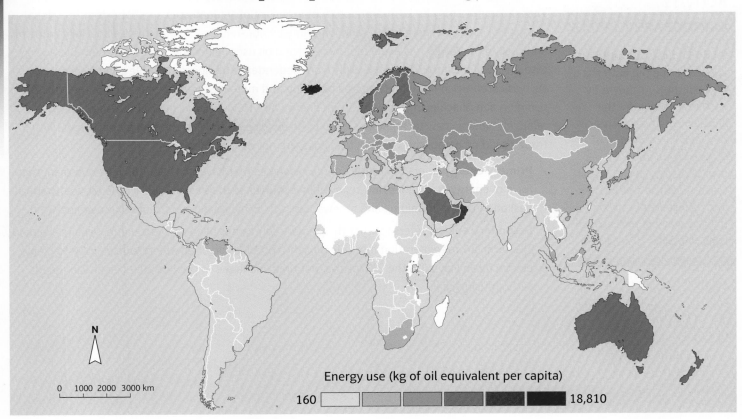

Figure 2.11: Energy use (kg of oil equivalent per capita), 2013

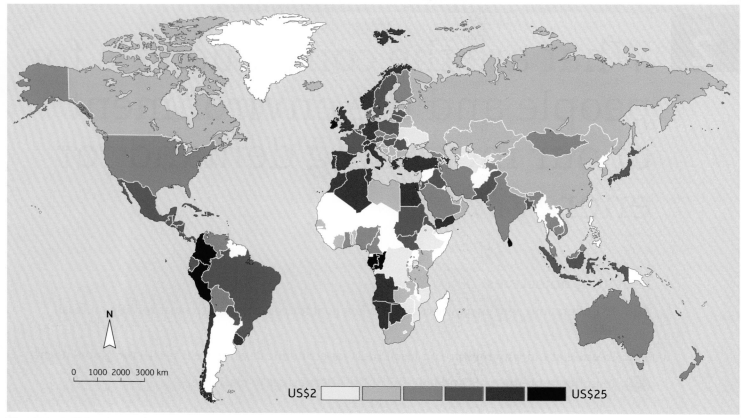

Figure 2.12: GDP per unit of energy use (PPP US$ per kg of oil equivalent), 2015

The **energy mix** of a country is the proportion of each primary energy resource it uses in a year. These resources may be both produced within the country (domestic) and imported. The global energy mix has been dominated for a long time by non-renewable fossil fuels, especially crude oil and coal. In the future it is expected that coal and oil use will decline, and that renewable energy resources will increase. This move away from dependence on non-renewable energy resources, such as carbon-rich fossil fuels, to less polluting and more sustainable energy resources, has the common aim of achieving a lower carbon energy mix. However, this energy transition may be relatively slow, as new technologies need to be developed and current **energy pathways**, with trade based on established geopolitical links, are not easy to break.

- **Primary energy** refers to natural energy resources that have not been converted into another form of energy. It includes non-renewable resources such as fossil fuels (crude oil, coal, natural gas) and nuclear (uranium), and renewables such as hydro, biomass/biofuels, solar and wind.

- **Secondary energy** refers to what the primary source has been converted into, usually electricity. The term 'power generation mix' refers to the combination of energy resources used to create electricity. In May 2016 the UK, for the first time, managed to provide all its electricity needs without using coal.

Yellow = Players, Orange = Attitudes and actions, Purple = Futures and uncertainties

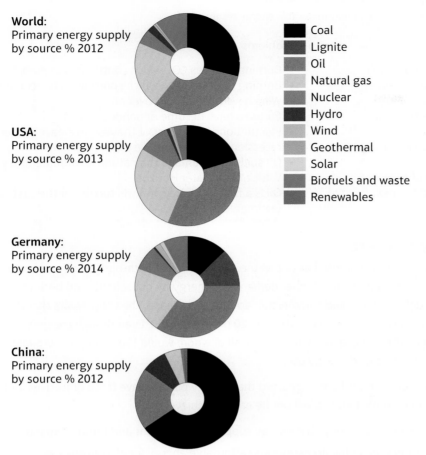

World:
Primary energy supply
by source % 2012

USA:
Primary energy supply
by source % 2013

Germany:
Primary energy supply
by source % 2014

China:
Primary energy supply
by source % 2012

- Coal
- Lignite
- Oil
- Natural gas
- Nuclear
- Hydro
- Wind
- Geothermal
- Solar
- Biofuels and waste
- Renewables

Figure 2.13: Energy mixes of the world, the USA, Germany and China, 2012–14

Access to energy resources

The energy mix of a country depends on several factors:

1. The availability of primary energy resources within the country, including levels of technology to extract and use them.
2. The accessibility of primary energy resources from outside the country.
3. The real or perceived energy needs of the country, based on lifestyles, economic development and climate.
4. Changing energy consumption patterns, perhaps linked to population or economic growth.
5. National and regional policies that affect energy production and consumption, such as legislation related to the natural environment (climate change targets) or the social environment (unemployment).
6. Cultural and historical legacies regarding energy use and geopolitical links.
7. The financial costs of each energy option.

As a result of the region's specific geological circumstances, the Middle East has the world's largest oil reserves and the largest production. Russia and the USA are also major oil producers, partly because of the high demand from their industries and transport. Some developing countries such as Venezuela, Mexico and Nigeria also have significant oil reserves; revenue from oil exports can help their economic development. The biggest trading pathways are into Europe and the USA, the largest consumers of oil due to the many vehicles and industries and many wealthy consumers who can afford to use the resource. Future problems exist over drilling in deeper water, such as the Arctic Ocean, and fracking (where gases and oil are forced out of rock layers); both are controversial because of possible negative environmental impacts such as oil spills and ground subsidence (see Book 1, page 249).

ACTIVITY

GRAPHICAL SKILLS

Study Figure 2.13. Compare the energy mixes of the featured countries. Include reference to the balance of non-renewables and renewables and the proportion of fossil fuel use.

Table 2.4: The benefits and problems of crude oil exploitation (a non-renewable energy resource)

Benefits	Problems
• Can be sold to bring wealth to a country and people • Can support industry, even in an economic recession • Easy to transport (e.g. pipeline or tanker) • Large worldwide demand and thus a very tradable commodity • Makes people's lives easier, especially allowing fast air and land transport • Has brought a culture based on freedom of movement • Money from oil can be invested in finding the next flexible energy resource	• Burning releases CO_2, which contributes to global warming • Burning releases NOx, which contributes to acid rain • Wars have been fought over oil • Cultures have become dependent on oil • Fluctuations in the cost of the resource have caused recession and inflation • Oil spills may damage the natural environment (e.g. Exxon Valdez, Deepwater Horizon) • Oil is a finite resource and will run out in the 21st century (estimated 2061)

Energy players

Population growth is regarded as one of the main causes of anthropogenic greenhouse emissions, because more people mean a higher demand for energy by households and businesses, and transport, deforestation and commercial farming all increase. The UN predicts that the world population will reach 9 billion by 2030 and 10 billion by 2050. Even though the growth rate is slowing, billions of people are being added, all of whom would like an improved quality of life which often leads to increased energy use.

The World Energy Council has suggested that energy players have three objectives, but recognises that these may conflict and so will not be easy to achieve:

- **Energy security:** ensuring that energy supply meets current and future demand.
- **Energy equity:** ensuring accessible and affordable energy for all countries.
- **Environmental sustainability:** ensuring efficient use of energy and use of renewable sources, so reducing pollution and moving towards lower greenhouse gas emissions.

During the 'oil age', **energy security** has been threatened by geopolitical tension, mainly in the Middle East since the 1973 Arab-Israeli war, more recent wars in Iraq and elsewhere, tensions between the USA and Iran, and the Arab uprising in North Africa and the Middle East which created political factions and terrorism. This attracted the involvement of the big energy players such as the USA and Russia, who wish to protect the energy pathways and supplies (Table 2.5).

Figure 2.14: Gazprom gas operations and the Trans-Alaskan oil pipeline (jointly run by BP, ConocoPhillips and ExxonMobil)

Yellow = Players, Orange = Attitudes and actions, Purple = Futures and uncertainties

TNCs that supply energy wish to keep trade flowing through the established pathways (Figure 2.14) and to control prices. Governments gain the revenues from state-owned energy TNCs such as Gazprom (Russia), Petronas (Malaysia) and Saudi Aramco (Saudi Arabia), and in most countries governments impose high taxes on energy use, and use the revenues to help develop the country. There is a conflict with environmental groups concerned about damage to natural ecosystems and the slow progress that large energy TNCs are making towards 'green energy'. While the number of major oil spills from tankers has decreased since double hulls were introduced in the 1990s, there are still risks of spills from pipelines in **fragile environments** and from extraction activities in deeper waters, including the Arctic Ocean. Governments are sometimes torn between ensuring a constant supply of energy and environmental considerations.

Many governments in Europe have been changing their energy mixes towards renewables, partly to meet ambitious CO_2 reduction targets. The UK government has supported solar and wind energy, albeit inconsistently. Germany has invested greatly in renewables since a policy change in 1990 and then again since the Fukushima nuclear disaster of 2011 (Figure 2.13). When China decided to develop its industries along more 'western' economic lines, China's need for energy jumped and the government sought resources in Africa. This increased global energy prices, because they were competing to buy the available supplies. When China's growth slowed significantly in 2015, energy prices fell, demonstrating their volatility. A key group of energy players is the scientists and engineers, whose research and development into alternative energy resources and more efficient technologies can change future energy mixes and security.

CASE STUDY: State-controlled energy companies

In order to improve energy security, a government may take over a company in order to control production or to gain maximum revenues. One example is Coal India Ltd, which is the world's largest coal producer, based in Kolkata, India. India's coal mining industry was nationalised in 1972/73, and in 2011 it became an official public service utility. It has 80 mining areas and is responsible for 5.9 per cent of the world's coal production and 80 per cent of India's coal production. Increased efficiency has seen a decline in workers, from 510,600 in 2002 to 352,300 in 2014, but an increase in production over the same period from 260.69 million tonnes to 494.24 m tonnes. India is the world's third-largest consumer of coal behind China and the USA, using it in the steel industry, in thermal power stations and in other industries associated with an emerging industrial economy. The government gained from the US$12.9 billion revenues in 2013/14.

Another example is the Russian company Gazprom, created in 1989 and put under state control in 2000. One of the largest producers of natural gas in the world, it has reserves in Siberia and the Ural and Volga regions of Russia. Gas is a major Russian export, helping its trade balance and earning revenues to help the government develop the country. An extensive pipeline network moves the gas to its domestic and international customers. Europe is an important customer, especially Germany, which imported 40 billion m³ of natural gas in 2013. The Ukraine also used 44.8 billion m³ in 2011. However, political conflict and terrorism in the Ukraine and Turkey have made the pipeline pathways vulnerable. Russia and Ukraine's dispute over gas supplies in 2005 increased tensions, and was part of the reason for the territorial conflict from 2013. In 2014 Gazprom's sales and profits declined, partly as a result of the EU and US sanctions imposed in protest at Russia's military involvement in the Ukraine, and also the falling value of the rouble; sales of natural gas to the Ukraine fell by 65 per cent.

ACTIVITY

1. Analyse the roles of the different energy players.

2. Suggest which player or players have the most important role in determining energy security in the future. Explain your choice(s).

Extension

Indonesia joined OPEC in 1962 but suspended its membership in 2009. It then reactivated its membership on 1 January 2016. Investigate the reasons why Indonesia changed its membership: were they political, economic, cultural or environmental, or a mixture of these? Evaluate the strength of these influences.

CASE STUDY: The Organisation of Petroleum Exporting Countries (OPEC)

OPEC is an intergovernmental organisation (IGO) set up in 1960 to coordinate member countries' oil policies. In 2016 there were 14 member countries in three continents, with Saudi Arabia the most important player. OPEC's aim is to create a stable income for oil-producing nations by controlling output and prices when selling to oil-consuming nations. In 2003 the OPEC General Secretary, reflecting on OPEC's 21st-century role, stated that oil would be important for decades and that the organisation was dedicated to ensuring that it was available to all countries through a stable oil market. It achieves this by increasing supplies from some members when supplies are disrupted from others, such as in Venezuela (strike), Iraq (wars) and Nigeria (civil conflict). OPEC controls over 40 per cent of the world's oil supply, and member countries have nationalised their oil so that they are able to control output and apply export quotas on member countries to reduce supply if prices go too low.

OPEC's mission is to help its members, but also the whole world; between 2000 and 2020 there will be a predicted investment of US$209 billion to expand production and form closer ties with non-OPEC oil and energy organisations. Some regard OPEC as a cartel that fixes prices and dictates the rules of supply: in the everyday commercial world such operations would be illegal, but OPEC decisions are national government actions, and oil is so important to world economies that no action has been taken. However, the oil price has fluctuated a great deal, which suggests that OPEC does not have complete control of the world market for oil, especially during times of world recession or increased demand from emerging countries. Consumption patterns also seem to operate independently of price (Table 2.5).

ACTIVITY

NUMERICAL AND STATISTICAL SKILLS

Table 2.5 provides data on the price of oil and consumption trends. As well as actual values, the table includes an index: this uses 1955 as a common starting point for both sets of data, which allows trends to be seen more easily, especially as the data has contrasting units.

a. Calculate the four missing index values. To do this, multiply the value for the year by 100 and divide by the value of the start year.

b. Use the index values to create a suitable graph to show the trends in prices and consumption between 1955 and 2015. Compare the trends shown by your graph, perhaps using best-fit lines.

c. Suggest a suitable statistical technique that would help to prove whether there is any correlation between oil price and consumption. What would your null and alternative hypotheses be for this statistical exercise?

Table 2.5: Global oil consumption and prices, 1955–2015

Year	Average oil price (US$ per barrel, 2014 values)	Index	Oil consumption (millions of barrels a day)	Index
1955	17.05	100	15.377	100
1960	15.17	88.97	21.471	139.63
1965	13.49	79.12	30.806	200.34
1970	10.97		45.348	294.91
1975	50.74	297.60	54.327	353.30
1980	105.81		61.233	398.21
1985	60.64	355.66	59.247	385.30
1990	42.97	252.02	66.737	
1995	26.43	155.01	70.322	457.32
2000	39.17	229.74	76.868	499.89
2005	66.09	387.62	84.411	548.94
2010	86.31	506.22	87.867	571.42
2015	50.75	297.65	93.700	

Yellow = Players, Orange = Attitudes and actions, Purple = Futures and uncertainties

CASE STUDY: National governments – the Danish energy model

Denmark has a diverse energy mix, with a significant proportion of combined heat and power (including district heating) and wind energy. Wind turbines produce about 40 per cent of the country's electricity. On a very windy evening, 9 July 2015, Denmark produced 116 per cent of its national electricity needs from wind turbines and was able to export energy (1,030 MW) to neighbouring countries. Since 1990 Denmark has reduced its greenhouse gas emissions by 30 per cent, and plans to reach 40 per cent by 2020. The Danish government has made agreements with Norway, Sweden and Germany (NordPool) to ensure energy security through transfer pathways from different sources, such as Norway's hydroelectric power. Denmark is also passing on its expertise and experience to the governments of China, Mexico, South Africa and Vietnam.

The Danish government believes in a holistic regulatory approach to improving energy supplies and reducing reliance on fossil fuels, using building codes, energy efficiency, investment in renewables, climate change laws, tax incentives and energy savings in the public sector (Table 2.6). All political parties are committed to this long-term energy policy, which it is estimated will cost US$530 million by 2020. Denmark's targets are to:

- generate half of its electricity from wind power by 2020
- eliminate coal from power generation by 2030
- produce all electricity and heat supply from renewables by 2035
- be a fossil-fuel-free society by 2050 (Danish Energy Agency 2015).

Table 2.6: Denmark primary energy production (petajoules), 1990–2015

Resource	1990	2000	2005	2010	2012	2013	2014	2015
Crude oil	255.96	764.53	796.22	522.73	429.14	373.36	349.63	330.66
Natural gas	115.97	310.31	392.87	307.42	216	179.27	173.65	172.57
Waste, non-renewable	6.97	13.68	17.01	17.15	16.8	16.86	17.42	17.49
Renewable energy	45.46	76.02	105.58	131.31	130.07	135.25	138.97	142.82
Total (PJ)	424.36	1164.53	1311.68	978.61	792.01	704.75	679.68	663.55

Table 2.7: Denmark observed energy consumption (petajoules), 1990–2015

Resource	1990	2000	2005	2010	2012	2013	2014	2015
Oil	343.47	369.57	348.3	315.84	287.87	278.22	273.14	282.65
Natural gas	76.1	186.27	187.54	184.98	146.49	138.09	119.25	120.97
Coal	254.84	165.92	154.99	163.7	106.24	135.62	107.49	74.15
Waste, non-renewable	6.97	13.68	17.01	17.15	16.8	16.86	17.42	17.49
Renewable energy	45.46	78.5	121.88	167.94	180.2	186.57	191.75	195.72
Net imports of electricity	25.37	2.39	4.93	-4.09	18.77	3.89	10.28	21.28
Total (PJ)	752.33	816.48	834.79	845.7	756.54	759.4	719.47	712.42

Extension

Identify when there were significant changes to the oil price shown in Table 2.5 and your graph. Use the internet to investigate the reasons for these changes and add these to your notes, perhaps by annotating your graph. Evaluate the relative importance of these reasons.

ACTIVITY

NUMERICAL AND STATISTICAL SKILLS

1. Study Tables 2.6 and 2.7.

   a. Assess the ability of Denmark to meet the targets set by its energy policy.

   b. Using Excel or similar software, compare the production and consumption data by energy source for Denmark between 1990 and 2015 using numerical or graphical methods.

Fossil fuels and economic development
Fossil fuels: supply and demand

Fossil fuels are found only where geological conditions in the past favoured their formation and entrapment. The presence of coal was important in driving the Industrial Revolution and determining the location of traditional heavy industry. From the mid-20th century, countries with oil reserves benefited from the wealth resulting from this valuable resource. During the last quarter of the 20th century and into the 21st century, production of oil, coal and natural gas steadily increased, with a dip in coal production in 2014 (Table 2.8). Emerging economies such as India and China have been responsible for increased global energy consumption since 2000, while in 2014 EU energy consumption fell to its lowest level since 1985.

This was the situation in 2014:

- Oil accounted for 32.6 per cent of global energy consumption, but was in slow decline (refining was at 80 per cent capacity), although production and refining remained high in the USA and the Middle East. The oil trade grew mainly because of China, which in 2013 had overtaken the USA as the world's largest importer of oil.

- Natural gas accounted for 23.7 per cent of primary energy consumption, with the USA having the world's largest increase, but with declines in Russia, Netherlands and the EU as a whole. Trade fell in 2014, including pipeline shipments, which fell by 6.2 per cent – the largest decline ever recorded.

- Coal's share of global primary energy consumption declined to 30 per cent, mainly through slower Chinese economic growth. The Ukraine and the UK had significant decreases in consumption, while India posted the largest increase (11.1 per cent).

Table 2.8: Oil data for 2014

World region	% of proven reserves*	% of production	% of consumption
Africa	7.6	9.3	4.3
Asia Pacific	2.5	9.4	33.9
Europe and Eurasia	9.1	19.8	20.4
Middle East	47.7	31.7	9.3
North America	13.7	20.5	24.3
South and Central America	19.4	9.3	7.8
Top countries	Venezuela (17.5) Saudi Arabia (15.7) Canada (10.2) Iran (9.3) Iraq (8.8) Russia (6.1) Kuwait (6.0) UAE (5.8)	Saudi Arabia (12.9) Russia (12.7) USA (12.3) Canada (5.0) China (5.0) Iran (4.0) UAE (4.0) Iraq (3.8) Kuwait (3.6) Venezuela (3.3)	USA (19.9) China (12.4) Japan (4.7) India (4.3) Russia (3.5) Brazil (3.4) Saudi Arabia (3.4) Germany (2.6) South Korea (2.6) Canada (2.4)

* **Reserves** include Alberta (Canada) tar sands and Orinoco Belt (Venezuela) tar sands

Yellow = Players, Orange = Attitudes and actions, Purple = Futures and uncertainties

ACTIVITY

CARTOGRAPHIC SKILLS

1. Using the data in Table 2.9, construct a proportional flow-line map on an outline map or GIS image of the world. Make sure that the lines are arrows with a width scale proportional to the percentage of world oil trade (perhaps 1 mm = 1%).

Table 2.9: Largest exports of oil in 2014

Countries or world regions	% of total trade in oil
Russia (former Soviet Union) to Europe	10.63
Middle East to China	6.16
Canada to USA	6.02
Middle East to Japan	5.63
Middle East to India	4.34
Middle East to Europe	3.64
Middle East to USA	3.34
South and Central America to USA	2.84
West Africa to Europe	2.81
USA to South and Central America	2.28

2. Describe the pattern shown by your completed map.

3. Using Tables 2.8 and 2.9 and other information, assess the strength of the link between the level of economic development of a country and its use of oil and other fossil fuels.

Energy pathways

Energy resources inevitably need to be transported from their source areas to areas of demand; this takes place by pipeline (oil and gas) over land, by bulk carrier ship (coal, uranium), tanker ship (oil and LNG) by sea, or via underground electricity cables. Pathways between producers and consumers may be complex, because they must move through different natural and human environments (Figures 2.15 and 2.16). Natural obstacles include vast distances and difficult terrain such as the tundra in Alaska (Trans-Alaskan pipeline; see Figure 2.14) and Siberia (Trans-Siberian pipeline). Extracting oil from the tar sands of Alberta in Canada or from deep water in the Gulf of Mexico – especially in hurricane season – also brings challenges. Human obstacles, including technical problems such as pipeline leaks, could temporarily disrupt pathways; supplies might run out, for example North Sea gas; or supplies might be diverted for greater profit, such as to China. Political tensions and disagreements may lead to pathways being blocked (at 'choke points'). These choke points are subject to change, but many fossil fuel resources are in unstable locations such as the Middle East and Russia, the South China Sea and the Red Sea. Examples include the Iraq wars (1990s), Somalian pirate activity, and Russian/Ukraine disputes (from 2004). Embargoes and sanctions also disrupt production and supplies, such as from Iran and Russia. Pathway disruption can have socio-economic and political consequences, such as recession and job losses as well as energy shortages affecting lifestyles, and even armed conflicts to secure pathways and resource locations. These pathways should become less important as energy mixes change, because many renewable resources are ubiquitous.

Extension

Investigate the effects that coronal mass ejections (CMEs) may have on electricity grids. For example, in 2012 the US Federal Emergency Management Agency (FEMA) completed a study.

Figure 2.15: A selection of oil pipelines to Europe

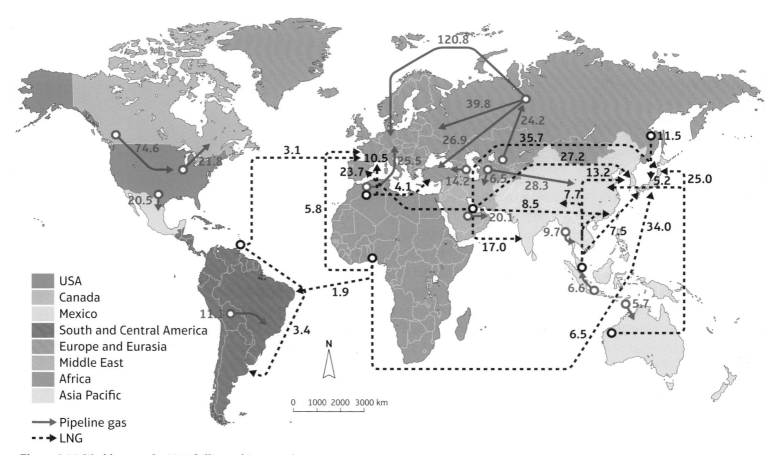

USA
Canada
Mexico
South and Central America
Europe and Eurasia
Middle East
Africa
Asia Pacific

→ Pipeline gas
⇢ LNG

Figure 2.16: World gas trade, 2014 (billion cubic metres)

Yellow = Players, Orange = Attitudes and actions, Purple = Futures and uncertainties

ACTIVITY

1. Study Figures 2.15 and 2.16.

   a. Identify locations where pathways could be or have been disrupted (i.e. the 'choke points').

   b. Explain the likely reasons for these disruptions, perhaps using historical evidence.

   c. Suggest what the players need to do in order to secure these pathways for the near future.

Unconventional fossil fuel resources

As fossil fuel reserves, notably oil, will soon be used up, especially as developed and emerging economies increase consumption, the search for other sources has intensified. This includes exploring deeper ocean areas such as the Gulf of Mexico, offshore Brazil and the Arctic Ocean, tar sands in Canada, and shale gas in the USA.

Deep-water oil is found in the Gulf of Mexico, where one of the largest and deepest oilfields is Atlantis (sea depth = 2,150 m). Although the number of new oil/gas platforms in the Gulf of Mexico has decreased, they are getting larger and being located in deeper water. In 2010 only 14 rigs were in deep water, but by 2014 there were 63 (Figure 2.17). Drilling in deep water is not easy and there are hurricanes as well as long distances to shore for undersea pipelines; despite this the USA's 2012–17 leasing plan extended the Gulf of Mexico area and also opened up the Alaskan Arctic, with the aim of reducing imports. BP's Atlantis platform (2007–22) produces 200,000 barrels of oil and 5.1 million m³ of gas a day. Since the Deepwater Horizon incident (2010) highlighted the risks involved with deep-sea drilling, new regulations and technological improvements have been made,

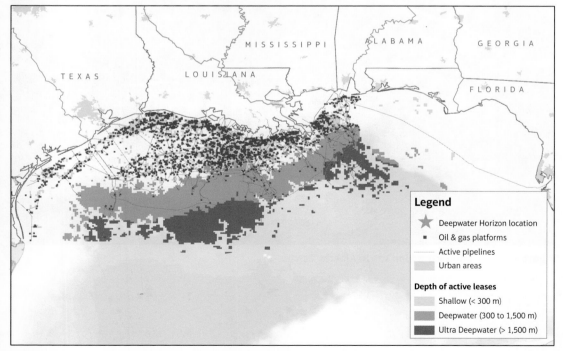

Figure 2.17: Deep-sea oil production and pipelines in the Gulf of Mexico

> **A level exam-style question**
>
> Study Figure 2.15. Suggest why the energy pathways shown may be prone to disruption. (6 marks)
>
> **Guidance**
> In 'suggest' questions, you should use the information to give you clues and prompts for ideas. You are not expected to know the area or place, although you may have studied it. You must provide a reasoned explanation of why pipelines or shipping routes may be disrupted, justifying your reasons (e.g. political or military conflict, piracy, extreme weather or policies to control supply).

> **Extension**
>
> Investigate the 2010 Deepwater Horizon incident from different perspectives. What effects did it have on the different players and on the natural environment?

Extension

There is a wealth of information about shale gas and fracking on the British Geological Survey (BGS) and Department of Energy and Climate Change (DECC) websites, such as the BGS's 2010 *Shale Gas Report* and the DECC's 2016 updated *Guidance on Fracking*. Investigate these websites to add to your understanding and notes.

such as remotely operated underwater robots that can seal a leak within 45 seconds. However, environmental groups believe that cost cutting (due to recession and low oil prices), along with complacency and untested technologies, could lead to future large oil spills.

Tar sands are a mixture of clay, sand, water and bitumen (very viscous oil). The oil is too thick to be pumped from the ground. Instead, it must be taken from an open pit, or strip-mined. To recover the oil it must be separated from the sands using very hot water diluted with lighter hydrocarbons. When oil prices are high – at least US$120 a barrel – it becomes economical to extract oil from the sands. The largest reserves and production are in Alberta, Canada (Figure 2.18) and in Venezuela (Orinoco Belt). Venezuela has more oil reserves than Saudi Arabia, which in 2012 raised its average GDP per capita (PPP) to US$12,772 and allowed the government to keep petrol prices very low.

Issues with tar sands include the large amounts of energy needed (heating and pumping) even before the oil is used; this results in an estimated contribution to global warming three times higher than conventional oil. The mining process leaves scars on the landscape – which are refilled with the sands once the oil has been extracted. There are also impacts on local wildlife and people, leaks into water supplies from tailings ponds, and infringements of indigenous peoples' treaty rights over fishing and hunting grounds.

Figure 2.18: Tar sands extraction site in Alberta, Canada

Yellow = Players, Orange = Attitudes and actions, Purple = Futures and uncertainties

Oil shale contains solid bituminous material (kerogen) that formed when silt and organic matter were heated and pressurised under water, but not enough to turn it into oil (see Figure 2.3). Oil shale can be mined, but must be heated to a high temperature to release the oil; this is expensive and releases greenhouse gases. The USA has large reserves in the Green River Formation of rocks in Colorado, Utah and Wyoming (perhaps 800 billion barrels), but currently there is little commercial development anywhere in the world, although in April 2016 it was reported that an Australian company would be expanding its oil shale exploration in Alaska. Environmental impacts include disturbance of land and vegetation cover, disposal of the waste after processing (over 1 tonne of waste rock for every barrel of oil), over-use of water resources (two barrels of water for every barrel of oil produced), and air and water pollution. However, the Royal Dutch Shell Company has developed a plan to heat shale underground while surrounded by a 'freeze wall', so that the kerogen seeps out into drilled holes for collection.

Shale gas is natural gas (mostly methane) trapped inside impermeable shale rocks, so it cannot be extracted by normal drilling. Instead, the rock must be broken to free the gas, which is done by hydraulic fracturing, more commonly called 'fracking'. This involves forcing water mixed with chemicals into the shale rock so that the rock splits apart and any gas flows into a prepared well where it is concentrated enough to recover. Fracking can involve horizontal as well as vertical drilling, which reduces the impact on the ground surface, but many drill sites are needed because the gas is dispersed. Other negatives include lowered local groundwater levels, possible chemical contamination of groundwater and surface water, methane gas leaks (not all the gas can be captured) adding to the greenhouse effect, and a risk of minor earth tremors and ground subsidence from altering the rock underground. Shale gas is still a fossil fuel, releasing carbon dioxide when combusted. Positives include increasing the energy reserves of a country and reducing the need for imports, and it is a flexible energy source; for example in 2015 in the USA natural gas overtook coal as the main generator of electricity (Figure 2.19). In addition, the carbon footprint of shale gas is about half that of coal and lower than liquefied natural gas (LNG), which is transported around the world.

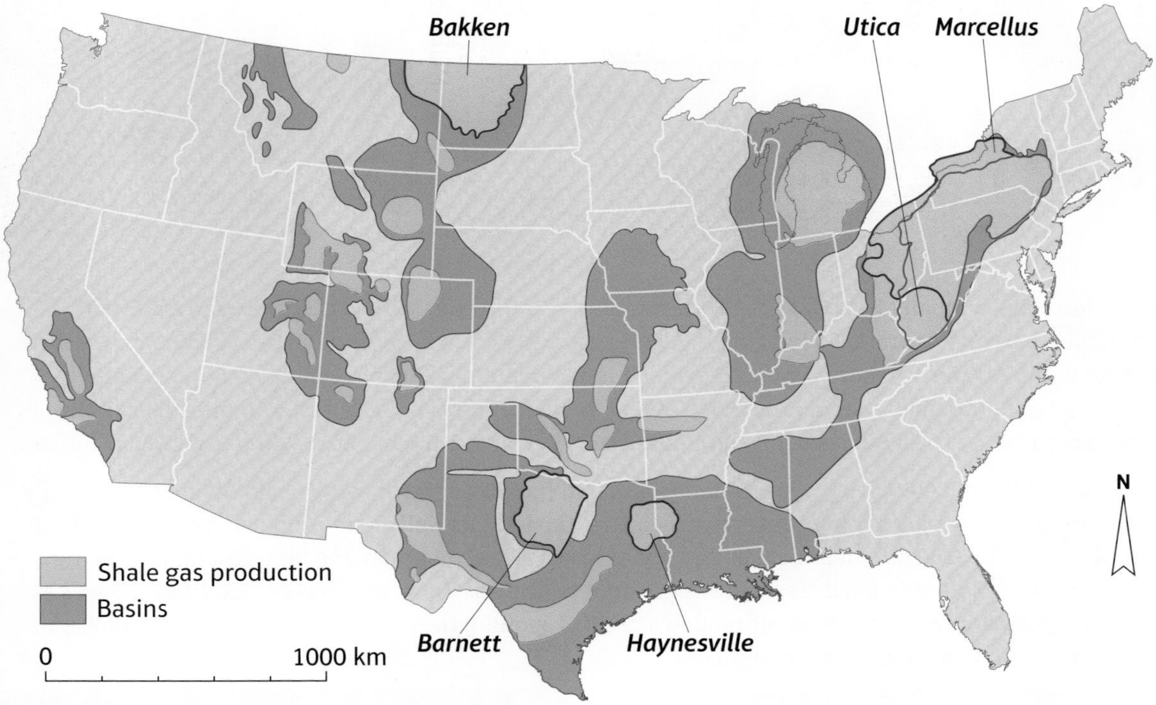

Extension

- Investigate the development and issues surrounding the development of shale gas in the USA (see Figure 2.19) and make notes to produce a case study.

- Read the case studies about tar sands in Alberta and deep-sea oil in Brazil. Research the internet to update your notes on the current situations.

Figure 2.19: USA shale gas areas

CASE STUDIES: Canadian tar sands and Brazilian deep-water oil

Tar sands in Canada

In Alberta, Canada, tar sand reserves of 166 billion barrels of oil exist in three areas covering 142,200 km²: Athabasca, Peace River and Cold Lake. There is surface mining over 3.4 per cent of the area, mostly in Athabasca, and other reserves have to be extracted by drilling. By 2013, 895 km² had been disturbed, including 220 km² of tailings ponds. Production increased from 0.1 million barrels a day in 1980 to 2.3 million in 2014, and is predicted to reach 4 million by 2030. It is estimated that between 1999 and 2013 US$201 billion was invested in developing tar sands, but US$4 trillion is expected to go into the Canadian economy between 2015 and 2035, with US$1.2 trillion in taxes and royalties to the government, and 151,000 direct jobs (2013) have been created.

The Canadian tar sands operations are regulated by national and provincial bodies and regulations, such as Joint Oil Sands Monitoring, the Alberta Environmental Monitoring, Evaluation and Reporting Agency and the Specified Gas Emitters Regulations – which set a target of a 20 per cent reduction in greenhouse gas emissions in Alberta by 2017. There is also environmental monitoring by the Wood Buffalo Environmental Association – which checks air, water and human health – and the Lower Athabasca Regional Plan – which protects air and water quality and sets land aside for conservation. Pollution remains a concern: in 2013 8.5 per cent of Canada's greenhouse gas emissions came from tar sands operations, with a predicted increase to 14 per cent by 2030, although there is a plan to capture and store carbon emissions (the Alberta Carbon Trunk Line). The tailings ponds contain toxic chemicals and have killed migrating waterfowl, and concern remains about leakages into groundwater and rivers and effects on the health of local people. An Oil Sands Community Alliance has been set up to provide services and facilities for local people, especially First Nation (indigenous) communities, but many regarded the US$18 million donated in 2013 by the oil business as too little.

Deep-water oil in Brazil

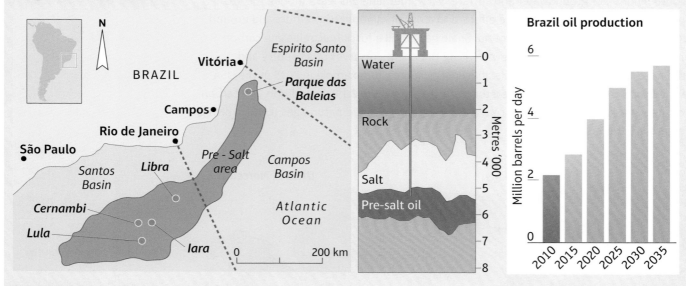

Figure 2.20: Brazilian deep-water oil production

With help from China, the Brazilian state company Petrobras began developing a deep-water oilfield discovered in 2006, with estimated reserves of 50 to 80 billion barrels (Figure 2.20). Pilot wells began producing oil in 2010, and in 2015 production reached 885,000 barrels a day. The oilfield is 200–300 km offshore, where the seabed is at a depth of 2,000 m. Drilling conditions are therefore difficult, with the addition of flammable gases and rock salt layers that flex. Costs are high; for example a special fleet of storage and offloading ships was required at US$2 billion each, and investments between 2014 and 2018 will reach US$221 billion.

At the present time only 13 per cent of Brazil's electricity is generated from fossil fuels, and the oil would help diversify its energy mix and provide greater energy security in case drought reduces hydroelectric capacity, as well as creating jobs and income from exports. However, by 2016 Petrobras was US$100 billion in debt as a result of lower global oil prices, the high cost of developing the oilfield and political corruption that resulted in job losses and the arrest of politicians. New construction activities, such as a refinery at Comperj, were delayed. There are concerns about oil-related activities spoiling the environment (Guanabara Bay) and causing oil spills, and also about the maintenance of safety equipment.

Yellow = Players, Orange = Attitudes and actions, Purple = Futures and uncertainties

Alternative energy resources
Renewable and recyclable energy

In 2014 the IPCC concluded that, in order to drastically reduce enhanced climate change, the share of energy production that is accounted for by renewable energy sources needs to be trebled, and must dominate world energy supplies by 2050. Renewables accounted for 3 per cent of global energy production in 2014, and continue to increase their share as action is taken to reduce dependence on fossil fuels and to mitigate climate change through reduced CO_2 emissions. Renewables represent 6 per cent of electricity production, especially in China, which accounted for all of the global increase in HEP production in 2014. Wind energy continues slowly to expand, but solar energy increased by 38.2 per cent in 2014 (Figure 2.21). But the development of renewable and recyclable energy is uncertain: for example nuclear energy is an increasing source of energy in South Korea, China and France, but decreasing in Japan, Belgium, Germany and the UK; and hydroelectric energy is suffering from climate uncertainties; it had an increasing share of world energy production (6.8 per cent) in 2014, but drought conditions reduced outputs in Brazil and Turkey.

The greatest **solar energy** potential is in equatorial areas where sunlight and heat are most focused and intense, and also in desert areas where skies are clear, for example the Saharan and Great American deserts (Table 2.10). Areas with potential include inland temperate regions that are not as affected by cloud cover, such as Germany, and high-altitude areas where sunlight is more intense, such as the Tibetan plateau. Heat from the Sun can be used to heat water, or photoelectric cells can generate electricity directly. Germany has invested the most in solar energy, producing nearly 25,000 gigawatts of electricity. It can do this because it is a wealthy country able to invest in research and development and has a strong Green Party. It is aiming to cut CO_2 emissions, and the 2011 Fukushima nuclear disaster in Japan (caused by a tsunami) encouraged them to develop alternatives.

Table 2.10: The benefits and problems of developing solar energy

Benefits	Problems
• Safe, clean and non-polluting once made and installed. • Renewable, so a sustainable source of energy. • Can be used by poorer countries. • Links well with other sources of energy. • Flexible and modular, so can be used on roofs of buildings or developed into a solar power station.	• Not enough research and development, especially into storage methods. • Electricity produced is initially more expensive than from conventional power stations. • Not very effective in cloudy climates or polar latitudes. • Energy still needs to be stored for later use.

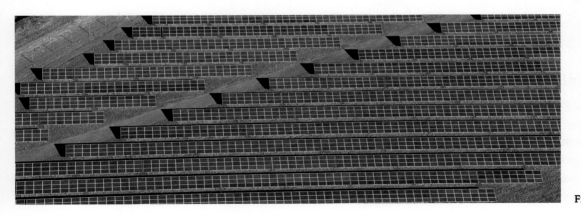

Figure 2.21: A solar farm in the UK

ACTIVITY

1. Compare the benefits and problems of oil and solar as energy sources.

2. Produce a profile of wind energy and nuclear power including characteristics, locations, and benefits and problems.

Extension

Study the latest Sankey energy-flow diagram for the UK (see the DECC website). Analyse the use of energy, imports and exports, energy losses, and the role of renewables such as bioenergy.

CASE STUDY: The changing UK energy mix

Figure 2.22 shows the changing energy mix of the UK. There has been a decline in the role of coal, while oil has remained constant; natural gas grew until about 2005 and then declined; and nuclear grew until the late 1990s and then steadily declined. Renewables have grown the most, but still account for only a small proportion of total energy use. Overall energy use, especially of fossil fuels, is falling and in 2014 reached its lowest level for about 50 years (coal use was the lowest for 150 years). There are several reasons for the decline of coal, mostly linked to carbon emissions targets but also to the loss of traditional UK industries such as steel, problems with coal-fired power stations, changing power stations such as Drax to biomass, and the costs of installing pollution-preventing technologies in older coal-fired power stations.

Despite the 6 per cent decline in total energy use in 2014, the UK economy grew by 2.8 per cent, proving that economic growth does not depend on greater energy use. In 2014:

- it was a warm year, which helped the UK reduce CO_2 emissions by 10 per cent; this warming trend could provide the negative feedback for even lower energy use in the future

- 7 per cent of energy was from renewable sources, which shows that the UK target of 15 per cent by 2020 may be achievable

- onshore and offshore wind energy had the largest renewables share (9.5 per cent) of the UK electric power mix, followed by biomass and landfill gas (6.8 per cent), with solar providing only 1.2 per cent – less than hydroelectric power.

However, fossil fuels still accounted for 85 per cent of total energy use and 60 per cent of electricity production in 2014. The consumption of energy by transport, based on oil and accounting for 38 per cent of UK energy consumption in 2014, is a large obstacle to changing the energy mix further.

UK primary energy use by source

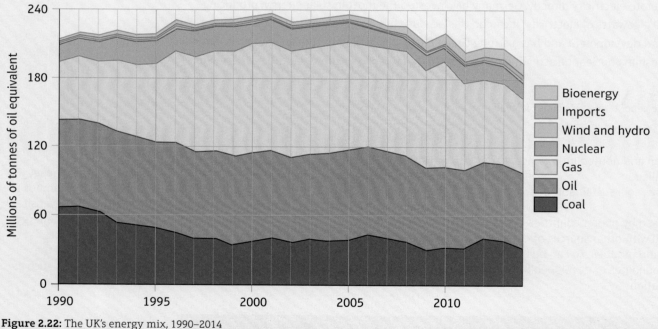

Figure 2.22: The UK's energy mix, 1990–2014

The growth of biofuels

In 2014 the IPCC identified bioenergy as having a crucial role in the mitigation of climate change through efficient biomass-to-bioenergy systems. However, biofuel is more suitable for small-scale rather than large-scale operations, and the resulting land-use changes will have an impact on **carbon fluxes** between soil, vegetation and atmosphere, on food security where food crops are replaced by biomass crops, on water resources (by changing the water cycle), on conservation of the environment, and on the livelihoods of local people.

Yellow = Players, Orange = Attitudes and actions, Purple = Futures and uncertainties

CASE STUDY: Biofuels in Brazil

Sugar cane has been grown in Brazil for nearly 500 years. Originally it was grown to produce sugar, but today Brazil is the world's biggest producer of ethanol from sugar cane. Most cars and light vehicles in Brazil use a mix of ethanol and petrol, with bio-refineries producing the equivalent of 930,000 barrels of oil per day, worth US$50 billion a year, providing 1.34 million direct jobs and 16 per cent of the domestic energy supply. Since as far back as 1931 it has been compulsory to mix ethanol with petrol, the exact mix depending on sugar yields. When crude oil prices increased dramatically in the 1970s, the government encouraged the production of ethanol; during the 1980s recession the ethanol business was streamlined and remained important.

In the 1990s the government gradually removed sugar cane subsidies, but ensured strong links between sugar cane producers, ethanol producers and ethanol distributors. Hydrous ethanol was developed, and in the 2000s cars were built to use various mixtures of fuel ('flex-fuel' engines). Ethanol production expanded as a result of greater efficiency and productivity rather than an increase in land area, so it is not linked to deforestation of tropical rainforest – and sugar cane grows best in other climates. Electricity is also produced from bagasse, the waste plant material that remains after sugar extraction. There are 370 cogeneration energy systems in sugar and ethanol mills, providing 7.2 per cent of the electricity-generating capacity of Brazil in 2012 (Figure 2.23). The ethanol industry provides nearly 11 jobs for every tonne of oil produced, and its workers are better paid and their jobs more formalised than those of other agricultural workers.

Ethanol struggled to compete when the government reduced taxes on fossil fuels to try to control inflation – biofuel plants were bankrupted, and fields were returned to growing food crops. However, in 2015 the government raised the minimum amount of ethanol in the fuel mix to 27 per cent and removed petrol subsidies. In Brazil, 6 million flex vehicles and 3 million others can run on hydrous ethanol, and biodiesel use is increasing (soybean oil). Sugar cane provides between eight and ten times more energy than the inputs to grow it, and reduces greenhouse gases by 90 per cent. In 2015 the price of a litre of ethanol in Brasilia, the capital, was 2.594 reals (US$0.72 or 50 pence). Domestic consumption of ethanol in 2016 was about 30 billion litres, with about 1.35 million litres exported, mainly to the USA and South Korea.

NOTE: Anhydrous ethanol has a purity of 99 per cent: hydrous ethanol has 93–96 per cent purity, which has proved efficient in vehicles.

Figure 2.23: An ethanol energy plant in São Paulo State, Brazil

ACTIVITY

Evaluate the potential of biofuels to help developed and developing countries achieve energy security and reduce carbon dioxide emissions.

The use of biofuels continues to grow slowly, mainly in the USA, Brazil, Indonesia and Argentina. In the UK, bioenergy accounted for 6.8 per cent of electricity production in 2014. Extensive areas in Africa are suitable for bio-ethanol (starch-based) or bio-diesel (oilseed) crops, but there are concerns about using land that is needed for food crops and water for people. Pressures from governments and companies could also lead to changes in land ownership (tenure), reduce the livelihoods of women, and take grasslands away from pastoralists or migrant farmers who need open access. Biofuel crops grown in marginal farming areas could also stress the natural environment and increase forest loss as people are forced to find new land, and deforestation would offset any carbon reduction benefits gained from switching to biofuels, so it would not be **carbon-neutral**.

More research is needed. For example the Jatropha plant has been suggested as an oilseed crop for Africa, but oil yields are variable and unconfirmed, with little information on impacts. In tropical South and Central America, palm oil is a significant biofuel crop, but is strongly linked to deforestation, especially in Colombia and Ecuador. In Peru, 72 per cent of new palm oil plantations were in forested areas, accounting for 1.3 per cent of the country's deforestation between 2000 and 2010. Biofuels can help Africa decrease carbon emissions; 24 per cent of non-electrical energy in 2012 was from biofuels and waste, compared with 16 per cent for the world as a whole, but biodiversity may be lost, food prices may increase and water resources may be stretched. Biofuels are climate-dependent, and future temperature and rainfall changes are not yet known with any certainty.

Radical technologies

If energy sources could be found to replace fossil fuels, greenhouse gas emissions would fall dramatically. Technically there are several possibilities and there may be more radical technologies to come. Some are geo-engineering ideas but they do not need to be large – for example, micro-hydro and solar cookers work well in developing countries.

Hydrogen fuel cells

This alternative energy source could replace petrol for transport or natural gas for heating. The term 'hydrogen economy' describes the possible widespread use of this energy source. Hydrogen has an instant appeal because, whether it is combusted to produce heat or used in a fuel cell to produce electricity, the only waste product is water. One of its best potential uses is in electric cars or public vehicles that use fuel cells converting hydrogen, a gas that can be obtained from a variety of sources. Fuel cells are more efficient than a petrol or diesel engine. Toyota developed a car with a fuel cell stack (the Mirai) and a range of 270 miles, which went on sale in California in 2015. However, hydrogen is not found in a pure form and has to be separated from other compounds such as water, biomass, ethanol or natural gas (methane). Processes to separate it require large amounts of energy and may emit large quantities of greenhouse gases. Hydrogen is an energy carrier or a way of storing energy, rather than a primary energy source, and hydrogen tanks need to be strong enough to withstand impacts.

Electric vehicles

The distances a purely electric vehicle can travel are relatively short before lengthy recharging is needed, so further technological development is required. Range varies from 62 to 340 miles, depending on battery capacity and linked technologies. Some communities, such as BedZED in the London borough of Sutton, have public charging points, but these are often scattered; according to Zap-Map (April 2016) there were 3,919 public charging locations in the UK to serve over 60,000 registered electric vehicles. London had 19.7 per cent of the charging points, and other cities have

Yellow = Players, Orange = Attitudes and actions, Purple = Futures and uncertainties

large concentrations; the fewest charging points are in rural regions of Wales (3 per cent) and the east of England (3.5 per cent).

Electric vehicles may be best suited to urban environments, helping cities reduce air pollution levels. Hybrid vehicles, using a petrol or diesel engine in conjunction with an electric engine, have engine management systems that decide which is the most economical to use during a journey. Tesla has manufactured electric vehicles since 2008; in 2016 it announced its third model and within three days had received 276,000 orders, with first deliveries planned for the end of 2017. However, it is not known if the small company can meet expectations.

The benefits of electric vehicles include zero carbon emissions and virtually no noise pollution, as well as being cheap to run. However, they can be more expensive to buy and they are so quiet that some people are concerned about collisions with pedestrians. Their biggest disadvantage is how the electricity is created for charging the batteries, because the carbon emissions of an electric vehicle depend on the energy profile of the country in which it is being used, so electric cars are more eco-friendly in Paraguay (hydroelectric) and Iceland (geothermal) than in India or Australia (both coal).

Carbon capture and storage

A complete carbon capture and storage system (CCS) collects CO_2 emissions from fixed points such as industrial and power plants, then transports the gas and injects it (in compressed form) into a suitable geological structure (over 800 m below ground). The storage is then monitored to ensure safety and no releases into the atmosphere. In 2014 Canada opened the first coal-fired power plant with CCS at Boundary Dam (Figure 2.24), at a cost of US$1.3 billion. It will reduce emissions by 90 per cent by pumping CO_2 underground and selling it to an oil company (Cenovus) for 'priming' nearby oilfields; this helped make the scheme economical. CCS can be combined with bioenergy (BECCS) to capture CO_2 produced during bioenergy production such as ethanol; this ensures that there is a net removal of CO_2 from the atmosphere, as all CO_2 produced during farming is stored as well.

The 2014 IPCC report recognised the uncertain availability and deployment of this geo-engineering technology – in 2013 there were only two small BECCS operations – but regarded it as essential mitigation action to limit temperature increase to under 2°C by the end of the century. The IPCC believes that power generation without CCS must be phased out by 2100, and expects great progress after 2050. As well as lower pollution and climate benefits, CCS could extend the use of fossil fuels and encourage greater efficiency. However, concerns include CO_2 leakage affecting human health, underground pressure causing small earthquakes, and increased water usage affecting natural environments.

Figure 2.24: The CCS system at Boundary Dam, Saskatchewan, Canada

ACTIVITY

Evaluate the costs and benefits of using radical technologies to help solve future energy demands.

A level exam-style question

Explain why renewable energy sources have costs in terms of their contribution to energy security. (8 marks)

Guidance

Remember that 'costs' here means problems and issues, not just economic costs. Energy security is ensuring that in the future, regions and countries can supply businesses, industries and homes with their energy needs. Renewables have some disadvantages in this respect, and this is what you need to identify through examples.

Nuclear fusion

Another radical technology is nuclear fusion, where two or more atomic nuclei join together to make a new larger nucleus, releasing energy in the process. The advantages of this energy source are that it is clean, with no greenhouse gas emissions or radioactivity, and it can use common elements. Nuclear fusion is still a long way from becoming a reality, but 35 countries are working together at the 500 MW International Thermonuclear Experimental Reactor (ITER) being constructed at Aix-en-Provence in France (Figure 2.25). In February 2016 a Chinese research team reported that it had managed to sustain a superheated plasma gas for over one-and-a-half minutes in a ring-shaped reactor using high-powered magnets. While this did not create usable energy, many regard it as a critical technical step forward. The German government has allocated over US$1.46 billion for research.

Figure 2.25: The new experimental reactor (ITER) in France

Nanotechnology

Another advanced approach uses nanotechnology to make solar fuels, which could replace fossil fuels. Solar fuels could be made using the action of sunlight on simple substances, a photochemical reaction or artificial photosynthesis – for example splitting water to make hydrogen, or using microorganisms or enzymes to 'harvest' light energy.

Yellow = Players, Orange = Attitudes and actions, Purple = Futures and uncertainties

How are the carbon and water cycles linked to the global climate system?

Learning objectives

6.7 To understand how human activity threatens biological carbon cycles and the water cycle.

6.8 To understand the implications for human wellbeing from the degradation of the water and carbon cycles.

6.9 To understand how further planetary warming may release large amounts of stored carbon, requiring responses from different players at all scales.

Threats to the carbon and water cycles
The growing demand for resources

The 2015 annual report of the Global Footprint Network showed that it is beyond the capacity of the planet to satisfy current human demand for ecosystem services. Collecting accurate data on all variables is almost impossible, but the Network's Ecological Footprint showed that the average forest carbon **sequestration** per year is 0.73 tonnes of carbon per hectare. Another calculation is $I = P \times A \times C \times T$, where I is impact, P is population size, A is affluence of the population, C is consumer behaviour (C_1 = food supply in calories and C_2 = crop production volume) and T is technology. When applied to food production, for example, it shows that population and affluence increase impacts as more food is grown; technology lowers impacts as it makes farming more efficient; and consumer behaviour may increase or decrease impacts through more or less consumption or

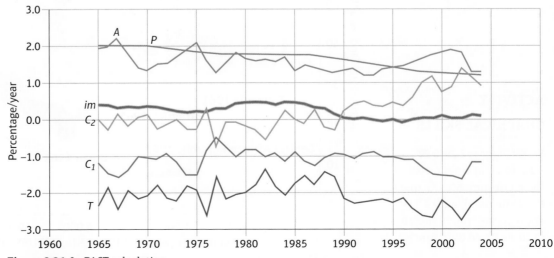

Figure 2.26: ImPACT calculation

through changing the ratio of food to non-food crops. Study the impact (*im*) line in Figure 2.26; when the line is below zero, cropland is released for other uses such as biofuels or reforestation.

Between 1990 and 2015 world energy use increased by 54 per cent, while the population increased by 36 per cent. This shows that while population growth may increase energy use, it is economic development and prosperity that account for most of the increased demand. Developed countries use half of the world's energy, and North America has the highest use per capita – more than ten times that of sub-Saharan Africa or South Asia. Many developing countries rely on wood or other biomass (such as charcoal or animal dung), which creates pressures on forests and impacts on human health.

Because the world's population is expected to increase to 9.2 billion by 2050 (UN estimate), especially in South, East and South-East Asia, more food and resources will be required and more energy will be used. In 2015 an estimated 1.4 billion people suffered from hunger or malnutrition even though absolute poverty fell, so there is a need to increase food production in some world regions. There is also a need to cater for the tastes of a wealthier population (Figure 2.26): the Food and Agricultural Organisation (FAO) expects world meat consumption (37.4 kg/person in 2000) to increase to 52 kg/person per year by 2050. Meat production is land and water intensive and requires more land for fodder crops, diverting farmland from staple crop production and degrading water and soil. World per capita food consumption (kcal/person/day) increased by 8.9 per cent between 1990 and 2015, but the FAO predicts that it will increase by only 3.5 per cent between 2015 and 2030. There are contrasting trends around the world, with the largest increases in food consumption between 1990 and 2015 in East Asia (+20.1 per cent), sub-Saharan Africa (+14.1 per cent) and Latin America and the Caribbean (+11.9 per cent). But the FAO predicts a slowing of food consumption in all world regions except sub-Saharan Africa (+7.2 per cent) and South Asia (+7.0 per cent).

Extension

Investigate the Sustainable Development Goals (SDGs) for 2015 to 2030. Consider carefully how SDGs 2, 6, 7, 12, 13 and 15 will help solve the carbon and water cycle issues caused by human activities.

ACTIVITY

GRAPHICAL SKILLS

Study Figures 2.26, 2.27, 2.32. To what extent does the environmental Kuznets Curve model (Figure 2.31) apply to the trends shown in these three diagrams?

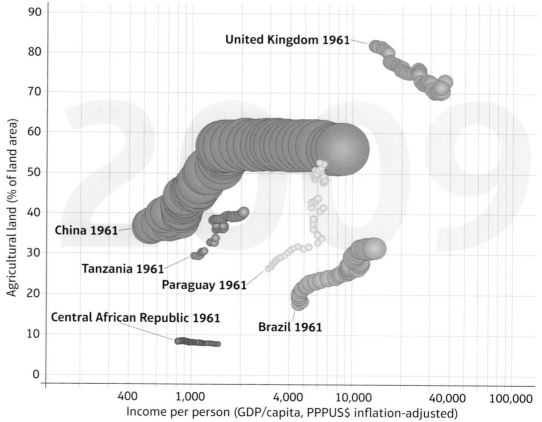

Figure 2.27: Comparison of crop area and GDP of selected countries

Yellow = Players, Orange = Attitudes and actions, Purple = Futures and uncertainties

Table 2.11: Resource and emission data for world regions

World region	Energy use change 1997–2013 (in %)	Deforestation rate av. annual % 2000–15 (minus figures mean forest gain)	Land protected difference 1996–2014 (in %)	CO_2 emissions change 1996–2011 (in %)
East Asia and Pacific	25.9	−0.17	11.1	175.6
Europe and Central Asia	159.1	−0.10	6.4	95.3
Latin America and the Caribbean	138.8	0.36	8.8	47.1
The Middle East and North Africa	485.6	−0.88	7.8	137.4
North America	47.3	−0.06	−0.3	−21.1
South Asia	−1.3	−0.37	0	107.0
Sub-Saharan Africa	107.1	0.48	7.3	58.2

Table 2.12: Freshwater resource use by world region

World region	Freshwater withdrawals (% of internal water resources)	Freshwater withdrawals for agriculture (% of total withdrawn)		Freshwater withdrawals for industry (% of total withdrawn)		Freshwater withdrawals for domestic (% of total withdrawn)	
	2014	1998	2014	1998	2014	1998	2014
East Asia and Pacific	10.8	80	72	14	16	6	12
Europe and Central Asia	7.5	63	36	26	34	11	30
Latin America and the Caribbean	2.4	73	72	9	11	18	17
The Middle East and North Africa	133.6	89	85	4	6	7	9
North America	9.1	30	38	59	48	11	14
South Asia	51.6	93	91	2	2	5	7
Sub-Saharan Africa	3.0	87	80	4	5	9	15

In 2013 the IPCC estimated that 24 per cent of greenhouse gas emissions came from agriculture and land-use change, such as biomass burning, deforestation and methane emissions from livestock farming and wet rice cultivation. Between 1995 and 2005 the global forest area decreased by 80 million ha, while arable land area increased by 24 million ha. At current rates the United Nations Environment Programme (UNEP) expects that between 320 and 850 million ha of natural grasslands, savannah or forest could be lost to crops between 2005 and 2050, and that this needs to be stopped by 2020. Emissions are decreasing in developed countries as the amount of farmland decreases; farming area is stable in the poorest and emerging countries, but is increasing in developing countries (Table 2.11; Figure 2.27). The FAO believe that changes to agriculture in

ACTIVITY

1. Describe and explain the patterns and trends shown in Tables 2.11 and 2.12.

2. Suggest where the future land and water issues are going to be.

Synoptic links

Sea-level rise is threatening coastal areas with increased flooding and erosion rates as waves reach new coastal land. Island communities may disappear.
Higher carbonic acid concentrations will increase chemical weathering at the coast, especially where there are coral reefs or limestone rocks. (See Book 1, page 155.)

Some sea-level rise is due to the changes to the cryosphere, with ablation being greater than accumulation. IPCC research shows that snow cover in the northern hemisphere is decreasing at a rate of 1.6 per cent per decade, with glaciers retreating and the Greenland ice sheet losing mass. (See Book 1, page 101.)

ACTIVITY

Explain how human activities may cause the health of marine ecosystems to reach a critical threshold.

sub-Saharan Africa could increase deforestation and carbon emissions, even with efficient changes, and increasing meat production could increase methane emissions in India and China. The IPCC recommends a switch to low-emission food crops, changing croplands to forests or bioenergy crops, reducing food waste (30 to 40 per cent is lost from 'farm to fork'), changing diets (plants not meat) and reducing food consumption.

Degraded soils, particularly organic soils that have lost organic matter and moisture, are not capable of storing much carbon (see Figure 2.9). When soils dry out through lack of vegetation cover, drainage or evaporation from increased temperatures, they emit rather than store greenhouse gases. It is estimated that soils can store three times more carbon than vegetation; the balance is decided by inputs of dead plant material and outputs from decomposition and mineralisation. Land management is important, especially in areas of moist, peat-rich soil: undisturbed waterlogged peat soils, which cover 3 per cent of the world's land area, store large amounts of carbon – up to 25 per cent of the global soil carbon store – but drainage has increased decomposition, respiration, fire risk and emissions of CO_2 and N_2O, especially in Asia and Europe. Degraded soils can be restored through managing fertility and water conservation, along with afforestation and longer fallow periods (conservation tillage), all of which help retain soil moisture and organic matter.

Ocean acidification

Acidification of the oceans is increasing as the sea absorbs more CO_2 from the atmosphere. Since the Industrial Revolution, pH has fallen by 0.1 (to 8.1), which on the log scale is a 26 per cent increase in acidity. Oceans are able to store 50 times more inorganic carbon than the atmosphere, creating a weak H_2CO_3 acid by releasing hydrogen ions, and are an important carbon sink for anthropogenic CO_2 emissions. In 2014 the IPCC reported that the oceans had absorbed about 30 per cent of anthropogenic CO_2 emissions, but absorption rates vary because of ocean oscillations, such as El Niño, and sea temperatures. The largest pH reduction has been in the North Atlantic and the smallest in the subtropical waters of the southern Pacific.

It is not easy to separate the effects of acidification from those of sea-level rise and seawater temperature increase. Sea levels have risen at an accelerating rate: by 1.7 mm/yr between 1901 and 2010, and by 3.2 mm/yr between 1993 and 2010. This increase is due to thermal expansion, glacier mass loss, ice sheet loss and increased water storage on land. Regional differences in sea-level rise are due to ocean currents and planetary winds pushing water towards certain coasts, such as the western Pacific. Oceans stored 90 per cent of heat energy between 1971 and 2010, with the upper 75 m warming by 0.11°C per decade during this period, and now deeper layers are also warming. Salinity patterns are changing – salty areas are getting saltier and fresher areas are becoming less salty – and evaporation and precipitation patterns over oceans are changing, as well as potentially affecting the movements of ocean currents and the thermohaline circulation.

The physiology and behaviour patterns of marine organisms are expected to change, perhaps at a **tipping point**. Marine plants, phytoplankton and microalgae may benefit from increased photosynthesis, and crustaceans and some fish may be less sensitive, but higher acidity may affect the ability of marine organisms to build shells and skeletons, so creating thinner or smaller shells in molluscs and reducing coral reef-building ability. Acidity may also reduce the availability of minerals, and there may be a build-up of CO_2 concentrations in fish, squid and mussels. Larval oyster fatalities have already been noted in the north-eastern Pacific, and fleshy algae and barnacles are replacing mussels. It is feared that the impact on cold- and warm-water corals will be negative, reducing habitats for other marine life in the ecosystem, leading to a decline in biodiversity and **ecosystem productivity** (the cascade effect).

Yellow = Players, Orange = Attitudes and actions, Purple = Futures and uncertainties

Shifting climates

Climate change models suggest that global weather patterns may be shifting, resulting in permanent changes that make some world regions wetter and others drier. Some scientists regard the tropical rainforest in Amazonia as a key element in offsetting anthropogenic carbon emissions, but are worried that a climate tipping point may be reached where the temperature and precipitation levels become unfavourable for tree growth, and the biome changes to savannah grassland instead.

ACTIVITY

Study Figures 2.28, 2.29 and 2.30. Describe and explain the patterns of potential change shown for droughts and floods. Which areas are most at risk from future water shortages or floods?

CASE STUDY: Droughts in Amazonia, South America

Drought events are a feature of the Amazonian climate, along with extreme flood events after exceptional rainfall. Since 1911–12, droughts have occurred about once every ten years, but the 2005 mega-drought was followed five years later by another mega-drought. Between 1995 and 2005 less water had been available for rainforest plants, which introduced stress into the ecosystem and made the 2005 drought worse. In 2005 about 70 million hectares of pristine mature forest in south-west Amazonia was damaged, amounting to 30 per cent of the total area, with 5 per cent severely affected. Visible changes could be seen in the forest canopy layer, with tree fall and dieback of branches, and NASA satellite data showed reduced moisture volumes and biomass. Half the area affected had problems for several years after 2005, despite higher rainfall, and so by the 2010 drought, trees had not recovered and were even more vulnerable. In 2010 nearly half of Amazonia was affected, 20 per cent severely, covering a large part of the 2005 affected area.

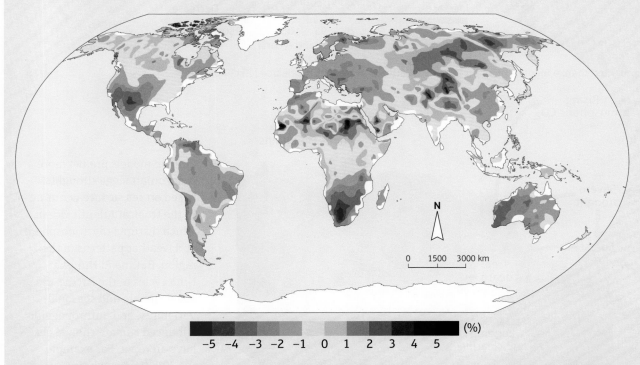

Figure 2.28: Mean projected change for soil moisture 2016–35 relative to 1986–2005 mean (top 10 cm): drought likelihood

Research by UNEP and the Global Ecosystems Monitoring (GEM) network showed that during these severe droughts the trees absorbed less CO_2, with photosynthesis slowing by about 10 per cent over a six-month period, despite growth continuing at the same rates as unaffected trees. It appears that drought-affected plants may prioritise growth rates – the race to sunlight over health. During the droughts, emissions of CO_2 increased as a result of more frequent wildfires and decomposing dead wood as more trees died. Using infrared gas analyses, it was estimated that the 2005 drought emitted 5 billion tonnes of CO_2 and the 2010 drought 8 billion tonnes – which was more than the CO_2 absorbed. This is worrying, as it suggests that the Amazonian tropical rainforest may in the future cease to operate as a carbon sink and instead become a carbon source, accelerating global warming, climate change and further change to the biome (an example of positive feedback).

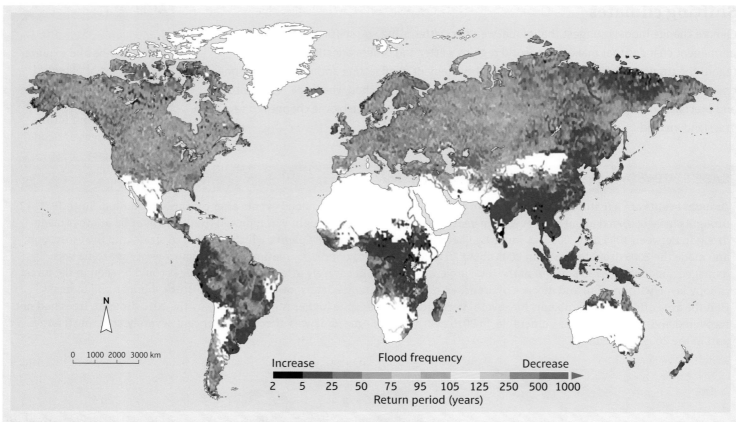

Figure 2.29: Mean projected change in total runoff for 2016–35 relative to 1986–2005 mean: flood likelihood

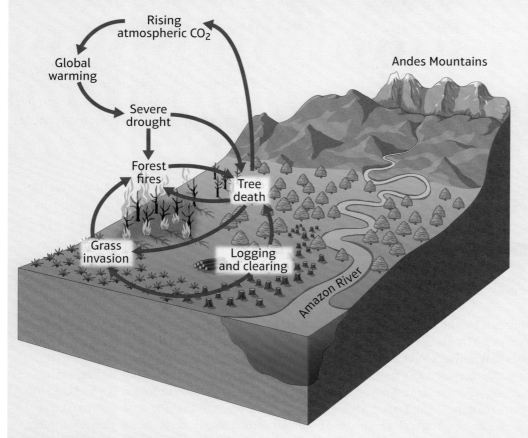

Figure 2.30: Links and feedbacks in Amazonia

Climate models predict more Amazonian mega-droughts, based on sea surface warming in the tropical Atlantic Ocean, which disrupts the atmosphere and weather patterns in the region. But the El Niño/La Niña cycles in the Pacific, the usual climate and seasonal variability, and anthropogenic-enhanced climate change also have an input. For example, the 2014/15 drought in south-east Brazil was due to a blocking anticyclone (high-pressure area), and it is not known whether this was a result of climate change.

Yellow = Players, Orange = Attitudes and actions, Purple = Futures and uncertainties

ACTIVITY

TECHNOLOGY/ICT SKILLS

1. Use the internet to find satellite images of an area that has experienced significant deforestation. Find at least two images of the same area from different years.

   a. Using a scale grid, calculate the area (in km² or as a %) that has been deforested over the time period.

   b. Explain why deforestation has been taking place in this area.

   c. Explain the implications of deforestation for the people living in this area.

Climates may shift if the thermohaline circulation changes. This could happen if a large input of freshwater occurs due to greater precipitation or melting glaciers and ice sheets (such as in Greenland) or melting sea ice (such as in the Arctic Ocean). Salinity levels would be reduced, altering seawater density, for example in the North Atlantic; this would prevent the Gulf Stream current reaching as far north, and would reduce the warming influence of the sea on north-western Europe. However, the exact climate change in NW Europe is difficult to predict because atmospheric temperatures will be increasing due to global warming, possibly at a greater rate than any decrease from a cooler sea.

Implications for human wellbeing
Forest loss

Landsat satellite data for 2000 to 2012 showed the loss of 2.3 million km² of the world's forests, and 0.8 million km² gained. Tropical forests had a trend of losses every year, for example in Bolivia, Indonesia and Angola. Intensive forestry in tropical climates caused the highest rates of change, while boreal forests had losses from both fire and forestry. Tropical rainforests – accounting for 32 per cent of the global loss, half of this in South America – have a major influence on regional and global climates and the carbon cycle (carbon storage) as well as on the water cycle through interception, absorption and evapotranspiration, affecting water supplies for people.

Forest loss from human activities has been a key concern since about 1970 and, while deforestation rates have slowed in some areas such as Brazil, it is still taking place. Predicting future forest loss (or gain) is problematic: past surges of deforestation have made way for soya farms, palm oil plantations and hydroelectric power dams. Policy changes may have an effect: Brazil's policies show how a country can create a 'turning point', as shown in the Kuznets Curve (Figure 2.31).

Fire is one of the main causes of forest loss, especially along drier forest edges. As a drier forest decays and is destroyed by fire, emissions of solid particles (aerosols) and wind-blown soil result in impaired air quality, and human respiratory problems increase. During the 2005 Amazonia drought, Capixaba in Acre State, Brazil, experienced a 181 per cent increase in hospitalisations, and in 2010, Fatima in Tocantins State had a 267 per cent increase (especially of under-fives). Drier soils and the lack of protection from vegetation cover increase soil erosion, which again may affect food supplies in the longer term.

The environmental Kuznets Curve is based on economic principles. It suggests that as a country develops, damage to the natural environment will at first increase, as resources are exploited and technologies cause pollution and degradation, but that over time, as technology becomes more efficient, fewer resources are used, new resources are created and pollution levels fall (Figure 2.30). All countries have removed natural forests to create farmland or living space or for fuel; this is occurring now in emerging countries such as Indonesia (largest increase in forest loss 2000 to 2012 – over 20,000 km² per year in 2011/12) and developing countries (Figure 2.32). In many developed countries, afforestation is now taking place, for a mixture of resource reasons and environmental reasons (see Tables 2.11 and 2.12). Eurasian boreal forests had the greatest expansion from 2000 to 2012 as a result of forestry management, reduced farmland and recovery after wildfires, while Russia had the largest global forest loss between 2000 and 2012 – 365,015 km².

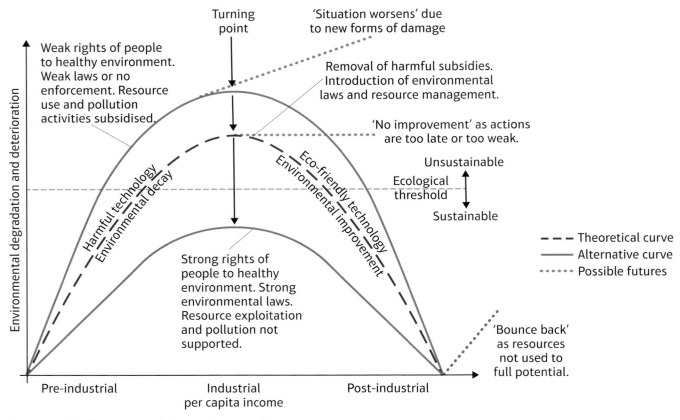

Figure 2.31: The Environmental Kuznets Curve model

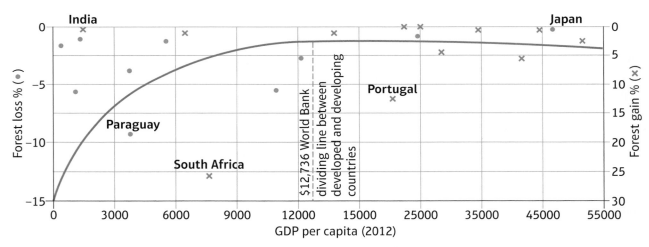

Figure 2.32: A scattergraph comparing GDP per capita and forest gain or loss for selected countries (2012)

Yellow = Players, Orange = Attitudes and actions, Purple = Futures and uncertainties

Increased temperature and evaporation rates

Changes to precipitation patterns due to climate change are complex, and more research is needed to separate natural variability from long-term change. One possibility is that the warming of the oceans and the lower atmosphere will result in a smaller temperature gradient within the troposphere, which would create stability, weaken planetary winds and limit increases in precipitation. However, higher temperatures will increase evaporation, leading to increased precipitation at the Intertropical Convergence Zone (ITCZ) and at polar latitudes.

CASE STUDY: Changes in the Arctic water cycle

The 2014 IPCC report identified changes to the Arctic water cycle:

- The largest global increase in temperature is in the Arctic, especially in winter (*very high confidence*).
- Higher temperatures and evaporation rates are drying up Arctic ponds.
- The greatest humidity increases are in northern high latitudes.
- The precipitation and evaporation balance will change (*medium confidence*) (Figure 2.33).
- Higher air temperatures will reduce the duration and extent of ice and snow cover (*very high confidence*) and increase snow and ice ablation.
- Ablation of glacial ice, such as in Greenland, will alter runoff ('freshwater flux') and create more local river-ice floods.
- Permafrost thawing rates will increase (*very high confidence*) and the permafrost area will decrease (*high confidence*); there may be increased river discharges in winter.
- Annual decline in Arctic sea-ice area will become more rapid (*high confidence*)
- The Arctic Ocean could be ice free by 2037 (*medium confidence*), especially as oceans warm, and reduced sea ice will provide increased evaporation and snowfall.
- Arctic lakes will freeze later, with earlier ice break-up and thinner ice in winter.
- Between 1977 and 2007, 19 polar Arctic rivers had a 9.8 per cent discharge increase, with increasing spring snowmelt producing earlier peak discharges.
- Earlier annual peak river discharges will occur in the Russian Arctic because of higher temperatures and an earlier spring thaw.
- Eurasian Arctic river discharges will increase as a result of a poleward shift of moisture transport (Figure 2.33).

People in the Arctic may experience building and structure collapse as a result of permafrost thawing (*high confidence*); and stresses on wildlife may lead to food insecurity as ecosystems change and hunting areas move (*high confidence*). Alaskan people may be forced to relocate or be trapped in isolated communities by extreme weather and hazards such as avalanches or river floods (*high confidence*). Damaged infrastructure, drought or changing hydrology – for example sediments and pollutants in rivers – may limit freshwater supplies.

There is uncertain confidence in many predictions, because climate models lack valid data from which to project the cascade and feedback effects in Arctic systems, as well as a co-ordinated network of long-term measurement sites for monitoring and assessing climate change in these northern areas.

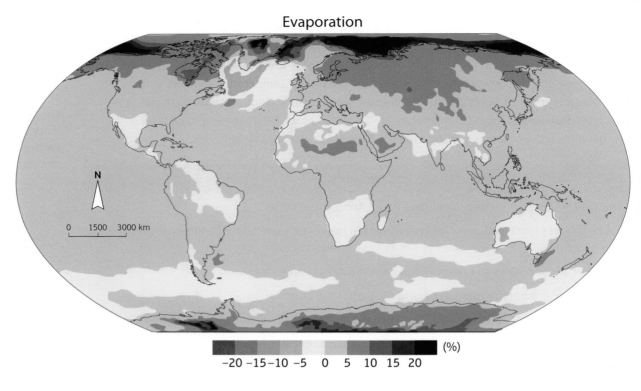

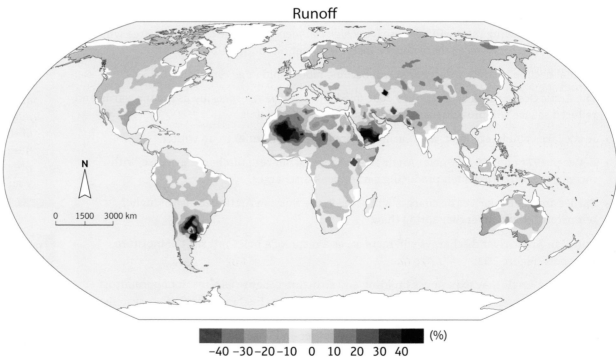

Figure 2.33: Predicted water cycle changes (runoff and evaporation) 2016–35

Threats to ocean health

Oceans have always supplied resources for people, and increasingly in recent decades food supplies for many developing countries with growing populations have been based on **marine resources**. However, ocean health is being affected by increasing temperature, acidity, salinity and possible changes to currents. Changes to coral reefs, the most biodiverse marine ecosystem, have an impact on ecosystem services, causing a loss of cultural and leisure opportunities and reducing the protection from storms that reefs offer coasts. The ability of countries to harvest marine

Yellow = Players, Orange = Attitudes and actions, Purple = Futures and uncertainties

resources through fishing could also change, especially if there is a poleward shift of fish species, suggested to be occurring at a rate of around 26 km per decade. Conflict between players may occur if Marine Protection Areas (MPA) end up in the wrong places, and territorial disputes over the new locations of fish stocks may arise.

CASE STUDY: CORAL REEFS

Warmer seas, stronger storms and acidification are damaging coral reefs. Bleaching is predicted to become severe for half of all coral reefs by 2050, with a transition to net erosion because of reduced calcification and coral mortality. Corals will be less effective at protecting coasts from storms and the resulting coastal erosion will affect millions of people living near the sea, especially in developing countries that cannot afford extensive hard-engineering defences. Marine productivity is decreasing, with fewer fish larvae, which is damaging fish stocks: in the Caribbean it is expected that fisheries will soon have a revenue loss, and in Pacific Island states a 20 per cent decline in reef fisheries by 2050 is predicted.

Over 100 countries benefit from the tourism and recreational value of coral reefs, which is worth US$9.6 billion according to the IPCC (2014). Damage to the reefs harms local economies in small island states and developing countries. In Viti Levu, Fiji, coral reef degradation will cause losses of US$5–14 million a year by 2050, due to reduced fisheries, habitats and tourism. Also in Viti Levu, human health is being affected as warmer seawater leads to fish eating ciguatoxic algae, which poisons people when they eat the fish; numbers affected are expected to increase to 700 times current levels.

Responses to large-scale carbon release
Natural and human factors and feedback mechanisms

Natural variations in climate are caused by Milankovitch cycles, Sun cycles and long-term ocean and atmospheric oscillations such as El Niño and jet streams. All of these create uncertainty when trying to predict future climate change. Similarly, human factors such as population growth, economic growth, land-use change and energy profiles may change in the future and alter anthropogenic greenhouse gas emissions.

Table 2.13: CO_2 levels from 1960 to 2015

Year	CO_2 annual mean level (ppm) (marine surface, Hawaii)	CO_2 annual growth rate (ppm/yr)
1960	316.91	0.54
1965	320.04	1.02
1970	325.68	1.06
1975	331.08	1.13
1980	338.68	1.73
1985	346.04	1.25
1990	354.35	1.19
1995	360.80	1.99
2000	369.52	1.62
2005	379.80	2.52
2010	389.85	2.42
2015	400.83	3.05

Synoptic link

Coastal areas are subject to storm surges, flooding and increased erosion because of climate change (see Book 1, Chapter 3). Territorial disputes between traditional and emerging superpowers, which mainly focus on resources such as energy and its pathways, may expand to include marine areas in the future – not just for deep-water oil as in the Arctic, but also for marine foods. Agreements protecting the oceans will require greater co-ordination and international legal status. (See Book 1, page 107.)

ACTIVITY

GRAPHICAL SKILLS

1. Study Table 2.13.

   a. Calculate the mean of the CO_2 annual mean level data.

   b. Using the data in the table, plot the values on a suitable graph, and add a line to show the mean value calculated in part (a).

   c. Calculate the rate of change from 1960 to 2015.

   d. Add a second vertical axis to your graph and plot the CO_2 annual growth rate data.

   e. Suggest how people should respond to the trends shown by the graph.

If emissions were stopped, some effects of greenhouse gases would immediately be reduced, but others could take centuries to be rebalanced by Earth's systems. Around 20 per cent of CO_2 could remain in the atmosphere for several hundred years; even an immediate drastic reduction would not enable the carbon cycle to return to a balance easily, as oceans would be saturated and vegetation unable to absorb more. If CH_4 and HCFCs emissions were reduced to zero, the atmosphere would return to normal concentrations within 100 years; a 30 per cent reduction would stabilise their warming effect. Nitrous oxide has a lifetime in the atmosphere of 110 years and would require a 50 per cent reduction to stabilise; zero emissions would return a natural balance in about 200 years. Any continuing emissions will lengthen the time period of anthropogenic warming.

In 2014 the IPCC defined a 'tipping point' as an abrupt, possibly irreversible, large-scale change over a few decades or less. The IPCC identified seven possible tipping points:

1. **Atlantic thermohaline circulation collapse:** Increased volumes of freshwater and density changes may alter currents in the Atlantic (but climate models do not agree).
2. **Seabed methane release:** Methane hydrates in seabed sediments (clathrates) could be destabilised by warming, releasing methane gas from a solid form (this is considered very unlikely).

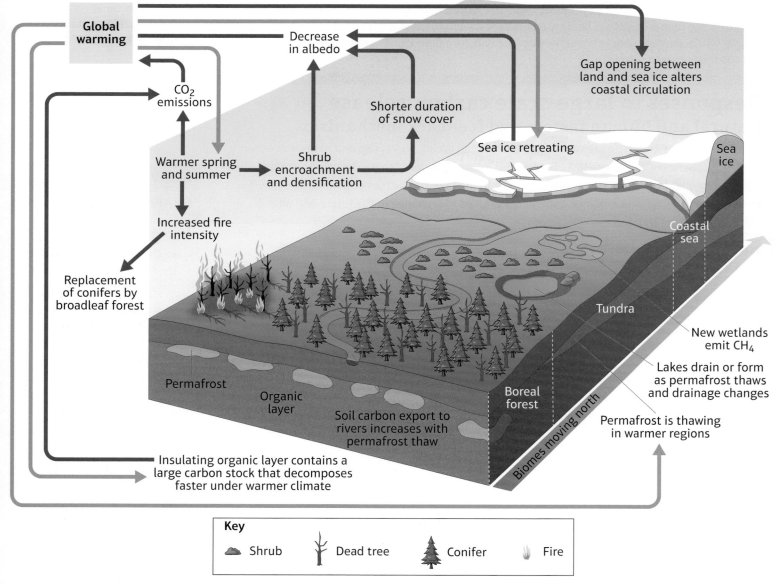

Figure 2.34: Shift of the boreal-tundra biome

Yellow = Players, Orange = Attitudes and actions, Purple = Futures and uncertainties

3. **Dieback of tropical rainforests:** Moist, complex forests, such as in Amazonia, could change to a less carbon-dense, drought- and fire-adapted ecosystem because of longer dry seasons (see Figure 2.28) (but a collapse over large areas is unpredictable).

4. **Dieback of boreal forests:** Arctic ecosystems are vulnerable to thawing permafrost, shrubs spreading into the tundra, an increased number of pests and fires in the boreal forest. Drought stress may cause a collapse of boreal forest since it cannot adapt, with southern margins turning into grasslands (but a large-scale collapse is unpredictable) (Figure 2.34).

5. **Arctic Ocean free of sea ice in summer:** Higher air temperatures and a warmer ocean melt sea ice (Figure 2.35) (considered likely). However, some research suggests that thinner ice and snow cover may increase albedo and cooling (but this negative feedback is not fully understood).

6. **Long-term droughts:** The subtropical dry zone moves poleward as large-scale atmospheric circulation modifies the Hadley cell over a long timescale (Figure 2.28) (however the frequency and duration of mega-droughts in the world is not predictable).

7. **Collapse of monsoon climate circulation:** More intense precipitation occurs during the monsoon wet season because of the transport of more evaporated moisture in warmer air (however a sudden collapse is not predictable).

Figure 2.35: Arctic sea ice extent and change 1980 (yellow line) and 2012

Two other possibilities may occur slowly: collapse of the Greenland or Antarctic ice sheets, and methane release from melting permafrost.

Around the world, species extinction may be abrupt in flat and high-altitude landscapes when a **critical threshold** is reached because flora and fauna have nowhere else to go. In oceans, warm-water coral reefs are unable to survive in cooler deeper water, and low-lying ecosystems flooded by sea-level rise could disappear completely. Any of these changes would have a cascade effect through food webs, affecting food sources.

Earth system models (ESMs) suggest that oceans will absorb more CO_2 from the atmosphere, but a concern is what happens when the ocean reaches saturation point. There is uncertainty about terrestrial ecosystem carbon uptake because of vulnerability to human activities, fire and respiration changes, which would return carbon to the atmosphere (positive feedback). Other positive feedbacks include reduced reflection (albedo) of solar radiation by the Earth's surface (see Figure 2.10) as snow and ice melt, with darker surfaces absorbing more heat energy, and methane releases.

Possible negative feedbacks include:

- increased plant growth as a result of higher CO_2 concentrations

- reduced CO_2 concentrations in the atmosphere slowing the rate of warming

- an increase in cloud cover due to higher evaporation rates which would reflect solar energy (see Figure 2.10)

A level exam-style question

Explain why there is uncertainty in the projections of future atmospheric CO_2 concentrations. (8 marks)

Guidance
You should be able to recall some of the IPCC predictions (via climate models) so that you can express the range. There are uncertainties in the science (feedback mechanisms, carbon fluxes and pathway changes), the effectiveness of international agreements (Kyoto and Paris), and future energy needs (linked to economic and population change).

- dust from drier climates entering the atmosphere and reducing the amount of incoming solar radiation.

Most feedbacks are not yet known – for example links between soil moisture and precipitation or between atmospheric chemistry and the biosphere – but overall it is expected that feedback between climate change and the carbon cycle will amplify global warming.

Adaptation strategies

All areas of human activity require carbon emission mitigation and adaptation strategies.

Solar radiation management

There is speculation about managing solar radiation and the carbon cycle through geo-engineering projects (see Table 2.14), which includes CCS and BECCS (see page 109).

Table 2.14: Geo-engineering methods to manage solar radiation

Geo-engineering method	Potential problems
Spray seawater into atmosphere: would provide hygroscopic nuclei for condensation to take place, creating bright white clouds that reflect Sun's energy into space.	May cause unpredictable weather changes, such as reduced precipitation and changes to regional temperatures. The albedo effect is not fully understood.
Add sulphur particles to stratosphere: the aerosol particles formed would reflect the Sun's energy back into space.	May cause unpredictable weather changes such as reduced evaporation. May affect stability of ozone layer and increase acid rain over large areas.
Trillions of wafer-thin discs in stationary orbit between the Earth and Sun: these would reduce the amount of the Sun's energy reaching the planet.	Very expensive, at least US$5 trillion. Unpredictable effect on whole atmosphere as heat budget would be changed. All countries would have to agree.

Countries will need to take practical action to adapt to climate change. Some populations may even become environmental refugees and move away from uninhabitable areas if they are flooded because of sea-level rise or prone to increased mega-droughts (see page 116).

Water conservation and management

The Lower Mekong River basin, which includes parts of Thailand, Laos, Cambodia and Vietnam, is home to 60 million people who are dependent on agriculture (rice) and fisheries. It is important to manage the river water in the face of climate change, but by 2016 there were no transboundary plans, and only 11 per cent of national plans were linked to projects in the Lower Mekong. In northern China, water-saving irrigation has been introduced in areas of huge demand in a drying climate, in order to adapt to water scarcity and food security issues. Between 2007 and 2009 the country saved up to 11.8 per cent of its previous water consumption.

Resilient agricultural systems

In northern China strategies of early or late planting can match climate change, using crops designed to withstand higher temperatures. For example, maize yields by 2050 could increase by 15.2 per cent. In northern Tibet, where alpine pastures are being degraded by higher temperatures and precipitation, a three-year demonstration plan made use of higher lake levels for irrigation during dry periods, which doubled grassland productivity and increased the number of plant species from 19 to 29.

Land-use planning

Areas of increased risk, such as from coastal or river flooding, can be zoned through urban planning and laws so that people and valuable property are not within them. Planning would also include the location and design of infrastructure to ensure its resilience to climate change factors such as stronger storms and higher water levels.

Yellow = Players, Orange = Attitudes and actions, Purple = Futures and uncertainties

Flood-risk management

Australia is adapting to increased flood risk by making all housing on floodplains more flood-resistant, with raised floors, stronger pile foundations and water resistant materials. Where possible, those living on floodplains are relocated, and an ecosystem approach is used to retain floodplains in urban areas and use wetlands as overspill areas. Reservoir levels will be closely monitored to anticipate inundations and release water at safe levels. However, issues include the high cost of relocating people, reduced property values in rezoned areas, and locals contesting the changes.

Rebalancing the carbon cycle

Energy targets and policies include the development of renewable energy sources, nuclear, carbon capture and storage (CCS), bio-energy with carbon capture and storage (BECCS), and switching from fossil fuels to renewables. Reduced emissions can also be achieved through increasing energy efficiency in supply, transmission and distribution pathways, as well as developing combined heat and power (cogeneration) and controlling methane emissions. In transport there could be a switch to low-carbon fuels and biofuels, more efficient technologies (for example, lighter materials), smaller vehicles, journey avoidance, car sharing and eco-driving practices. In addition, businesses and governments can make infrastructure improvements to improve journey efficiency and move freight from road to rail. In industry, emissions can be captured (CCS) or processed, and a switch made away from fossil fuel electricity to the use of BECCS or waste products. Consumers also have their part to play, by moving away from a 'throwaway' society, sharing journeys, reducing demand, using goods more intensively (as in 'reduce, reuse, recycle, recover/repair').

CASE STUDY: Germany's *Energiewende*

Even before the Japanese Fukushima nuclear disaster after the 2011 tsunami, Germany had been reducing the amount of electricity it produced by nuclear power and developing renewable energy, partly because of pressure from a strong Green Party. Now the country faces the challenge of reducing fossil-fuel use and nuclear energy at the same time. The final nuclear power station closure is scheduled for 2022, and it is planned that solar, wind and biomass energy will produce 80 per cent of Germany's electricity by 2050. The emissions target is to reach 70 per cent of the 1990 level by 2040. However, this policy is not without its challenges:

- There are fixed high consumer energy prices and subsidies (US$18 billion in 2013) to help establish renewable energies (for example in Bavaria).

- Higher energy costs make businesses and industries less competitive in a globalised world and so the government has subsidised these, but this may conflict with EU laws on unfair competition.

- The policy has created a state-planned (top-down) energy industry and economy.

- Solar and wind energy are dependent on weather conditions, and this unreliability needs back-up technology such as battery storage, which is still under development.

- Energy security in the short term comes from conventional power stations burning lignite (brown coal), which produces the highest emissions of any energy source. Since many jobs rely on lignite mining, political difficulties may arise when production is cut (for example in North-Rhine Westphalia).

The 2016 Climate Change Performance Index, based on 58 countries which together emit over 90 per cent of world CO_2 emissions, showed that no country is performing well enough to reduce the impact of their emissions overall. However, good performers included Denmark, the UK and Sweden, with European countries also doing well (Figure 2.36). The poorest-performing country was Saudi Arabia, with Japan, Australia and South Korea also performing poorly. This index considers emissions, use of renewable energy, energy efficiency and government policies on climate change.

Extension

Investigate the extent to which 'crowdsourcing' is a way of informing people about how to save energy and reduce carbon emissions.

Performance
- Very good
- Good
- Moderate
- Poor
- Very poor
- Not included in assessment

At present there are no countries in the 'very good' category

0 1000 2000 3000 km

N

Figure 2.36: The Climate Change Performance Index 2016

Mitigation methods are progressing better in some countries than others, although the 2015 Paris Agreement may lead to greater urgency and action in the near future. Mitigation measures include:

- **Carbon taxation:** Government taxes can encourage people and businesses to reduce carbon emissions by using less fuel or electricity. Specific taxes include congestion and emissions charges, road tolls and taxes on vehicles and vehicle fuels. Construction and industry can be charged carbon taxes to encourage them to use environmentally friendly materials and minimise waste and emissions. Farming can have taxes on fertilisers, especially nitrogen (a greenhouse gas). In the UK, vehicle fuels are highly taxed; in 2016 taxes accounted for over half the cost of petrol and diesel. London has congestion and emissions charging zones, but at present no tax is payable on a car in the UK emitting less than 100 g/km of CO_2.

- **Renewable switching:** UK policy is based on switching away from fossil fuels to low-carbon energy sources. The UK established a target of 15 per cent of energy consumption by 2020 to be from renewables, including onshore and offshore wind, biomass electricity, biomass heat, air and ground heat pumps, and renewable transport. Projected savings in greenhouse emissions in 2020 could be 134 $MtCO_2$, with over 65 per cent of these savings from renewable switching (Figure 2.37).

- **Energy efficiency:** Technological improvements mean that vehicle engines are more efficient. For example, between 1970 and 2008 the fuel efficiency of the average passenger car in the USA improved by 57 per cent. Hybrid vehicles such as the 1,200 Routemaster buses in London are extending efficiency and reducing emissions. In the EU, household electrical items now use less electricity, being rated on an A-to-G scale with A+++ being the most efficient. The UK government estimates that by 2020 its energy-efficiency policy should save the energy use of 9 million households, or the output of 19 power stations.

Yellow = Players, Orange = Attitudes and actions, Purple = Futures and uncertainties

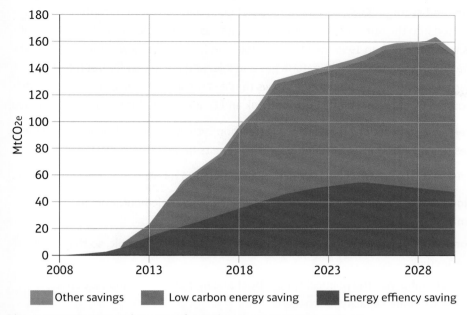

Figure 2.37: UK projected savings of greenhouse gas emissions, 2010–30

Legend: Other savings | Low carbon energy saving | Energy effiency saving

- **Afforestation:** Forests are an important carbon store (see page 50), and planting more trees will absorb more CO_2. In addition to schemes such as REDD (Reduced Emissions from Deforestation and Forest Degradation), which pay forest owners not to cut down forests, afforestation is the conversion of land that has not been forested for at least 50 years. In Mapanda district in Tanzania, over 3,500 ha of eucalyptus, pine and fruit trees have been planted on degraded grassland and on maturity will be used by local people. It is estimated that between 2000 and 2020 this afforestation will have removed an average of over 166,000 tonnes of CO_2 per year from the atmosphere.

CASE STUDY: The Paris Agreement, 2015

The 1997 Kyoto Protocol set CO_2 emission reduction targets for industrialised nations to meet. It was not a truly global agreement, as it omitted emerging and developing countries on the grounds that they had not created the pollution. However, with the severity of climate change apparent around the world and very high emissions from emerging countries such as China and India, it was soon clearly necessary to involve everyone in solving the global problem. In 2015 at a conference in Paris, 195 countries promised to reduce greenhouse gas emissions to almost zero by 2065; countries will monitor themselves and report progress every five years. The richer nations pledged around US$100 billion a year by 2020 to help poorer countries adapt to climate change and switch away from fossil fuels. The overall aim is to limit temperature increase to 1.5°C above pre-industrial levels, as many believe that 2°C would create damaging change (see Table 2.2).

However, there is little to force countries to meet the targets, and progress reporting may not be accurate. Perhaps only pressures from within countries (a bottom-up approach), arising from severe air pollution and a high risk of coastal flooding, such as in China, will force governments to take action. However, science suggests that the target is unrealistic, with 1°C of warming being reported at the start of 2016. In addition to the huge reduction in fossil-fuel use it will also be necessary to remove carbon from the atmosphere, which is expensive – the International Energy Agency estimates it will cost about US$16.5 trillion by 2030.

ACTIVITY

1. Consider the range of methods available to players for rebalancing the carbon cycle.

   a. Evaluate these methods by considering their effectiveness in limiting the average global temperature increase to 1.5°C above pre-industrial levels.

   b. State which method you believe is the most important, and why.

Extension

Investigate the Living Planet Index, the Ecological Footprint, the Safe Operating Space model, the Happy Planet Index, the Environmental Sustainability Index and the Climate Change Performance Index. Use the information that these provide to assess:

- the progress being made to adapt and mitigate climate change

- the attitudes of different groups towards action.

ACTIVITY

1. Investigate the World Bank's Climate Change Knowledge Portal. For a selection of countries, explore rainfall and temperature change over time, future projections by climate models, impacts on agriculture (land and yields), water, forest area and natural hazards (floods and droughts). Make notes to be able to compare changes and impacts.

2. Investigate the World Bank's climate adaptation country profiles for the countries you selected in 1. Make notes on the vulnerability of the countries and the actions they are taking.

- **Carbon capture and storage (CCS):** Large producers of CO_2 could be forced to capture the gas before it is emitted into the atmosphere, compressing and storing it underground in disused oil and gas fields (see page 109). In 1996 the Sleipner project in the North Sea was the world's first CCS scheme, taking unwanted CO_2 from natural gas and injecting it 3 km underground into a porous and permeable rock layer. Monitoring shows that it is securely confined. There are plans to transfer captured CO_2 from mainland UK to the Auk and Goldeneye fields under the North Sea.

Global-scale agreements and actions to rebalance the carbon cycle have proved slow to implement because countries put their own economic and political interests first (Figure 2.38).

The IPCC (2014) suggests that three spheres of action are needed to rebalance the carbon cycle:

- **Practical actions** that are observable and measurable, including implementation, monitoring and evaluation

- **Political actions** to set up systems and structures, such as legal and economic, so that practical action can be effective

- **Personal actions** to influence the political sphere by changing behaviours and responses, including systems to create capacity for change.

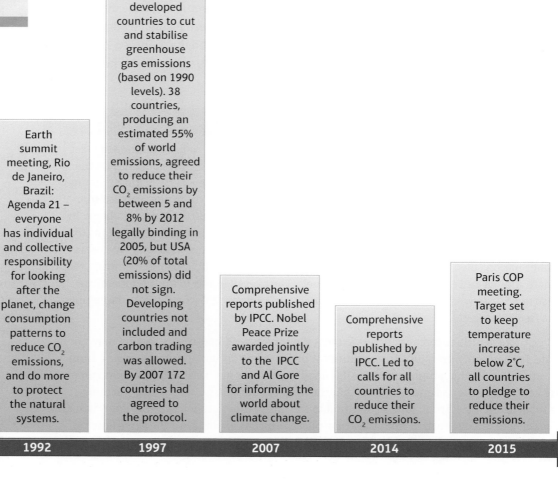

Figure 2.38: Timeline of global action on climate change

Yellow = Players, Orange = Attitudes and actions, Purple = Futures and uncertainties

Summary: Knowledge check

Through reading this chapter and by completing the tasks and activities, as well as your wider reading, you should have learned the following and be able to demonstrate your knowledge and understanding of the carbon cycle and energy security (Topic 6).

a. Explain the interactions within the biogeochemical carbon cycle, including stores and fluxes.

b. Describe how geological processes store carbon and also release it into the atmosphere.

c. Explain how processes within the ocean store and move carbon.

d. Explain how processes on land store and move carbon, including biological carbon.

e. Describe and explain the natural greenhouse effect.

f. Explain how natural systems and human activities affect the balance of carbon pathways and stores.

g. Explain the global pattern of energy consumption and the role of energy players.

h. Describe and explain energy pathways and their role in providing energy security.

i. Explain the role of unconventional fossil fuels and renewable and recyclable energy in future energy mixes.

j. Describe and explain the costs and benefits of using biofuels and radical technologies to provide energy security and reduce carbon emissions.

k. Describe and explain how human activities may affect the health of terrestrial and marine ecosystems.

l. Describe and explain the threats to human wellbeing of changes to the health of the natural environment and its systems.

m. Explain why there is uncertainty about future changes as a result of higher levels of greenhouse gases.

n. Describe and explain ways in which people may adapt to climate change.

o. Describe and explain ways in which people may mitigate climate change and rebalance the carbon cycle.

As well as checking your knowledge and understanding through these questions, you also need to be able to make links and explain the following ideas:

- The Earth has a long geological history in which the natural carbon cycle has operated in conjunction with other systems such as the rock cycle. The carbon cycle consists of stores and fluxes operating on different timescales and through a variety of processes, including exchanges between atmosphere and oceans, and volcanic activity.

- The biosphere has an important role to play in the sequestration and transfer of carbon, both on land (terrestrial) and in the sea (marine). Numerous small, shorter-timescale processes are involved, such as decomposition, respiration and photosynthesis, as well as large-scale processes such as the thermohaline circulation.

- The natural balance of the Earth's systems is largely based on energy received from the Sun, which determines atmospheric movement and the pattern of precipitation and temperature, which in turn affect biome location and photosynthesis. Human activities are disturbing the balance between climate, ecosystems and the water cycle by adding more greenhouse gases to the atmosphere.

- Countries wish to develop and therefore consume energy, especially fossil fuels, to provide for businesses, industries and people. Securing energy for the future is a key aim, and many players are aiming to ensure that this happens, including OPEC, energy TNCs and consumers.

- Sources of energy, especially fossil fuels, are often not located in the areas of greatest demand, and so they have to be transported along a pathway. These pathways are not always secure due to physical and human factors, which can have consequences for fragile environments.

- Research and development is being carried on into unconventional fossil fuels, renewable and recyclable energy, biofuels and radical technologies. All of these alternatives to traditional fossil fuels have advantages and disadvantages, but some may enable a fossil-fuel-free energy mix in the future.

- Many human activities threaten the natural carbon and water cycles, changing stores and fluxes, and adding even more CO_2 into the atmosphere. Higher CO_2 concentrations increase the acidity of oceans and affect marine ecosystems, as well as increasing drought and floods.

- Degradation of carbon and water cycles has impacts on the wellbeing of people, threatening marine and terrestrial food supplies, income from tourism and other businesses, and protection from flooding – both coastal and riparian.

- Actions are needed to prevent further changes and tipping points, but achieving global action is not easy and takes many decades. Although the IPCC has assembled global evidence in various reports, there is still uncertainty about what will happen. Adaptation and mitigation measures are required, ranging from geo-engineering to vehicle-sharing.

Preparing for your A level exams

Sample answer with comments

Assess the importance of renewable energy sources to achieving a carbon-neutral energy mix.
(12 marks)

There is a range of renewable energy sources including offshore and onshore wind, solar, and nuclear and these are being promoted in a number of countries such as the UK, Denmark and Germany as ways of reducing greenhouse gas emissions such as CO_2. However, can these supply sufficient energy to meet the demands of the future?

The energy mix of the UK has changed over time, with a decrease in the use of fossil fuels and an increase in the use of renewables. Similarly, Denmark has seen a dramatic fall in primary energy production from crude oil, from about 800 petajoules in 2005 to just over 300 in 2015, while renewable energy has increased from about 50 petajoules in 1990 to just below 150 in 2015. At the same time, Denmark's energy consumption has changed so that renewables are now second behind oil, well above natural gas and coal. It is the fossil fuels that when burned release CO_2, so reducing them helps reduce carbon emissions.

On a world scale the primary energy supply in 2012 was mostly from the three main fossil fuels, with biofuels and waste being the largest of the other energy sources, mainly due to their use and importance in developing countries where poor people cannot afford the expensive fossil fuels or the technology to develop other renewables. In China, hydroelectric power is significant and wind energy is developing but these are very small contributors to the primary energy mix of the country.

Solar power is found in several forms from solar panels on roofs, solar farms and solar power stations. In Germany this form of renewable energy is an important part of their Energiewende and is helping to reduce greenhouse gas emissions.

Nuclear energy may be carbon neutral but it has its own long-lasting source of pollution in radiation, as illustrated by the Fukushima incident in 2011, which continues to cause problems in Japan, and has prompted countries such as Germany to plan to close all their nuclear power plants by 2022. So renewable energy resources at the moment can only partially help create a carbon-neutral energy mix because countries cannot make the switch on a large enough scale to replace all fossil fuels, and renewables themselves have problems.

This introduction adequately sets out the themes, including a hint in the last sentence of assessing importance, but the idea of 'carbon neutral' must be the central idea throughout the answer – not energy security.

Appropriate examples are used, but the assessing is basic and needs further development.

Rather than focusing on the command word 'assess', the candidate continues to attempt to use examples, but more facts were needed. For example, what are the difficulties involved with replacing oil and coal in China with renewables and therefore how important can they become?

There is a good range of knowledge in this answer and a basic point relevant to the question in this paragraph.

This conclusion refers to the question and makes a main 'assess' point, but there is only a limited explanation within broad statements.

Verdict

This is an average answer, with some application of knowledge and understanding with appropriate connections to the question themes. But the interpretation of the question is only partial: for example, in assessing the importance of renewables the candidate should have considered their role in transport and electricity production (secondary energy) as well as primary energy sources. Greenhouse gas levels could have been considered because, even as renewables have been increasing, the contribution of CO_2 from fossil fuels and industry use has still been increasing. So this answer is unbalanced.

Preparing for your A level exams

Sample answer with comments

Evaluate the extent to which geological processes control the carbon cycle. (20 marks)

The carbon cycle involves many processes. Some work on a relatively short timescale, such as the biological processes, while others work on a very long timescale, such as the geological processes. At times these processes combine in the biogeochemical carbon cycle with fluxes between the different parts. In this way a natural balance is reached, with several controls including the geological processes.

This introduction is packed with terminology and it outlines a possible structure for an answer. However, it lacks incisive points, such as what is meant by 'geological processes'.

In the carbon cycle most carbon is stored within sedimentary rocks. This takes a long time, millions of years through the biogeochemical processes in the ocean, where carbon absorbed from the atmosphere is used by marine creatures to make their shells or skeletons. For example, coral polyps create the base on which successive generations live. Over time, dead sea creatures sink to the seabed, creating a calcareous ooze which eventually gets buried by other sediments. This creates a process of diagenesis and limestone rocks are created. In some places biological matter is also buried and turned into fossil fuels.

This shows a thorough understanding of the long-term geological cycle with good use of terminology, but it could have said how much carbon is stored in sedimentary rocks and fossil fuels. A comparison with other stores would enable an evaluation of roles in the carbon cycle to be made.

Other geological processes involve tectonic activity where greenhouse gases are released by out-gassing where there is volcanic activity, such as at constructive plate boundaries, subduction zone volcanoes and hot spots. Much of this is intermittent. Biological processes can sequester carbon wherever there is plant life carrying out photosynthesis, taking it from the atmosphere and storing it for shorter timescales than the geological processes. People have also disrupted the carbon cycle by increasing emissions of greenhouse gases, especially CO_2.

There is a high level of understanding, but no supporting case study evidence or facts related to how much carbon moves or is stored. A factual comparison was needed in order to allow a full and objective evaluation.

Crustal rocks contain carbon because sedimentary rocks, especially limestone, are formed when calcium carbonate has been deposited on the seabed. During the Carboniferous geological period the formation of coal and crude oil created a long-term sink, for example oil contains about 85% carbon.

There are many carbon stores, such as in the soil, vegetation, freshwater, ocean layers and sedimentary rocks, including fossil fuels. Over time there is some movement between these stores, and the geological stores are largest and so control the carbon cycle over geological time. However, their influence over a shorter period of time is limited to weathering and out-gassing by volcanoes, and the settling of new calcareous ooze on the bottom of oceans.

The conclusion attempts to answer the question and does so in general terms with two evaluation points, but the candidate also needed to consider the alternative processes involved in the carbon cycle.

The implication in this paragraph is that geological processes are very important over a long time scale, but it needed to be expanded to explain how this controls the carbon cycle.

Verdict

This answer is at the upper end of average, but could have been even better. The processes are covered well, demonstrating understanding throughout, but knowledge is weaker – for example there is no data comparing stores or fluxes, which limits the opportunities for a full evaluation. An interpretation is supported only by understanding and not evidence. For example, sedimentary rocks contain about 83,000,000 PgC, with the next largest store being the intermediate and deep ocean with 37,100 PgC, and so geological processes are extremely important to controlling the long-term carbon cycle. However, in the short term, photosynthesis and respiration have the largest fluxes (123 PgC/yr and 118.7 PgC/yr respectively), greater than volcanic activity (0.1 PgC/yr), so have more control of the carbon cycle than geological processes.

Read the following extract carefully and study Figure A. Think about how all the geographical ideas link together or overlap. Answer the questions posed at the end of the article. This extract is from *The Economist* online, source dated 10 January 2015.

THE TWILIGHT OF THE RESOURCE CURSE?

Over the past decade Africa was among the world's fastest-growing continents – its average annual rate was more than 5% – buoyed in part by improved governance and economic reforms. Commodity prices were also high. In previous cycles African economies have crashed when the prices of minerals, oil and other commodities have fallen. In 1998–99, during an oil-price fall, Nigeria's naira lost 80% of its value. African currencies again took a beating during a period of turmoil in commodity markets in 2009. Since last year the price of oil has fallen by half and many metals such as copper and iron ore have also dropped sharply. With commodity prices plunging, will the usual pattern repeat itself?

In some economies large drops in commodity prices have led to currency falls. At least ten African currencies dropped by more than 10% in 2014. But there have been few catastrophic depreciations. This suggests that investors do not see lower commodity prices as a kiss of death. Ghana's currency, the cedi, was the continent's worst-performing currency in 2014, having lost 26% against the dollar. But it tumbled not because investors fret about the impact of lower commodity prices. In fact, Ghana is by African standards not especially commodity-dependent. Rather, it has in recent years run a lax fiscal policy. In 2013 its budget deficit hit 10% of GDP.

The mall, not the mine

One reason currencies have been robust may be because economic growth is starting to come from other places. Manufacturing output in the continent is expanding as quickly as the rest of the economy. Growth is even faster in services, which expanded at an average rate of 2.6% per person across Africa between 1996 and 2011. Tourism, in particular, has boomed: the number of foreign visitors doubled and receipts tripled between 2000 and 2012. Many countries, including Ethiopia, Ghana, Kenya, Mozambique and Nigeria, have recently revised their estimates of GDP to account for their growing non-resource sectors.

Despite falling commodity prices, the outlook also seems favourable. Wonks at the World Bank reckon that Sub-Saharan Africa's economy will expand by about 5% this year. Telecommunications, transportation and finance are all expected to spur economic growth. What explains Africa's increasing economic diversification? A big pickup in investment helps. That has arisen partly because governments have worked hard to make life better for investors. The World Bank's annual 'Doing Business' report revealed that in 2013/14 sub-Saharan Africa did more to improve regulation than any other region. Mauritius is 28th on the bank's list of the easiest places to do business. Rwanda, which 20 years ago suffered a terrible genocide, is now deemed friendlier to investors than Italy.

After two decades of poor performance, Africa's total investment as a percentage of GDP increased after 2000. Foreign direct investment (FDI) into Africa rose by 5% in 2012 and 10% in 2013, despite global stagnation. Ten years ago almost all FDI went to resource-rich African economies; resource-poor economies received very little. Resource-rich countries still receive more FDI in absolute terms; but resource-poor economies outpace them when investment is measured as a share of GDP. Foreign investors from other African countries are especially keen on non-commodity industries: nearly a third of their investments are in financial services.

The most resource-intensive economies are working hard to diversify. For the past three years growth in Nigeria, Africa's biggest economy, has exceeded 5%. You might think its growth is being powered by oil exports. Nigeria has Africa's second-largest reserves, it is the fifth-largest exporter and, according to the IMF, oil accounts for 95% of all exports. But in recent years the Nigerian oil industry has stagnated. Growth has instead come from things like mobile phones, construction and banks. Services now represent 60% of GDP. Angola is similar. It is Africa's second-largest oil producer and the stuff makes up the vast majority of exports. But its 5.1% expansion in 2013 came mainly from things such as manufacturing and construction. In 2013 fishing expanded by 10%, and agriculture by 9%. About a third of government revenue now comes from non-oil sources, compared with almost nothing a decade ago, economists at Standard Bank reckon. In Botswana the percentage of GDP made up by the mining and quarrying of goodies like diamonds, gold and copper has fallen from 46% in 2006 to 35% in 2011, according to the 'African Economic Outlook'. Other countries that are successfully diversifying are Rwanda – which has thriving banks and business-services firms – and Zambia, which although still copper-dependent has posted growth of 12% a year in financial services. Congo-Brazzaville, where oil makes up 80% of exports, is seeing rapid growth in construction and transport. That may be further fuelled by the All-Africa Games, which are to be held this year in the capital, Brazzaville.

Better fiscal policy also plays an important role. Commodity markets are volatile; government spending smooths out the booms and busts. Until a few years ago, nearly all African economies spent freely when their economies were hot, only to rein in spending when things cooled down. That is the opposite of what most economists would advise a finance minister to do. But in recent years, according to a report from the World Bank published on January 7th, fiscal policies in many African countries have become more sensible. These days a fair number of African economies save money during the good times, in order to spend it in the bad ones. There is still a long way to go. Africa is still the continent most dependent on commodity exports. Countries such as Tanzania and Nigeria want to develop giant gasfields which, while boosting the economy in the short term, could tie them more closely to commodity cycles. Some worry that investment in infrastructure will fall

as mining companies retrench. Even so, there is reason to think the 'resource curse' is losing its power. Despite turmoil in commodity markets, Africa is still one of the world's fastest-growing regions. With better education systems, investment in infrastructure and sensible regulatory reforms, the continent could completely break the spell that has held it back so often in the past.

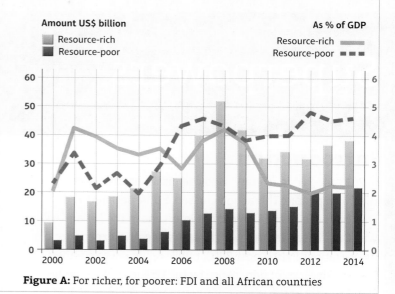

Figure A: For richer, for poorer: FDI and all African countries

ACADEMIC SKILLS

1. How reliable and unbiased are the sources of information used in this article?

2. Explain why percentage growth rates may be misleading when assessing the development of countries.

3. Study Figure A. Assess the strengths and weaknesses of this graph, especially the extent to which it clearly shows the correlation between foreign direct investment (FDI) and resource-rich/resource-poor African countries.

4. Suggest what other information and data would be useful to being able to make a more accurate judgement on the end of the 'resource curse'.

ACADEMIC QUESTIONS

5. a. Explain the links created between industrialised nations and resource-rich African countries'.

   b. Why has this sometimes been described as the 'resource curse'?

6. Explain the role of African countries' government policies and regulations, and foreign direct investment in assisting the development of African countries.

7. Assess the extent to which the article offers evidence that African countries are progressing through Rostow's stage model of economic development.

8. Suggest which other improvements African countries need to make in order to achieve sustainable economic development.

9. Evaluate the following statement: 'Investors do not see lower commodity prices as a kiss of death for African economies.'

Human systems and geopolitics

Superpowers

Introduction

In 2013 the Syrian military used chemical weapons to attack their own people. This action crossed a USA-declared 'red line', but even so the USA did not intervene in the Syrian Civil War. One reason for this was the USA's negative experience of its operations in Iraq post-2003; public opinion was another. Perhaps decades of dominance by the USA in conflict arenas – and control of the world order by a single country – could be coming to an end?

This is not the only event suggesting that the USA's status as the sole global superpower in the 21st century is being challenged. Votes by the UN Security Council on taking action against fundamentalist terrorism in the Middle East have been subject to vetoes and abstentions by China and Russia. When, in 2014, Russian troops entered Ukraine to annex the Crimean Peninsula, certain assumptions that power in a globalised world might be exercised through 'cultural influence' instead of 'military action' were shown to be premature, and modern geopolitical theory was turned upside down. Even when, in 2015, China constructed a runway on artificial reefs in the Spratly Islands in the South China Sea, reinforcing its controversial claim over the archipelago and its associated mineral wealth and trade routes, the strong economic links between China and the USA made conflict unlikely.

These changes are the by-products of tremendous shifts in global power over the last century. Europe's former colonial empires may have fragmented, but the legacy of mineral extraction and trade, arbitrary border creation and two world wars may help to explain many contemporary tensions around the world. Cultural divides intersect with economic inequality and scarcity of resources such as food, water and energy. The problems have been particularly acute in the Middle East, which has the highest population growth rate in the world as well as most of the world's oil. Decisions by a few players have a disproportionate effect on the global economy, politics and the natural environment, and competing spheres of influence create huge uncertainty when defining 'power' today.

In this topic

After studying this chapter, you will be able to discuss and explain the ideas and concepts contained within the following enquiry questions, and provide information on relevant located examples:

- What are superpowers and how have they changed over time?
- What are the impacts of superpowers on the global economy, political systems and the environment?
- What spheres of influence are contested by superpowers and what are the implications of this?

Figure 3.1: The Statue of Liberty, New York, USA. To what extent do the Statue of Liberty and the New York skyline symbolise the superpower status of the USA?

Synoptic links

The role and interaction of key global players are crucial for understanding this topic, whether that be individuals, TNCs, intergovernmental organisations such as the EU and ASEAN, nation-state governments or the United Nations. In your studies of globalisation you will have learned that some players are more powerful than others, and that their influence changes over time. Judgements need to be made about the continued importance of countries like the USA and UK within NATO, and whether they can or should continue to act as global police. Chapter 4 considers whether there is a time and place for military intervention in cases of human rights abuses. Every player has contrasting attitudes and ideologies: some are willing to work together on critical issues; others are not. When this comes to demonstrating action and leadership over global issues such as climate change, water and energy security, the stakes are high. Tremendous complexity and significant tension in some parts of the world leave considerable uncertainty about how a world that is not dominated by one country might operate.

Useful knowledge and understanding

During your previous studies of Geography (GCSE and AS Level), you will have learned about some of the ideas and concepts covered in this chapter, such as:

- The characteristics of developed and emerging countries.

- The causes of global inequalities, including the environment, colonialism, and neocolonialism.

- Rostow's modernisation theory and Frank's dependency theory.

- How TNCs contribute to globalisation to the benefit of some countries but not others.

- The links between TNCs and intergovernmental organisations.

- Trading and trade routes.

- The national, regional and global context that countries find themselves in.

- Impacts of economic development and globalisation on the environment, both locally and globally.

- Changing geopolitical relationships between emerging countries and the EU/USA.

- The global patterns of megacities and future projections of urbanisation.

- The uneven distribution of oil supply and consumption, and how prices are affected by international relations.

This chapter will reinforce this learning but also modify and extend your knowledge and understanding of geopolitics and international relationships between the emerging and the developed world.

Skills covered within this topic

- Construction of power indices using complex data sets, including ranking and scaling.

- Mapping of past, present and future spheres of influence and alliances using world maps.

- Use of graphs of world trade growth using linear and logarithmic scales.

- Mapping of emissions and resource consumption using proportional symbols.

- Plotting of the changing location of the world's economic centre of gravity on world maps.

- Analysis of future gross domestic product (GDP) using data from different sources.

What are superpowers and how have they changed over time?

Learning objectives

7.1 To understand that geopolitical power stems from a range of human and physical characteristics of superpowers.

7.2 To understand why patterns of power change over time and can be uni-, bi- or multipolar.

7.3 To understand how emerging powers vary in their influence on people and the physical environment, which can change rapidly over time.

Geopolitical power
The characteristics of superpowers

A **superpower** is a nation, or group of nations, with a leading position in international politics. From the mid-19th century to the early 20th century the UK was arguably the world's superpower, having successfully created a global empire with strong trading links, which it then defended against challenges by other European countries. However, following the intervention of the USA in two world wars, that country started to emerge as a superpower, challenged by Russia during the Cold War. The USA established sole authority during the Cold War, but the long-term legacy of change, especially through the globalisation of freedom and democracy, meant that within each continent a number of countries can now claim to have significant influence. These **regional** powers include countries such as Saudi Arabia, South Africa, Chile, Australia, India and, within Europe, Germany and the UK. As a resurgent China benefits from its own unique twist on capitalism, some see its wealth and influence as a future challenge to the USA's status as the primary global economic superpower. Table 3.1 shows the different ways superpowers can be defined. In 2015 *Time Magazine* identified that:

- The USA's per capita GDP was US$53,000 compared to China's US$6,000 (note that there are different ways to measure GDP, some of which suggest the gap is much smaller)

- 80 per cent of all financial transactions and 87 per cent of foreign currency market transactions are in US dollars

- The USA's military spending is four to five times that of China, accounting for 37 per cent of global military spending

- The USA is the most favoured destination for migration – 45 million people living the USA were born in a foreign country, four times that of the next-highest country

- The USA hands out the most money in the world in financial assistance (US$33 billion), with the UK being second (US$19 billion)

- 16 of the top 20 universities in the world are in the USA.

Yellow = Players, Orange = Attitudes and actions, Purple = Futures and uncertainties

Table 3.1: The characteristics of superpowers

Economic	High GDP and high levels of trade, including influence over global trade Home to many TNCs Hard currency held in reserve by other countries
Political	Permanent seat on the UN Security Council, together with powerful allies Many multilateral agreements
Military	High expenditure, largest amount of hardware and personnel, including nuclear weapons Could command global military control Unparalleled intelligence networks Exporters of technology
Cultural	Long-standing tradition and rich cultural history or way of life voluntarily enjoyed by many around the world, for example music and fashion
Demographic	Significant percentage of global population Attracts skilled migrants and other workers
Access to resources	Able to export and control the supply of valuable commodities, for example oil, or able to secure the resources it needs. On the other hand, multiple resources make a country less dependent on others (e.g. energy security) Occupying a world location that enables it to command influence

Maths tip

Normalisation is a process of scaling data in different datasets; this ensures that each set has the same variance (between 0 and 1), enabling a direct comparison and reducing the impact of outliers when comparing the data.

In Excel use this formula:
=(**A1**-MIN($A$1:$A$12))/ (MAX($A$1:$A$12)-MIN($A$1:$A$12))

Without Excel:

- for each column, calculate the difference between an individual country's factor and the lowest score for all the countries

- divide this result by the highest score for all the countries.

Table 3.2: Data for a 'soft'-power index score

	Gender equality	Normalised gender score	Labour participation	Normalised labour participation score	Education spending (% of GDP)	Normalised education spending score	Combination of index scale scores	Ranking
Norway	1.00	0.89	68.70	0.04	6.60	1.00	1.93	12
UK	0.97	0.74	68.70	0.04	6.00	0.87	1.65	7
USA	1.00	0.89	68.90	0.05	5.20	0.70	1.64	5
Japan	0.96	0.72	70.40	0.12	3.80	0.39	1.23	4
China	0.96	0.71	67.80	0.00	3.80	0.39	1.10	2
Russia	1.02	1.00	71.70	0.18	4.10	0.46	1.64	6
Ethiopia	0.84	0.14	89.30	1.00	4.70	0.59	1.73	10
Philippines	0.98	0.80	79.70	0.55	3.40	0.30	1.65	8
Mozambique	0.88	0.34	82.80	0.70	5.00	0.65	1.69	9
Ivory Coast	0.81	0.00	81.40	0.63	4.60	0.57	1.20	3
Malaysia	0.95	0.66	75.50	0.36	5.90	0.85	1.87	11
Georgia	0.96	0.73	75.10	0.34	2.00	0.00	1.07	1

TECHNOLOGY/ICT SKILLS

1. A skill you need to develop is the ability to calculate power indexes using data from multiple data sets. There are two parts to this process: scaling and ranking.

Use a database such as those in the United Nations Development Report or the World Bank website:

a. Choose 6–8 different indices (factors) for a set of 10–15 countries. (Table 3.2 shows indices chosen to assess 'soft' power.)

b. Normalise the data in each column by changing the raw data for each factor into a figure between 0 and 1 (this is called scaling). See the Maths tip for how to do this in Excel.

c. Calculate the total of index scale scores for each country (see Table 3.2). This total can be referred to as a 'power index': the higher the score, the more power a country has.

d. Place your chosen countries into rank order according to how much power they have. Are there any surprises in this order? Explain why some countries are higher or lower than you would have expected. For example, in Table 3.2 Norway ranks highest.

How superpowers maintain power

A country can gain or maintain power through mechanisms that are broadly classified as 'hard' or 'soft' power. Writing about the latter, Antonio Gramsci (an Italian Marxist and geopolitical theorist) was inspired by how Mussolini had maintained power in Italy in the 1930s. The Italian people's willingness to accept the government's values kept Mussolini in power without the use of force. Gramsci described this as a form of 'cultural **hegemony**', or what Professor Joseph Nye of Harvard University has described as non-coercive, **soft power** that attracts and co-opts the views and agreement of other countries. Successful use of soft power explains why the UK, while no longer a global superpower, continues to exert considerable influence around the world. This soft power has three main features:

- **History** – families from all over the world send their children to study at British universities, particularly those in London, Oxford and Cambridge. The cultural and other relationships established through the British Empire live on through the Commonwealth. The British common-law legal approach (case law), where judges decide cases on the basis of previous judicial outcomes, and other aspects of our legal system, are widely modelled around the world, while its neutrality, transparency and continued development (which are attractive to companies) have facilitated the powerful growth of international finance in the City of London.

- **Culture** – the BBC is a major international broadcaster and besides a rich literary, artistic and musical legacy, English is the most widely spoken language after Mandarin. The 2012 Olympics reasserted Britain's capability to host major international events and the opening ceremony showcased the country's many contributions to the world, not least the invention of the internet. The UK also exports knowledge management in the form of international consultancy firms such as PriceWaterhouse Coopers.

- **Diplomacy** – the UK has one of the largest networks of embassies and high commissions. British diplomats are widely respected and Britain has been hugely influential in imposing economic sanctions, for example on Russia after its involvement in Ukraine. Henry Kissinger, the former US Secretary of State, suggested that throughout European history Britain has always sought to support weaker countries against takeover by stronger ones, thereby maintaining crucial balances of power across the continent.

The failure of the USA to react to human rights abuses in countries such as Syria, in 2013–16, has called into question the effectiveness of soft power. The historian Niall Ferguson asserts that superpowers should stand astride the world like a 'Colossus', recognising that **hard power**, in the form of military force and economic change, is vital. Figure 3.1 shows the Statue of Liberty, a deliberate reformulation of the Ancient Colossus of Rhodes, emphasising the 'freedom' that is possible in the USA because of its hard power around the world. There are two major aspects to hard power:

- **Military power** – the USA exercised hard power by confronting the Taliban and bringing about the death of Osama bin Laden. It responded to Kuwait's request for military help in the 1991 Gulf War, following that country's invasion by Iraq, and subsequently removed Saddam Hussein from power in Iraq in 2003. The Afghanistan War was prompted by the 2001 terrorist attack on the World Trade Centre by Al-Qaeda, an extremist organisation.

- **Economic power** – although USA and Chinese GDP totals are similar, 2015 per capita income in the USA was four times that of China. The USA remains the largest trading partner for many countries, exporting high-value goods (for example military aircraft) and global brands (for example, Apple). The USA has dominance in innovation and intellectual property, such as patents. The World Bank, the IMF and the World Trade Organization (WTO) are all vital economic tools for spreading Western influence.

Yellow = Players, Orange = Attitudes and actions, Purple = Futures and uncertainties

Although hard power may be necessary for establishing dominance, the consequences of US and British operations in the Middle East have damaged the reputations of both countries, and both appear to have become increasingly reliant on soft power to maintain their influence.

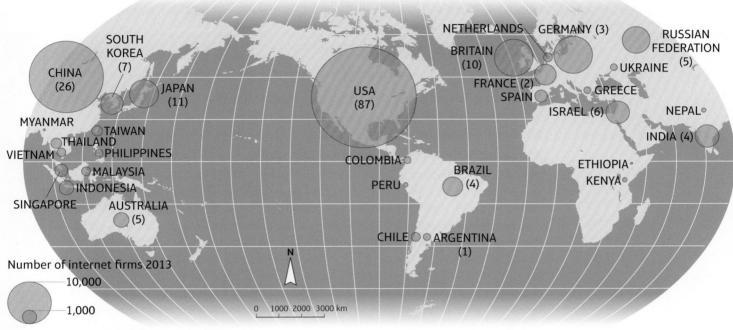

Figure 3.2: Economic dominance by technology (number of internet firms)

Superpowers and geostrategic theory

Geostrategic theory is not new. For example, the Romans and other ancient civilisations believed that their success depended on a strategic understanding of physical geography. During the period of European empire-building, 'completing the world's blank maps' was an incentive for exploration and sometimes conflict. In Britain, Halford Mackinder (1861–1947), the first Reader of Geography at Oxford University, believed that, to exert influence, it was crucial for a country to control strategic areas of land. After the First World War, Nazi Germany believed that occupying other countries was a logical and vital way to ensure *Lebensraum*, or 'living space'. This association of **geopolitics** with physical geography (a form of environmental determinism) in Nazi Germany and the colonial powers meant that the term 'geopolitics' was avoided for many decades.

After the Second World War, the USA adopted a policy of containment, to ensure that Russian ideology (communism) would not spread by force or influence countries recovering from the war. Both Russia and the USA poured money into countries on the boundaries between 'West' and 'East' – such as Germany, Afghanistan, the Caribbean region, Korea and Malaysia. This increased aid and support aimed to encourage countries to resist communism. The indirect battles and conflicts that resulted are described as the Cold War.

In the 21st century, countries try to maintain power in many different ways. Klaus Dodds (Professor of Geopolitics at Royal Holloway, University of London) says that 'geopolitics is as much about high profile and dramatic (for example the Obama administration's policy to pursue high-profile terrorist targets in Pakistan and Yemen) as it is about the everyday, the banal, and the apparently mundane (for example children reciting the pledge of allegiance in classrooms all over the United

ACTIVITY

Explain the advantages and disadvantages for a superpower when exercising influence through hard power and soft power.

States)'. The UK Government, faced with a possible conflict between nationalistic reactions to migration and the rise of extremist ideology, has introduced counter-terrorism strategies such as 'Prevent' to ensure that interactions online or face to face are shaped by 'British values' (such as freedom of expression and democracy). At the same time, films and other media (for example the two James Bond films *Skyfall* and *Spectre*) challenge the relevance of military capability and intelligence networks in the face of global media and the internet.

CASE STUDY: Mackinder's geographical pivot theory

Halford Mackinder believed that the world was divided into three components (Figure 3.3):

1. The World Island, comprising Europe, Asia and Africa – the largest and most wealthy combination of continents

2. Offshore islands, including the British Isles and Japan

3. Outlying islands, including North and South America and Australia

At the centre of the World Island was the Heartland – which Mackinder described as the Pivot area because it contained 50 per cent of the world's resources. Controlling this area secured control of the World Island and, in turn, the rest of the world. When Mackinder was writing, the Heartland was controlled by Russia but could be invaded by Germany or Japan and an alliance. Previous invasions had not succeeded because land transport technology meant that conflict inland could not be sustained (for example, Napoleon's ill-fated invasion of Russia in 1812 had succumbed to lack of reinforcements and supplies, and the Russian winter). Britain's naval power had been able to maintain control of coastal waters, but Mackinder was keen to ensure that Britain was prepared for a world where improved technology would make inland conflict and invasion more likely.

Extension

Planes leaving Dubai can reach any other location in the world non-stop. It has become a major hub airport, threatening the global dominance of Heathrow and the traditional European airlines.

In a world of global air transport, to what extent has the UAE become a modern geographical pivot? You should research data on other world airport hubs to be able to offer a comparison in your answer.

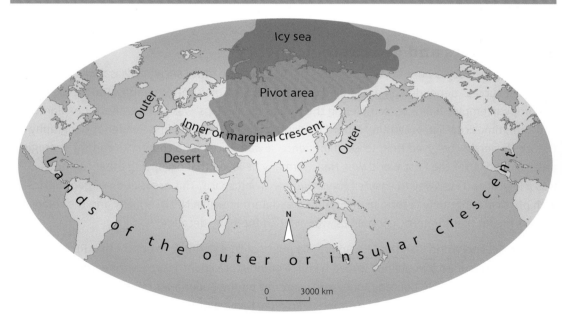

Figure 3.3: The Geographical Pivot, according to Professor Halford Mackinder in 1904

Patterns of power
Colonial (direct) control

At its height, the British Empire extended over about a quarter of the world's land area and ruled a fifth of its population. Its origins can be traced back to the late 1400s, when England was competing with Spain and Portugal to explore the world, Christopher Columbus discovered

Yellow = Players, Orange = Attitudes and actions, Purple = Futures and uncertainties

America, and Sir Francis Drake defended claims to West Africa. Early colonial actions included settlements in Ireland by English and Scottish Protestants and the establishment of settlements in the Caribbean and North America, along with the slave trade between west and central Africa and the Americas. Rivalry with other European powers led to the Anglo-Dutch Wars in the 1600s, and a long history of war with Spain and then later France.

Like other European countries, Britain established trading companies to finance voyages to search for valuable commodities, such as spices from the East Indies and India. These raw materials were brought back to British cities such as Liverpool, Bristol, Hull and London, driving the Industrial Revolution. In 1875 Britain bought the largest shareholding in the Suez Canal and subsequently occupied Egypt. The Empire grew to include New Zealand, Australia, India and Ceylon (now Sri Lanka), as well as large expanses of west, east and southern Africa. For much of the 1800s Britain was unchallenged by any other superpower.

The British Empire worked by direct colonial control. The steamship and telegraph were new technologies developed to help maintain the empire. The so-called 'All Red Line', an early precursor of the internet, consisted of a network of underwater telegraph cables, for example under the Atlantic Ocean from the UK to North America. British cultural values and the legal system, together with the English language, sports such as cricket, football and rugby, as well as British inventions such as railways, were introduced around the world, facilitating the growth of more complex trade networks and links.

Britain's policy of 'splendid isolation' during the **imperial era** meant that, although it had almost total global control, it played little part in European politics except for maintaining the balance of power and participating in the 'Scramble for Africa' in the late nineteenth century – which divided land arbitrarily along lines of latitude and major physical geographical features. The British army helped defeat China in the Opium Wars to ensure that Britain would enjoy favourable trade arrangements with China. Occasional rebellions were put down by force, followed by more direct rule; for example the Indian rebellion of 1857 concluded with Queen Victoria being crowned Empress of India.

By 1914 Britain's Empire was becoming overstretched and was facing competition from a rapidly industrialising Germany. Although by the end of the First World War Britain had gained control over

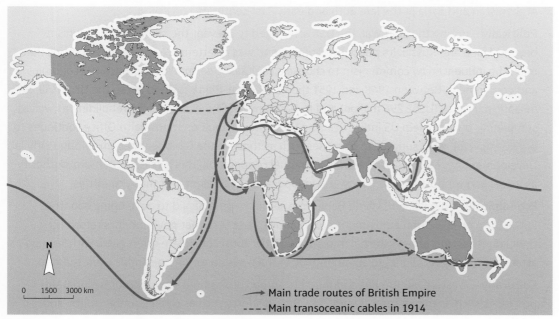

Figure 3.4: The British Empire (red) and its trade and communications routes

— Main trade routes of British Empire

---- Main transoceanic cables in 1914

0 1500 3000 km

additional territories, including parts of the Middle East (Palestine, Jordan and Iraq), the idea of empire was now being questioned:

- Increasing agitation in Ireland for home rule since the close of the 19th century had led to a guerrilla war against British rule and the eventual creation of the Irish Free State in 1922 and the separation of Northern Ireland, which remained part of the United Kingdom.

- Discontent in India over the killing of hundreds of Punjabis at the Amritsar Massacre (or Jallianwala Bagh Massacre) in 1919 led many in Britain to question the morality of colonialism.

- People in other countries also showed dissatisfaction and declared independence, including Egypt, Australia and South Africa.

The First World War saw the growth of US and Japanese naval power, challenging Britain's control of the seas, forcing the country to make choices regarding its international priorities.

These were the first signs of the world becoming multipolar. By the time Britain emerged from the Second World War, near-bankrupt and dependent on US support, the country was facing rising anti-colonialism around the world. One consequence of the dismantling of European empires was that the colonial boundaries – which often did not consider cultural frictions – became the borders of new countries. For example, Palestine was split into (Jewish) Israel and (Arab) Palestine, while (Hindu) India was separated from (Muslim) Pakistan and Bangladesh.

Post-war indirect control

Since the end of the Second World War and the beginning of the **neocolonial** period, countries have exercised power in multifaceted ways:

- **Militarily** – independence was not an easy process for some countries, and in some instances was accompanied by civil unrest and war, for example the Mau Mau uprising in Kenya, a communist insurrection in Malaya, the IRA terrorist campaign in Northern Ireland and a fierce guerrilla war in Rhodesia (now Zimbabwe) during the 1960s and 1970s following the unilateral declaration of independence by the minority white regime there. Britain continued to use military force to exercise influence in many countries. Britain, sometimes together with the USA, intervened militarily in many countries, including the Falkland Islands, Sierra Leone, Libya, Iraq, Afghanistan and Syria in the late 20th and early 21st centuries.

- **Politically** – during the Cold War, the USA attempted to prevent the spread of communism beyond China and Russia. Its policy of 'containment' was an attempt to persuade countries that might be influenced by communism to choose a capitalist free-market approach to economics and government. Between 1948 and 1951 the US Marshall Plan provided aid to the UK (US$3,297 million), to France (US$2,296 million) and to Germany (US$1,448 million) for rebuilding as well as stimulating trade to help US industries. The UK's Foreign and Commonwealth Office continues to provide development support to countries such as those in the Caribbean region through Official Development Assistance (ODA). As well as mutual cooperation, this gives the UK moral authority as it seeks to support racial equality and independent sovereignty for those countries.

- **Economically** – the IMF and World Bank were set up to provide aid to developing countries in the form of 'structural adjustment programmes' to ensure that governments reformed their countries into pro-Western democracies. Other forms of aid are often given with 'strings attached', forcing recipients to spend money in the ways donor organisations want them to.

- **Culturally** – Western culture has continued to spread around the world through globalisation processes such as the internet (Figure 3.2). British sports such as cricket, tennis, rugby and football have remained key aspects of culture in many former colonies. Western music,

Extension

- What is meant by the term 'neocolonialism'?

- Make a list of examples of neocolonial influences featured in this chapter section.

- Evaluate the extent to which the geopolitical problems of some developing countries are the result of external influences.

Yellow = Players, Orange = Attitudes and actions, Purple = Futures and uncertainties

books and architecture can be found around the world, with many TNCs operating globally. Increasingly, however, TNCs are emerging from countries such as Brazil, India, China and the Gulf States, and they are able to compete on a world scale, threatening the domination of US, European and Japanese TNCs (Figure 3.5).

Figure 3.5: Worldmapper diagram showing the proportion of scientific papers published in various territories in 2001 (colours group countries in to 12 regions, colour section is a rainbow based on the 2004 Human Development Report: Japan, purple, the richest; Africa, red, the poorest)

China's investment and trade with countries in Africa have grown rapidly. Today some 1 million Chinese are estimated to be living in Africa, many in areas with large amounts of raw materials. In 2010 80 per cent of Chinese imports were mineral products from Africa, and Figure 3.6 shows that China is Africa's top business partner. Chinese companies create jobs, upskill locals and spend

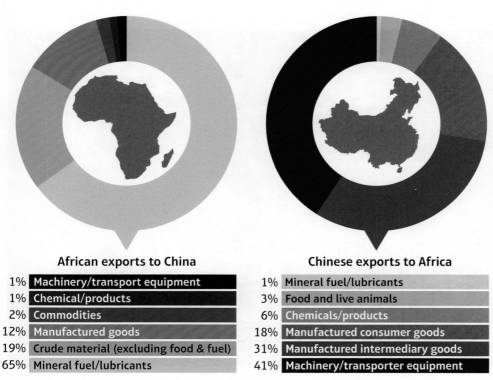

African exports to China

1%	Machinery/transport equipment
1%	Chemical/products
2%	Commodities
12%	Manufactured goods
19%	Crude material (excluding food & fuel)
65%	Mineral fuel/lubricants

Chinese exports to Africa

1%	Mineral fuel/lubricants
3%	Food and live animals
6%	Chemicals/products
18%	Manufactured consumer goods
31%	Manufactured intermediary goods
41%	Machinery/transporter equipment

Figure 3.6: Sino-African trade

ACTIVITY

CARTOGRAPHIC SKILLS

1. Describe the global pattern of research as indicated by scientific papers published, shown in Figure 3.5.

2. Suggest how changes in post-war political and economic power might have influenced the pattern.

Literacy tip

Remember all the descriptive geographical language from GCSE and AS when describing locations, such as north, east, south, west, clustered, scattered, peripheral, in the centre, the names of countries and continents, latitude and longitude, proximity to oceans and mountains, landlocked, and coastal.

money in the economy, stimulating further growth. China's influence is also spreading to less resource-rich countries such as Ethiopia. China would like to develop either Hargeisa in Somaliland or Djibouti as a major gateway to East Africa and the Red and Mediterranean Seas (via the Suez Canal). The relationship between African and Chinese leaders is strong: for example, China played a major peacemaker role in negotiations between North and South Sudan. However, unlike direct colonial rule, Chinese companies operate within rules of the countries; and corrupt Chinese workers have been expelled from Malawi and Tanzania.

Geopolitical stability and risk

The global pattern of power has changed from a **unipolar** (British Empire) to a **bipolar** world (USA competing with Russia), followed by a return to a unipolar world (just the USA). Evidence in the 21st century suggests that a **multipolar** world is developing, but as countries compete on a world scale for power and influence there is much instability and uncertainty.

CASE STUDY: Somalia and Somaliland

Somaliland is a region of Somalia that declared independence in 1991. The region was occupied for many centuries by the Ottoman Empire, but in 1888 it became a protectorate of the British Empire. Britain's main interest in the region was as a supply of meat to British military outposts in Yemen, while the rest of Somalia was ruled by Italy. Somaliland declared independence from Britain in 1960 and united with Somalia to form the Somali Republic.

In 1969 the army's leader, Muhammed Siad Barre, initiated a *coup d'état*. Somaliland resisted his military rule and was bombed by Barre's army, supported by Russian-built MIG aircraft (see Figure 3.7). Over the following years, US military troops were deployed to the country, while Russia constructed of one the longest runways in Africa (later identified by NASA as a potemntial emergency landing site for the Space Shuttle). Resistance to Barre from other ethnic groups in Somalia led to civil war in 1991 and the collapse of the government. The country is described as a failed state, while Somaliland remains one of two relatively stable autonomous regions within Somalia.

Figure 3.7: A Russian-built MiG fighter jet, shot down in Hargeisa, Somaliland

Yellow = Players, Orange = Attitudes and actions, Purple = Futures and uncertainties

Many residents of Somalia fled to European countries, particularly France and Britain. Somaliland has relied on this diaspora returning money through the *hawala* Islamic finance system. Although it has not officially recognised its sovereignty, the UK government is one of a few countries to attempt links with Somaliland, sending aid workers and diplomats. However, simultaneously dealing with Somalia has irritated Somaliland. There is frustration that while many African borders created by colonialism are considered legal, the fractured partitioning of Somalia during the Scramble for Africa, which ignored ethnic territories, remains 'illegal' in the eyes of the world. Mozambique, a former Portuguese colony, has been admitted to the Commonwealth despite no historic links to Britain, but Somaliland, a former protectorate, has not been allowed to join.

In 2009 Saudi Arabia lifted a ban on livestock exports from Somaliland, and imports from Vietnam and China are growing as the diaspora begins to return. Business growth has been strong enough to justify investment in fibre-optic broadband and mobile-phone networks. However, tension is growing between rich returnees buying up land and enjoying lavish lifestyles, and those from other Arab countries who are keen to capitalise on business opportunities. Meanwhile in Mogadishu, the capital of Somalia, ongoing terrorist attacks initiated by groups such as Al-Shabaab have reinforced the consequences of past colonial influences creating borders that don't fit the local geopolitical situation. The instability also led to frequent incidents of piracy from Somalia, whose coastline is adjacent to one of the world's busiest seaways.

The influence of the emerging powers
BRICS countries and the G20

A global shift and outsourcing of manufacturing has increased jobs, income and consumer spending in emerging and developing countries around the world, particularly in Asia. In 2009, the British Prime Minister Gordon Brown hosted a G20 summit in London, noting that recovery from a global recession required a wider group of countries to make decisions about global economic policy. Subsequent summits have attempted to reform the World Bank and the International Monetary Fund (IMF) and examine ways of tackling the demographic issues created by an ageing world population. There has been criticism that widening the group membership from 8 to 20 still excludes and under-represents the African continent, besides omitting some developed countries such as Norway, the world's seventh-largest contributor to UN development programmes. A wider group (G77) represents the interests of developing nations, which broadly includes all UN members except those in Europe, the Organisation for Economic Co-operation and Development (OECD), the Commonwealth of Independent States, and a few other countries. In 2014 Australia proposed the exclusion of Russia from the G20 following its military action in the Crimea, Ukraine, in addition to its ban from the G8.

Russia is considered a country with economic potential, along with Brazil, India, China and South Africa, collectively referred to as the BRICS countries. The term BRIC was first used in 2001 to represent the largest emerging economies at the time; the S was added in 2010. Their relatively small influence at the World Bank and the IMF prompted their first summit in Yekaterinburg in 2009. By July 2014 the BRICS announced that they would create two new financial institutions in order to increase their influence around the world:

- The New Development Bank (NDP) will compete with the IMF to finance infrastructure and other development projects, with a budget of US$50 billion.
- As a rival to the World Bank, US$100 billion will be made available through the 'Contingent Reserve Arrangement' (CRA).

These new institutions are intended to meet the needs of developing countries that experience frustration having to implement pro free-market reforms before being allowed access to funds from the World Bank. However, emerging countries have yet to sustain their growth, and they were badly affected by the 2008 recession: Brazil and South Africa's economies are under considerable threat, and China's economy, although still growing (at a much slower rate), is based on high levels of debt that could trigger a significant economic collapse.

The BRICS countries influence is based on their ability to purchase commodities and manufacture goods so that other countries are dependent on them. In 2011 the term MINT was first used, to refer to Mexico, Indonesia, Nigeria and Turkey, which are also showing signs of economic emergence (Figure 3.8). The growth of the BRICS and MINT countries suggests that a multipolar world is developing. Together these countries also play an important role in global environmental governance, not least through their contribution to global conferences such as the COP Climate Change summits, the latest of which was held in Paris in 2015.

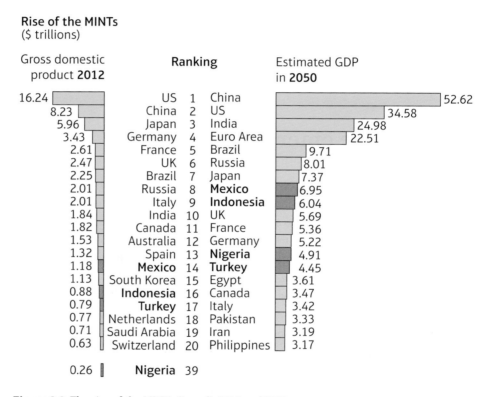

Rise of the MINTs
($ trillions)

Gross domestic product **2012**		Ranking		Estimated GDP in **2050**	
16.24	US	1	China		52.62
8.23	China	2	US		34.58
5.96	Japan	3	India		24.98
3.43	Germany	4	Euro Area		22.51
2.61	France	5	Brazil	9.71	
2.47	UK	6	Russia	8.01	
2.25	Brazil	7	Japan	7.37	
2.01	Russia	8	**Mexico**	6.95	
2.01	Italy	9	**Indonesia**	6.04	
1.84	India	10	UK	5.69	
1.82	Canada	11	France	5.36	
1.53	Australia	12	Germany	5.22	
1.32	Spain	13	**Nigeria**	4.91	
1.18	**Mexico**	14	**Turkey**	4.45	
1.13	South Korea	15	Egypt	3.61	
0.88	**Indonesia**	16	Canada	3.47	
0.79	**Turkey**	17	Italy	3.42	
0.77	Netherlands	18	Pakistan	3.33	
0.71	Saudi Arabia	19	Iran	3.19	
0.63	Switzerland	20	Philippines	3.17	
0.26	**Nigeria**	39			

Figure 3.8: The rise of the MINTs (in red), 2012 and 2050

The likely success of emerging countries

Overall, many regard the G20 as a useful global forum for action after the 2008 global financial crisis and ensuing recession. However, individual countries may act in more pragmatic ways in order to attain a secure financial future. The effectiveness of BRICS countries partly depends on whether developing countries look to their institutions (CRA and NDB) for alternative funding. The consequences of military incursions by China and Russia represent a significant threat to stability and the superpower balance. Table 3.3 summarises some of the key ways the BRICS countries and G20 countries might influence the rest of the world.

Yellow = Players, Orange = Attitudes and actions, Purple = Futures and uncertainties

Table 3.3: The main ways BRICS and the G20 influence the world

	BRICS	G20
Economic	• They are the only US$1 trillion economies outside the OECD. • But growth rates have slowed since 2013, and consumer spending is still low because of ageing populations.	• Minimal impact on financial markets because discussions about change happen over a longer time-period. • More money is available for the IMF, to help tackle global recession.
Political	• China has become a 'mega-trader', larger than imperial Britain: commodity prices have soared while manufacturing costs have shrunk.	• Newer members of the G20 tend not to keep agreements and commitments as seriously. • Countries have started to share financial information to fight tax evasion, and agreed to monitor one another's actions.
Military	• Increased military expenditure with incursions into South China Sea and NATO airspace, Ukraine and Georgia. • But direct conflict with NATO would still leave them outmatched.	• The G20 was divided over military action in Syria – Russia and China lead opposition against the USA. • Some countries are engaged in unilateral military action, for example Saudi Arabia in Yemen.
Cultural	• Cultural differences mean there is a lack of common understanding, which limits sharing of experiences.	• Indirectly support the spread of globalisation.
Demographic	• Large populations mean a huge labour market and flourishing universities with many science and engineering graduates. • However, working-age populations are starting to shrink, and not all countries are creating enough jobs.	• Focused on creating economic growth by encouraging private business to invest in infrastructure; will help tackle youth unemployment.
Environmental	• Historic rise in greenhouse-gas emissions: three of the top four polluters are BRICS countries. • Starting to lead the world in renewable energy production, for example solar panel production in China.	• Agreed a post-recession 'green stimulus package' worth US$1.1 trillion and commitment to remove fossil fuel subsidies. • Have agreed on a need to tackle climate change and global health issues, for example ebola; but without committing money or agreeing quantitative targets.

ACTIVITY

1. Study Table 3.3. For each type of influence shown, explain whether BRICS or the G20 have the greater influence.

2. Does their influence benefit just world regions or the whole world? Give reasons for your answer.

Development theories

Three theories attempt to explain the changes in development since 1900 and may suggest possible changes to superpower status in the future:

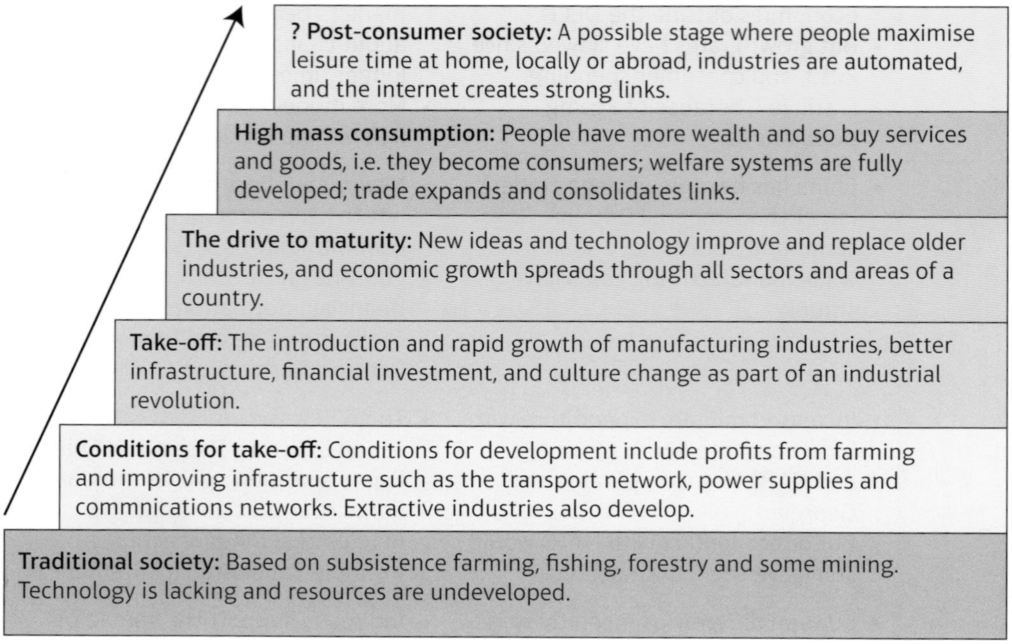

? Post-consumer society: A possible stage where people maximise leisure time at home, locally or abroad, industries are automated, and the internet creates strong links.
High mass consumption: People have more wealth and so buy services and goods, i.e. they become consumers; welfare systems are fully developed; trade expands and consolidates links.
The drive to maturity: New ideas and technology improve and replace older industries, and economic growth spreads through all sectors and areas of a country.
Take-off: The introduction and rapid growth of manufacturing industries, better infrastructure, financial investment, and culture change as part of an industrial revolution.
Conditions for take-off: Conditions for development include profits from farming and improving infrastructure such as the transport network, power supplies and commnications networks. Extractive industries also develop.
Traditional society: Based on subsistence farming, fishing, forestry and some mining. Technology is lacking and resources are undeveloped.

Figure 3.9: Rostow's five-step modernisation theory, with possible sixth step

- **Modernisation theory:** Developed by Walt Rostow in the 1960s, this five-step model is summarised in Figure 3.9. The model was used to explain the growth and dominance of the British Empire and the USA. In the post-war period, the government of the USA believed that if enough investment was made in developing countries, this would stimulate industrial change ('take-off') and they would be able to repay their loans. This would also lead to the economic growth necessary to reduce the influence of communism from Russia or China. This theory is sometimes associated with a neoliberal approach to the development gap and explains why some believe continued investment by a country's TNCs (for example Chinese) would lead to increased superpower status around the world.

Peripheral 'developing' countries
Political support
Raw materials
Debt repayments/ purchases
Brain drain
Core 'developed' countries
Manufactured goods
Aid
Polluting industry
Political and economic ideas

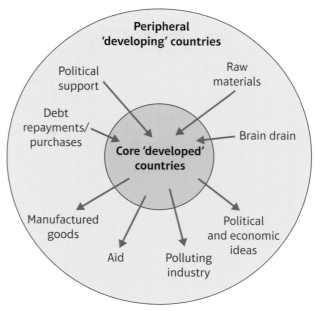

Figure 3.10: Frank's dependency theory

ACTIVITY

1. Use the information about the three theories to answer the following:

a. Which theory presents the most positive viewpoint and which the most negative viewpoint of the world? Give two reasons for each.

b. Which presents the most economic explanation, and which the most social explanation of change? Explain your choices.

c. Which one do you think best explains why the world was, until recently, a bipolar or unipolar world? Give three reasons for your choice.

d. Which theory do you think best explains why the world might become multipolar? Give three reasons for your choice.

2. According to the Kondratiev Economic Cycle model:

a. When would a UK government want to invest capital in major new infrastructure?

b. To recover from the 2008 recession, which economic areas should governments and private businesses be investing in? Why?

Yellow = Players, Orange = Attitudes and actions, Purple = Futures and uncertainties

- **Dependency theory:** Developed by Andre Frank in 1971, this theory is illustrated in Figure 3.10. Frank believed that TNC investment in developing countries led to the exploitation of skilled labour and cheap raw materials, as well as creating international debt. Frank described this unequal relationship as the 'development of underdevelopment', because poorer countries did not have the resources, technical skills or institutions that could help them resist exploitation. Some believe that the USA's influence over the WTO and IMF allow the country to benefit, to the detriment of developing countries. The implication of dependency theory is that some kind of revolutionary break would allow developing countries to have a voice; such a viewpoint would see the BRICS countries actions to establish a new development bank as an example of this kind of change.

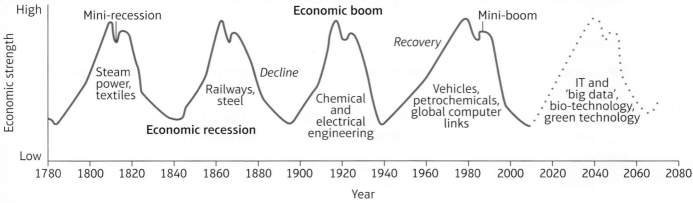

Figure 3.11: Kondratiev cycles, past, present and future

- **World systems theory:** Developed by Immanuel Wallerstein in 1974, this theory looks at change from a wider spatial and temporal perspective. Spatially, the world's global market is divided into three sections where countries compete politically and economically: a developed core, a developing periphery and the semi-periphery where changes and tensions might occur, for example in the BRICS and MINT countries. Temporally, the world economy moves in long-term (Kondratiev) cycles, in which global depressions follow major changes in production roughly every 50–60 years. As Figure 3.11 shows, the most recent depression was in 2008. Global finance had taken advantage of the internet, and investments occurred faster and were more widespread, with increasingly risky investments and loans being made that were ultimately unjustifiable. Although governments stepped in to support struggling banks and some businesses, the loss of jobs slowed the movement of money and consumer spending, causing an economic recession.

After a recession, economic growth normally recovers following a stimulus of new industries and increased spending. Often, a government will make public sector investments in expensive new infrastructure, which helps stimulate growth in the private sector (a multiplier effect). Highly entrepreneurial businesses take advantage of the new technology, which determines how economic growth takes place in the next wave; Figure 3.11 suggests this could be in the environment and biotechnology sectors. If governments are unable to provide investment at the bottom of a recession, sovereign wealth from other countries, such as resource-rich areas of the Gulf States, might provide the stimulus. The investment creates a new power dynamic where the lender gains influence. After 2008, China had an opportunity to assist European banks, but it did not. Political analysts such as Will Hutton concluded that China's internal challenges, such as an ageing population, threaten its rise to superpower status. The interactions between countries in the semi-periphery, and with the struggling established superpowers, will shape the pattern of power over the next few decades.

CHAPTER 3

What are the impacts of superpowers on the global economy, political systems and the physical environment?

Learning objectives

7.4 To understand why superpowers have a significant influence over the global economic system.

7.5 To understand the key role played by superpowers and emerging nations in international decision-making concerning people and the physical environment.

7.6 To understand why global concerns about the physical environment are disproportionately influenced by superpower actions.

The global economic system
Intergovernmental organisations

After the Second World War, the Allied countries believed that an international economic system with greater cooperation between countries was necessary to prevent a future war. The Bretton Woods conference held in New Hampshire in 1944 led to the rules and **intergovernmental organisations (IGOs)** that now dominate international decision-making. The most notable IGOs are the **International Monetary Fund** (see page 207), the **Organisation for Economic Co-operation and Development** (OECD) and the predecessors to the **World Bank** and the **World Trade Organization** (WTO) (see page 207). The OECD's self-stated mission is to improve the economic and social wellbeing of people world wide through the promotion of improvement policies. The World Bank provides financial and technical assistance to developing countries. It aims to end extreme poverty (those living on less than US$1.90/day) and to promote prosperity among the lower 40% of each country's inhabitants. The WTO manages the global rules of international trade; part of which includes world patents. Predominately controlled by western states one should remember they are indirect instruments through which the west exerts its influence.

Figure 3.12 shows that, although colonialism was an early precursor of globalised **free trade**, the opening up of markets after 1945 led to unprecedented levels of international trade. Typically, the biggest contributors to these IGOs are the USA, UK, Germany, France, Japan and Canada.

At the time of the Bretton Woods Agreement, Soviet Russia refused to ratify agreements to set up these organisations. After the West formed the predecessor to the OECD to distribute US aid to rebuild Europe through the Marshall Plan, Russia formed Comecon, an IGO intended to facilitate cooperation between socialist and communist countries. In response to raised levels of threat from

Yellow = Players, Orange = Attitudes and actions, Purple = Futures and uncertainties

the Soviet Union, Western Europe and North America created the NATO (North Atlantic Treaty Organisation) military alliance. While Comecon ceased to exist after the fall of the Berlin Wall in 1991, other IGOs continue to place key decisions about global issues in the hands of a relatively small number of leaders – as the following case study shows.

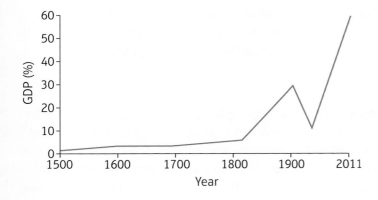

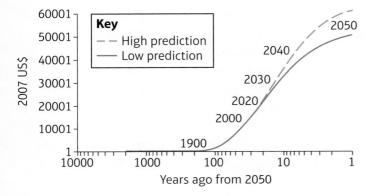

Figure 3.12: Globalisation over five centuries

CASE STUDY: The Côte d'Ivoire, West Africa

The IMF did not allow the Côte d'Ivoire (Ivory Coast) to receive aid until 2013, when its government agreed to set up commercial courts and allow free presidential elections. After economic reforms, US$4.4 billion of debt was cancelled, but cancellation of a further US$10 billion of debt depended on reform of the country's electricity sector to allow companies to react to changes in world energy prices. Investment in education and training has depended on the government creating a more competitive banking sector. The UN still bans diamond exports from the Côte d'Ivoire after the country used them to finance a civil war in 2005. After minimum wages agreements and investment in road infrastructure were achieved in 2011, the country was allowed to export cocoa again.

IGOs aim to promote free trade and capitalism in countries around the world. The policies they encourage or impose are intended to help countries go through Rostow's hypothesised five stages of modernisation theory that the USA and Europe have already experienced. Critics point to dependency theory to show how some policies, such as structural adjustment programmes, although intended to reduce poverty through loans, may erode national sovereignty and promote austerity programmes in those countries, sometimes undermining education, health and social programmes. Jubilee Debt Relief programmes established between 2000 and 2010 require countries to stay on track to continue with debt cancellation. These decisions demonstrate the influence superpower countries maintain through IGOs.

Extension

Investigate the current trading situation of the Cote d'Ivoire. Describe and explain the country's current trading profile.

An IGO of global significance is the Davos Group. The group originated in Switzerland where it meets annually at the World Economic Forum (WEF). About 2,500 people (business leaders, political leaders, intellectuals and journalists) are invited to Davos to discuss different themes, which have included 'educational technology', 'resilient dynamism', 'mastering the 4th industrial revolution' and 'income inequality'. The group is criticised for prioritising capitalism and globalisation at the expense of tackling poverty, as well as placing too much trust in individuals who are not always trusted in their own countries. However, bringing influential leaders together to exchange ideas and agree to cooperate is a vital way to tackle international problems.

TNCs and trade

Transnational companies invest large amounts of money in research and development in order to maintain their competitive edge. Each new product requires expertise, time, specialist equipment and materials. The academic research that underpins these developments is often funded by TNCs. For example, Queen's University in Belfast, which has one of the leading university engineering departments in the UK, maintains links with companies such as Bombardier, Jaguar Land Rover and Rolls-Royce.

Innovation and the 'knowledge economy' are driven by the global patents system, which protects intellectual property by issuing **patents**. Whether the product is artistic, commercial, financial, technological or even strands of DNA, the rights of the inventor are protected, usually for 20 years, during which time they will be able to make a profit from their discovery or invention. While patent law goes back many centuries, the system as currently organised by the WTO was established to give Western companies international protection in the face of global competition. Nevertheless, there is a lot of counterfeiting of popular brands and products – mainly US, Italian and French ones – particularly by China. Global trade in fake goods accounted for about 2.5 per cent of world imports in 2013, and 5 per cent of EU imports (OECD and EU Intellectual Property Office).

The pharmaceuticals industry is a controversial beneficiary of patenting laws: TNCs control access to medicinal drugs, and some will not invest in developing treatments for diseases in developing countries because they cannot guarantee a profit. There have also been cases where indigenous technical knowledge about the medicinal properties of local plants has become the intellectual property of companies. However, since the development of medicines is very costly, without patenting laws companies might not bother to do the research in the first place. The WTO has tried to help by permitting developing countries to import cheaper versions of essential medicines before patents expire. Sometimes, to improve its media image, a company runs charitable programmes; for example GlaxoSmithKline has teamed up with the Bill and Melinda Gates Foundation to distribute more than US$28 billion of the malaria vaccine Mosquirix.

In some countries, particularly China, TNCs operate under tight state controls, and some strategic industries (banking, energy, telecommunications and transport) are state-owned. Although private individuals may own a small number of the shares, TNCs operate much like government departments. The advantage of this is that governments can invest in long-term development, such as roads, railways, ports and renewable energy. This drives economic growth and allows countries like China to make **neocolonial** investments in other continents, or provide sovereign wealth funds for projects. The disadvantage of these state companies is that they can become corrupt, discourage private investment and innovation, and are vulnerable to accumulating large levels of debt as a result of inefficient operations. Following the privatisation policy of the 1980s and 1990s, few state-owned companies are left in the UK and the USA. The UK government has, however, rescued banks such as Northern Rock and the Royal Bank of Scotland, to save them

ACTIVITY

To what extent do TNCs represent the influence of the superpower countries where their headquarters are based?

Yellow = Players, Orange = Attitudes and actions, Purple = Futures and uncertainties

from collapse, but with the stated aim of re-privatising them. A few UK companies remain state-controlled in the public interest (such as the BBC), or where safety is too high a priority to be left subject to financial priorities (for example Network Rail, the National Health Service and the National Nuclear Laboratory).

TNCs exert a lot of power and influence. In 2015 HSBC announced that it might move its global headquarters out of the UK because the UK government was imposing a banking levy that would threaten its profits. In 2016, partly due to this threat, the UK government reduced the levy to a surcharge, which will still lower HSBC profits but not as much as the levy would have. Together with a slowing Chinese economy, this arguably persuaded HSBC to keep its headquarters in London.

'Western' cultural influence

Using the latest communications and transport technology, TNCs have spread Western culture to other parts of the world. Even during colonial times, modernisation was about more than just economic growth; then as now, traditional ethnic and religious values were viewed as barriers to profitable relationships between a company and its workers. By the 1990s a **Westernised** 'global culture' was beginning to emerge, dominated by consumerism, capitalism, wealth-creation, an Anglicised culture with English as the dominant language and selective absorption of other aspects of popular culture.

CASE STUDY: The cultural impact of Apple

Apple's development of the iPod, the iPhone and the iPad has transformed how people work, communicate, listen to music and interact socially. The quality of design is high and, until 2015, Apple was the largest company in the USA, worth US$724 billion, twice that of the second largest. Digital music, video and podcasts have completely changed the music industry, both on the high street and in the way artists produce music, creating an industry now worth US$6.85 billion. The iPhone made communication more dynamic through instant messaging and video conferencing. The iPad brought age groups together, providing them with, as Apple CEO Steve Jobs described, 'more intimacy than a laptop and so much more capability than a smartphone'. By 2015, Apple had sold more than 1 billion iOS devices (iPads and iPods).

The transformation of culture has been particularly strong in China, because of the large difference between traditional Chinese communist culture and that of the capitalist USA. Apple's brand image has perhaps had as much impact as the technology itself. For a young, outward-looking generation that wishes for symbols of freedom, the iPhone and the Apple logo have become reflections of taste and lifestyle and a status symbol.

Figure 3.13 shows that, in 2015, the UK was ranked highest in terms of soft-power score, and China was at the bottom of a 30-country index. Although the survey was by a London-based company, some reasons for the scores include the factors of diplomatic engagement (the number of missions and embassies overseas), the number of internationally chart-topping music albums, the foreign following of football teams and the quality of higher education. The UK performs well on all these factors. Some of the reasons for low scores in other countries involved factors such as restricted access to the internet, undemocratic government and gender gaps.

Synoptic link

Supplies of primary energy such as natural gas, and secondary energy such as electricity, are essential for a high quality of life and the success of businesses and industries, and some of the largest state-owned TNCs are those associated with energy, such as Gazprom in Russia. Energy has also become a focal point of power struggles in various regions of the world, not least the Middle East and Eastern Europe (see page 94). Different players such as TNCs trade and spread ideas about free markets, liberalisation and privatisation. These forms of economic and cultural globalisation help maintain the power and economic wealth of 'pro-Western' nations. (See page 240.)

ACTIVITY

1. Explain how pro-Western IGOs and players influence power in the global economic system.

2. Explain how the system of intellectual property law benefits the current superpowers.

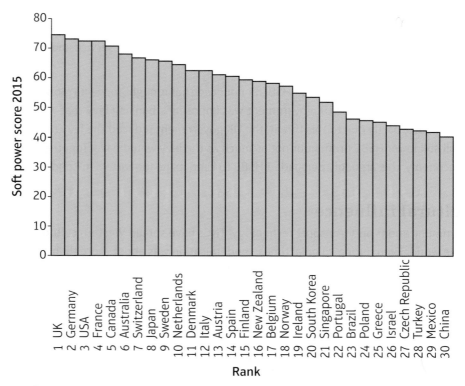

Figure 3.13: Soft-power index, 2015

International decision-making
Global action

In 1945, US President Franklin D. Roosevelt proposed a council he called 'The Four Policeman'. His vision was that the USA, the UK, the Soviet Union and the Republic of China would be responsible for guaranteeing peace around the world, sitting as four permanent members of the UN Security Council. The final 'version' of the United Nations, although different, included veto power for these four countries, underlining their significance as real or perceived superpowers. Their global actions have included the following:

- **Crisis response:** Britain contributed over US$1.5 billion of aid to the 2015/16 humanitarian crisis in Syria, making it the second-largest contributor behind the USA. Unlike other countries in the EU, the UK has focused its aid on humanitarian projects in Syria and aid for refugees in nearby countries. By focusing on quality of life in source countries, the UK hopes to prevent migrants being trafficked into the EU.

- **Conflict prevention:** The UK prioritises prevention of conflict in fragile states, using programmes like 'Official Development Assistance' and the Arab Partnership Programme to help countries in the Middle East and North Africa (MENA) region. On some occasions military action has been used to prevent conflict and, in 2014–15, Britain was part of a coalition action against Islamic State territory in Iraq. Britain has one of the most advanced military targeting systems in the world, so that loss of civilian life is minimised.

- **Climate change mitigation:** Figure 3.14 shows the reduction in greenhouse gas emissions resulting from the actions of countries after agreements at the regular Conference of the Parties (COP) under the United Nations Framework Convention on Climate Change (UNFCC). However, some data is subject to estimation, calculation, and incomplete knowledge as to long-term

Yellow = Players, Orange = Attitudes and actions, Purple = Futures and uncertainties

impacts. Some important actions have come about separately due to changes in superpower status, such as the collapse of the Soviet Union which helped to reduce the number of polluting factories and inefficient state farms, and the slowdown in China's economy and consequent energy use. Figure 3.14 also shows that a clear global policy (such as the 1989 Montreal Protocol on the stratospheric ozone layer) can have a significant impact. It has long been noted that successful mitigation of climate change requires the world's largest polluters (the USA and China) to take action to reduce emissions. The decision in 2014 by the USA and China to recognise the need for action was a significant shift in policy by both countries, perhaps in response to multipolar world influences.

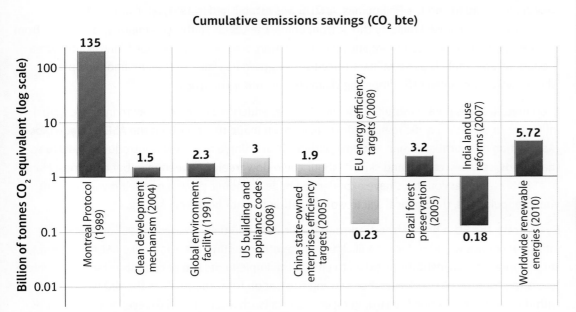

Figure 3.14: Greenhouse gas emissions savings, 1989–2010

Global geostrategy

Successful **geostrategy** requires countries to work together in alliances, both military and economic. France, the UK and the USA were founding members of the North Atlantic Treaty Organisation (NATO) in 1949. As members of this military alliance, they are committed to defending against military incursion into any member state. Membership has increased, with eastern European countries on Russia's border joining NATO after the collapse of the Soviet Union. In 2016 NATO was operating in Turkey to install Patriot Missile Systems to defend against factions in Syria. In Afghanistan, NATO forces are training Afghan security forces to combat fundamentalist terror threats from groups within the country. Following Russia's invasion of Ukraine territory in 2014, NATO became a world player again, as countries in the Caucasus (western Asia) may seek membership of NATO, and Russian military aggression has shown the role that a military alliance can have.

Synoptic link

There is a need for global-scale actions in the form of agreements about rebalancing the carbon cycle, future emissions and changing climate adaptation strategies. The consequences for the hydrological cycle and water supply systems due to climate change have implications for transboundary water agreements. (See page 66.)

Literacy tip

Be careful to know your protocols well and not muddle them; always ensure that you use them in the correct context. For example, the Montreal Protocol of 1989 brought about agreement for reducing chemical emissions (e.g. CFCs) that were found to be enlarging the natural holes in the stratospheric ozone layer over the Polar regions. This is not directly related to climate change and should not be muddled with the Kyoto Protocol of 1992, which agreed that developed countries should reduce their CO_2 emissions. The Helsinki Protocol was an international agreement to reduce SO_2 emissions that were causing 'acid rain' (dilute sulphuric acid), damaging aquatic wildlife in lakes and killing forests.

Synoptic link

Superpowers like the USA and the UK are seen as powerful players whose role is to act as 'global police', intervening, sometimes militarily, in situations where human rights are under threat. The actions of Russia and China, as well as those of the USA and UK, are often questioned, as their intervention can be perceived as beneficial or problematic from different points of view. Countries that question the role of the superpowers in this process can abstain or vote against military action in the United Nations. (See page 222.)

A similar military alliance in the Asia-Pacific region, ANZUS, operates between Australia, the USA, New Zealand, Japan and the Philippines. ANZUS was established in 1951, later widened in 1953–4. Recognising an increased military threat from China, the USA is shifting its military focus away from Europe. 1,000 Chinese missiles are still aimed at Taiwan, tension between North and South Korea is ongoing, and China is using military force to strengthen its claim on the uninhabited Spratly Islands, which are subject to competing claims by five other countries.

Economic alliances regard mutual trade and interdependence as the best way to increase global influence. In the Asia-Pacific region, efforts have been made to strengthen the ASEAN trade bloc (Association of SouthEast Asian Nations), within which a single market was introduced in 2016 so that goods, services, capital and skilled labour could flow freely between member countries. Many of ASEAN's member countries have experienced huge economic growth on their own, and, with young populations, the political will to work together has not been very strong.

By contrast, the EU is both a free-trade and political bloc, initially created in 1957 so that interdependence would reduce the likelihood of another continental war. Unlike ASEAN, a complex set of supranational institutions exists to ensure the implementation of common standards, which forces the countries to work together, causing frustration for some (such as the UK) when faced with the burden of supporting struggling economies (such as Spain and Greece). Some countries in Europe (such as Norway and Switzerland) prefer to opt in to the free-trade market but not to political union, and they continue to govern themselves.

Table 3.4 outlines the benefits and problems for economic alliances within the North American Free Trade Agreement bloc (NAFTA). TNCs such as Bombardier can benefit from operating within NAFTA, as they can manufacture in Mexico where labour is freely available and cheap, design in Canada where research and development are particularly strong, and export via the USA's high-quality infrastructure, free from Canada's strict rules, such as shipping movements on its Pacific coast.

Table 3.4: The benefits and problems of trade blocks, such as NAFTA

	Benefits	Problems
For all countries	• Region-wide transport infrastructure increases efficiency – for example rail network transporting components and goods to and from Mexico and Canada • Production and services in the cheapest and most efficient place (Mexico), which lowers prices for consumers	• Agreements are sometimes biased to rich-world interests, for example the USA has heavy border restrictions on goods and people • National policy can get in the way, for example environmental regulation in Canada prevented new ports • Subsidies can get in the way, for example fair trade
For developing countries (Mexico)	• Better market access • Better-paid jobs because more FDI, for example aircraft fuselages constructed in Mexico	• Bulk exports of some goods can force local producers out of business • Rich countries shop around and new agreements sometimes override old ones, for example TNCs locating in China

Yellow = Players, Orange = Attitudes and actions, Purple = Futures and uncertainties

The UN and global stability

One of the most significant IGOs in the world is the United Nations. It was established in 1945 and has grown from 26 countries in 1945 to 193 by 2016. The aim of the UN is to prevent a recurrence of global conflict, by focusing on establishing fundamental human rights and equal rights for both men and women in all nations. The UN is a forum for member nations to express opinions and grievances, propose actions to resolve disputes and ask for (or offer) assistance from other member states. In addition to hosting conferences on global issues (annual conferences on climate change began in 1991), superpowers and emerging countries can influence geopolitical stability through the various UN institutions:

- **The UN Security Council** can authorise and direct action to resolve conflict, either through economic sanctions (for example trade restrictions), insisting on the use of the International Court of Justice, or by authorising military intervention (for example UN peacekeeping forces). In 2016 there were 16 UN **peacekeeping missions** deployed around the world, made up of military personnel from every member state. Some of the largest contributors are India, Nigeria, Indonesia and Egypt. There are five permanent members of the UN Security Council – the USA, Russia, China, France and the UK. Other countries are elected for two-year periods.

- **The International Court of Justice** resolves legal questions and disputes brought to it by UN member states. It is based in The Hague in the Netherlands, and its nine judges are nominated by the UN General Assembly and the UN Security Council. They are all from different countries, distributed by geographical region, but the five permanent members of the Security Council always have a judge in the court. Recent decisions include settling a dispute between Bolivia and Chile: Bolivia is a landlocked country but has a navy. and the Chilean government was resisting giving Bolivia access to its coastline.

The relationship between the permanent members of the Security Council is often crucial for the success of the UN. Until recently China repeatedly abstained from votes, believing that even peacekeeping is interference, but it has started to engage with issues that help it achieve its international superpower ambitions. China's financial contribution has grown to help with peacekeeping, climate-change impacts and development projects. For example, China deployed 8,000 troops to South Sudan in October 2015 and has allocated US$6 billion to help meet the UN's new sustainable development goals. However, China still refuses to engage with the UN's Permanent Court of Arbitration, charged with ruling on the dispute between China and the Philippines over the interpretation of UNCLOS (the UN Convention of the Law of the Sea) in the disputed areas of the South China Sea.

A level exam-style question

Explain why military alliances are an important part of international decision-making. (6 marks)

Guidance

You should consider examples of military alliances within your answer, such as NATO or ANZUS. For example, being part of ANZUS offers the Philippines protection against any aggressive stance by China over the Spratly Islands. You should also identify how these alliances link to wider cooperation and the establishment of international power.

ACTIVITY

1. **a.** Explain the importance of the UN in creating global geopolitical stability.

   b. What obstacles exist to the efficient functioning of the UN?

2. Compare the influence of global players such as the UN with regional players such as the EU and ASEAN.

3. Do agreements such as NAFTA bring equal benefits to all countries, or reinforce the dominance of one?

Global environmental concerns
Environmental degradation

Synoptic link

Interventions by global IGOs (such as the UN) play a powerful role in both accelerating globalisation – through increasing the likelihood of foreign direct investment – and maintaining geopolitical stability by recognising the legitimate claims made by different cultural and ethnic groups for territory, such as in the case of South Sudan. (See page 275.)

Superpowers, established or emerging, consume large amounts of resources. These are often non-renewable and, while these countries may have large supplies of their own, such as oil and gas in Russia and rainforest products and iron ore in Brazil, they import other resources. With their growing populations, there is an increasing demand for resources (including food and water), which often affects the natural environment through degradation:

- **Landscape scarring** – Opencast mining removes vegetation and scars the landscape (see Figure 3.15). Most mines in Britain closed in the 20th century, but in 2015 Drakelands Pit in Devon became the first new mine to open since 1969, extracting tungsten. The price of tungsten had doubled because supply from China and the USA was running low. These operations also create local noise and air pollution, and if extraction of a mineral requires other chemicals, the waste may contaminate groundwater.

Figure 3.15: Open-cast mining in Devon

- **The built environment (fossil fuels)** – Coal is used in power stations, in some industries and occasionally in homes. Increasing car ownership means that greater supplies of oil are consumed. Industrial chimneys may emit pollutants high into the atmosphere, where they mix with rainfall to form acid rain. In 2003 more than 250 Chinese cities were affected by acid rain, created economic losses of US$13.3 billion. Residential chimneys emit pollutants closer to the ground and these pollutants become local dry deposition, often causing weathering of cement or limestone in buildings. China has become the world's largest emitter of CO_2 – 9.7 billion tonnes during 2014 – with the USA second, with emissions of 5.6 billion tonnes. Figure 3.16 shows that Chinese CO_2 emissions grew by only 1.2 per cent in 2014, compared to the earlier average annual growth rate of 6.7 per cent, but only because of economic slowdown.

- **Oil spills** – In 2010 BP's Deepwater Horizon drilling rig exploded in the Gulf of Mexico. About 450 million litres (100 million gallons) of oil escaped from the undersea wellhead, polluting the sea and nearby coastline. Chemicals in the spraying dispersant used to reduce the oil slick damaged marine life as well as other wildlife along the coast near the Mississippi delta.

Yellow = Players, Orange = Attitudes and actions, Purple = Futures and uncertainties

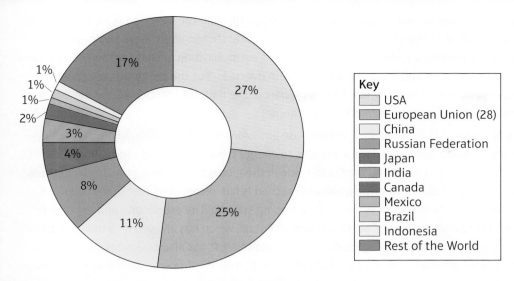

Figure 3.16: Cumulative CO_2 emissions 1850–2011 (% of world total)

Key
- USA
- European Union (28)
- China
- Russian Federation
- Japan
- India
- Canada
- Mexico
- Brazil
- Indonesia
- Rest of the World

- **Removal of forests (for food)** – Brazil's food production increased by 26 per cent between 2002 and 2012 (FAO), turning it from the world's largest importer to a major exporter. Forest has been cleared and the land converted to cropland and pasture (particularly for growing soya, used for cattle feed). Beef exports have increased by ten times and Brazil's cattle ranching is now the second-largest in the world, behind India. Other Brazilian exports include poultry, sugar cane, ethanol and soya bean. However, meat production, in particular, requires more intensive use of resources: approximately 15,000 litres of water are needed to produce a kilogram of beef, but only 1,250 litres for a kilo of maize or wheat. Agriculture causes 8–18 per cent of global greenhouse gas emissions, when deforestation to create pasture is included. In the 1990s Brazil deforested an area the size of Belgium, although deforestation rates fell significantly (by 70 per cent) in the early 21st century.

Willingness to act

The actions of superpowers to reduce environmental degradation set an example for other countries to follow, and have the benefit of stimulating the growth of eco-friendly technology. The USA and China agreed only in 2014 to start reducing their CO_2 emissions – by 2025 and 2030 respectively. Mass production of new technology (for example solar panels) can lower the unit price, making such technology more viable for consumers. However, superpowers and emerging countries often resist initiating changes that might damage their economies, so achieving global agreements on environmental issues can be difficult and takes a long time.

- **China's** use of resource is motivated by national pride. China had a disastrous famine in the 1950s, and never wants to experience a lack of food again. Like other developing and emerging economies, agreement to cut CO_2 emissions could limit China's economic growth. However, the post-2008 economic slowdown has drawn attention to the costs of environmental degradation (9 per cent of GDP in 2012) and the associated poor global reputation for pollution. Measures to protect the environment have started to be applied, and Chinese companies are now some of the biggest clean-energy firms in the world, with heavy investment in the production of solar panels.

Synoptic link

The attitudes of different countries towards energy and environmental issues, for example the USA compared to Denmark or Saudi Arabia, influence the supply and demand of fossil fuel energy. Threats to the sustainability of the carbon cycle include pressures for deforestation and increased use of fossil fuels. Therefore the power of countries has important implications for the environment and sustainability, as clearing or burning trees influences the greenhouse effect by both increasing atmospheric carbon and decreasing the potential for terrestrial photosynthesis, and combusting fossil fuels on a large scale is adding considerably to climate change. (See page 97.)

- In **the EU, f**or many decades, EU agricultural subsidies through the Common Agricultural Policy (CAP) encouraged farmers to grow food intensively, using chemical fertilisers and pesticides, which caused problems such as lost hedgerows and eutrophication of water resources. The same food could have been imported at lower cost from overseas, but the policy was to protect EU farmers, especially in France, where farming is part of the national identity, and where certain local foods still define town markets and restaurants.

- **Russia** has taken a lead in reducing greenhouse emissions by using nanotechnologies, energy-efficiency laws and other mandatory changes to energy consumption and production. As a result, it has a surplus of carbon credits to trade in the global carbon emissions trading scheme. Russia is keen to ensure that any agreement reached is fair, recognising that many countries could lose out economically in systems like a trading scheme. The production and export of natural gas provide a source of political power in Russia, which has allowed it to manipulate countries on its border, including those in the EU. However, it has also created political uncertainty, which has harmed its level of influence. Carbon dioxide emissions limits may also reduce the movement of fossil fuels and further reduce Russia's power.

- In **the USA** some people, including a few scientists, remain sceptical about the problems and consequences of climate change. Political debate has therefore lasted several decades and is an issue between right-wing and left-wing thinkers: the former resist government intervention and want a free market to allow businesses to grow without regulation, whereas the latter argue for the health of people and the environment. Reducing carbon dioxide emissions is a cost for companies, and would challenge energy businesses in particular, which makes them resistant to change.

Climate change is an example of a global 'tragedy-of-the-commons', because in the long term all countries potentially gain from reducing the effects of climate change. Unfortunately, any country that initiates change by reducing greenhouse gas emissions may suffer in the short term, as others 'free ride' on the benefits. This discourages countries taking action independently, which shows the importance of international agreements such as Kyoto in 1997 and Paris in 2015.

Resources and future consumption

Increasing wealth in emerging countries has increased demand for more high-tech goods, many of which depend on **'rare earth'** elements. Table 3.5 shows how some of the different rare earth elements are used in common products. Some estimates suggest the business of generating rare earth elements is worth US$4 billion a year, with the collective worth of companies that rely on them estimated to be US$5 trillion.

Table 3.5: How rare earth elements are used in everyday life

Element	How it is used
Neodymium	Wind turbines and hybrid cars
Lanthanum	Camera lenses and studio lightning
Cerium	Catalytic converters
Praseodymium	Aircraft engines
Gadolinium	X-rays and TV screens
Europium	Control rods in nuclear reactors

Yellow = Players, Orange = Attitudes and actions, Purple = Futures and uncertainties

Many rare earths have similar chemical properties because they are often found in the presence of radioactive thorium or uranium. A great deal of water, acid and electricity is used in the extraction processes to separate the ore from toxins, many of which are carcinogenic. Processing one tonne of rare earths can produce 2,000 tonnes of toxic waste, and if this waste mixes with surface water or groundwater there is a significant environmental impact. China produces 85 per cent of the global rare earths: in Baotou, a small village that produces over half of China's supply, 10 million tonnes of waste water per year is pumped into containment ponds from where it seeps into groundwater and drinking water sources. Livestock near the mines are also affected, decreasing the income of agricultural communities nearby. In recent years China has restricted the refining and eventual export of these products in order to keep their price high, but this has had the effect of expanding mining in other parts of the world because the higher price makes it economic to do so.

The food supply

Although the increase in the supply of staple food grains is undisputed, there are some unintended consequences. For example, the Green Revolution in India caused soil degradation and chemical runoff of excess fertilisers resulting in eutrophication. Although rice consumption has stabilised in India, consumption has increased by 50 per cent in sub-Saharan Africa and has grown steadily in the USA and the EU because of immigration and increasing awareness of the value of a healthier fibre-based diet, so environmental problems may continue or spread.

The water supply

There are concerns that industrialising countries will catastrophically overuse available water, for both drinking and farming staple grains. In India, where farmers have been supplied with solar-powered pumps, groundwater in some states is being used three times faster than it can be replenished. As glaciers continue to melt and river discharges eventually decrease due their source diminishing, climate change could make this problem worse. In other places like California, drought in the south-western USA also affects crop production. Pre-existing water rights along the Colorado and Sacramento Rivers mean that extraction has been difficult to redirect in a drought that has lasted since 2010. Many Californian farmers have used more groundwater, but the water table is dropping and ground subsidence has been noted. California is a very important farming area in the USA, producing two-thirds of the country's fruit and nuts for example. Consumer demand for vegetables, salads and nuts has increased rapidly as healthy diets have become a priority; this encourages agribusinesses to extract more water to increase yields.

The use of resources has increased over time due to larger population numbers, with more consumers needing energy, food and water for example. Since the independence of colonial countries many more have economically developed (emerging economies) with a larger range of industries using more raw materials and energy. People have also become relatively more wealthy and so demand more resources and a wider range of them, either directly or through the products that they buy.

Oil is an example of a resource that has been in increased demand. The 'oil age' is a term that shows human dependence on this non-renewable fossil fuel, especially for transport and also the creation of plastics for example. Some countries and TNCs have benefited from oil, such as Saudi Arabia and BP. There is a lot of trade and geopolitical manoeuvring between those countries that have oil resources, such as OPEC, USA and Russia. One prediction is that oil will run out in 2061, certainly 'peak oil' production has passed, this has put pressure on more remote areas such as the Arctic Ocean from oil exploitation.

A level exam-style question

Explain the extent to which superpowers influence action on global environmental concerns. (8 marks)

Guidance

Give reasons to support the idea that superpowers have a strong influence or that they have a weak influence, or that their influence varies according to the environmental issue. Developed industrialised countries cause environmental degradation, but can also provide global leadership in tackling the causes and consequences of environmental concerns, and they often have the technology to help.

CHAPTER 3

What spheres of influence are contested by superpowers and what are the implications of this?

Learning objectives

7.7 To understand how global influence is contested in a number of different economic, environmental and geographical spheres.

7.8 To understand the changing relationships between developing nations and superpowers and the consequences for people and the environment.

7.9 To understand why existing superpowers face ongoing economic restructuring, which challenges their power.

How global influence is contested
Disputes over resources

It is estimated that a quarter of the world's undiscovered oil and natural gas may be located in the Arctic Ocean. Receding polar ice is increasing the possibility of accessing these reserves. This is a potential area of conflict and there are already territorial boundary disputes (Figure 3.17). The debate has started on how to exploit these resources sustainably, and who has sovereignty over different parts of the seabed. The costs are high. The US Geological Survey estimates the cost of extraction to be about US$37 per barrel, in comparison to US$2 per barrel to extract oil in Saudi Arabia. Eventually, though, the price of oil may increase to a level where it becomes economically feasible to extract from the Arctic Ocean.

1 USA Continental Shelf
If the USA ratified the Law of the Sea treaty, it could claim territory here roughly half the size of Alaska.

2 Chukchi Sea
Shell has plans to explore here. But since Russia is claiming nearly half the Arctic Ocean, it may run into trouble.

3 Beaufort Sea
A 100-square-mile area in this body of water is said to be rich in oil and gas, but it's in dispute – so no one has bid on a drilling lease offered by both Canada and the USA.

4 Lomonosov Ridge
This giant undersea landmass extends from Russia to Greenland – and the two countries are fighting over it. In June, 2007 Russia said its scientists found evidence of a 70-billion-barrel deposit and claimed rights to the whole ridge.

Figure 3.17: Competing claims to the Arctic

Yellow = Players, Orange = Attitudes and actions, Purple = Futures and uncertainties

Under the United Nations Convention on the Law of the Sea (UNCLOS), countries can claim the right to exploit resources in an area up to 200 nautical miles beyond their coastline (the Exclusive Economic Zone, or EEZ). Figure 3.17 summarises those parts of the Arctic Ocean where EEZs overlap and are disputed. A Russian submarine controversially planted a flag on the Lomonosov Ridge, claiming it was directly connected to the Russian continental shelf. Russia has claimed nearly half of the Arctic (half a million square miles) and granted permits to its own companies to exploit it. Canada compared the move to an old-style colonial land grab and, together with the USA and Norway, does not recognise Russia's claim as legitimate. Figure 3.18 represents some of the key military developments that have raised geopolitical tensions in the area.

2002	Canada recommended military exercises in the Arctic that are now conducted annually.
2003	2003–2007: Norway built Fridtjof Nansen class frigates.
2007	Russian submarine planted a Russian flag at the North Pole and Russia restarted long-range Arctic bomber patrols.
2009	USA released its National Arctic Policy placing Arctic security as the number one priority.
	Denmark published plans to create both an Arctic military command and an Arctic Response Force.
2012	Chinese icebreaker, the Xuelong, navigated the Northern Sea Route for the first time.

Figure 3.18: Timeline of recent military developments in the Arctic

ACTIVITY

CARTOGRAPHIC SKILLS

1. Study Figure 3.17.

   a. Describe the competing claims to the Arctic Ocean basin.

   b. Explain how geopolitical processes have led to a contested Arctic (consider economic, political, geographic and environmental influences).

Energy development in the Arctic threatens the natural environment – a fragile ecosystem already under stress from climate change. The Arctic National Wildlife Refuge in Alaska, close to the route of the trans-Alaskan pipeline, is an area protected from resource exploration or exploitation. However, 88 per cent of Alaska's budget depends on oil revenues, and the US government is under considerable pressure to allow exploration. The development of fracking in parts of the USA has made it a bigger exporter of oil and gas than Saudi Arabia. A crucial factor may be the price of oil; if prices are low then the Arctic may be safe. Surprisingly, in November 2014 OPEC countries decided against reducing production, and this kept prices low compared with previous years. Although Figure 3.19 shows that this created serious economic concerns for some countries, such as Venezuela, Nigeria and Saudi Arabia, the intention was to reduce or negate the profit margin for USA oil producers. This may make production in marginal areas unviable and force big consumer countries like the USA to return to buying oil from Saudi Arabia and OPEC countries, so also returning power over this resource to them.

A level exam-style question

Use the information in Figure 3.19 to explain how tension can increase between countries over the acquisition of physical resources such as oil. (6 marks)

Guidance

Sometimes a six-mark 'explain' question might be linked to a source; but use your knowledge and understanding of tension over resources, as well as evidence from the figure. Figure 3.19 shows how the continued exportation of oil by Saudi Arabia has reduced the global oil price to below a viable limit for Venezuela and the USA, for example, which could lead to tension.

Synoptic link ⤴

Attitudes to resources, especially energy, sometimes result in political tension. Tension in the Arctic reflects competing demands over sovereignty, particularly between Russia and other nations, and these are challenges to IGOs such as UNCLOS and NATO, which must regulate decisions about disputed overlapping marine areas. The driving force would often seem to be linked to establishing energy security and, perhaps, profit. (See page 94.)

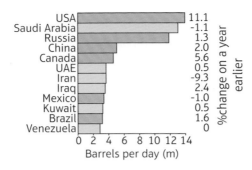

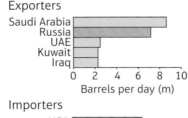

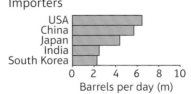

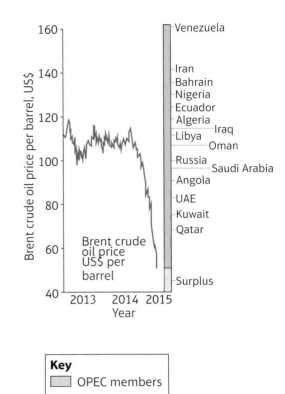

Figure 3.19: The affordability of the oil export price for OPEC countries (*The Economist,* 19 February 2015)

Intellectual property and counterfeiting

Counterfeiting can undermine the global system of intellectual property rights and patenting, straining trade relations and TNC investment. Counterfeited and pirated products are sold across the world. Based on goods seized by customs, in 2016 Bloomberg Business identified the chief suspect countries as China, Taiwan, Singapore, Malaysia, the Philippines, Thailand, the UAE, Russia and India. The OECD also identified some of these countries, as well as Turkey, Morocco, Pakistan and Egypt, from 2013 studies of goods seized by customs. Pharmaceutical products are among those that are counterfeited, particularly in Africa, while counterfeit electrical goods and music, movies and software are significant around the world. Counterfeiting costs G20 countries US$85 billion a year. Digitally pirated music, movies and software account for losses of up to US$75 billion.

The damage to brand image and, in some cases, the endangering of consumers through inferior products (for example electric shocks from generic mobile phone chargers) can have long-term effects on the revenues of leading TNCs. Many companies are unable to sustain investment in new technology, and this has discouraged innovation. Trade relationships have become strained; for

Yellow = Players, Orange = Attitudes and actions, Purple = Futures and uncertainties

example, in Nigeria there is growing resentment towards China as a result of fires having started due to inferior Chinese electrical goods that cannot cope with Nigeria's electricity supply, while Chinese factories have been undercutting Nigeria's textiles industry by producing counterfeit African fabric designs.

Counterfeiting has grown because global manufacturing has shifted to countries where intellectual property rights are poorly protected. The internet has made it easier to find the technological information required to make fake products, as well as to sell and distribute such products around the world. In addition, times of global recession can tempt companies to cut costs by using fake goods as part of their supply chain and this, in turn, harms legitimate businesses, raising unemployment levels. It is estimated that approximately 2.5 million jobs in developed countries have been lost due to counterfeiting around the world.

In 2011, 32 countries signed the Anti-Counterfeiting Trade Agreement, including the USA and EU, but the biggest counterfeiter, China, did not sign it. US and European companies – Tiffany and Gucci are two particularly high-profile ones – have filed lawsuits against prominent Chinese banks that are regarded as 'safe havens' for counterfeiters. The Chinese government regards bank secrecy as a matter of national sovereignty and has refused US State Department subpoenas for information about Chinese assets. However, as China develops, it is encouraging its companies to become more innovative, and is starting to take copyright and intellectual property law more seriously (Figure 3.20).

ACTIVITY

GRAPHICAL SKILLS

1. Study Figure 3.20. Compare the rate of patent applications between different superpowers and emerging countries.

2. Is intellectual property law a benefit or a problem for (a) superpowers, (b) emerging nations and (c) developing countries? Explain your answer.

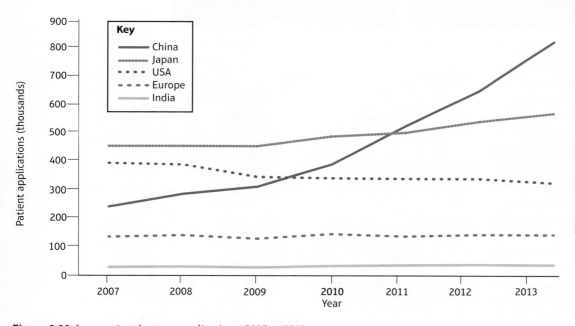

Figure 3.20: International patent applications, 2007 to 2013

Challenges to spheres of influence

The term **'sphere of influence'** dates back to colonial days, when it referred to areas of the world linked by the same culture, for example francophone countries around the globe. During the Cold War, when Soviet Russia and the West were trying to secure greater influence over countries on their periphery, the spatial extent of their level of control was described as a sphere of influence. For example, Western Europe, Oceania, Japan and South Korea were all part of the USA's sphere of influence, while Russia had Eastern Europe. The fractured consequences of the Cold War and post-war US dominance created new implications for people and economies on the edge of Russia and China, or countries that became isolated with the weakening power of the USA and Russia, such as Cuba.

CASE STUDY: Disputes over the South and East China Seas

Japan and China are in dispute over about eight islands (the Senkakus) in the East China Sea, currently administered by Japan. Sovereignty is critical to each country because the surrounding seas are rich fishing grounds and there are extensive gas and oil reserves under the seabed. The islands were transferred to US sovereignty in 1972, but evidence going back to the 1600s is being used to justify claims of ownership by China. In 2010 the Japanese coastguard rammed a Chinese fishing trawler, arresting the boat and crew. In 2012 Japanese activists staged a protest on the islands, provoking an angry reaction from Beijing.

In the South China Sea, China has established a small military presence on the Spratly Islands, which lie between Malaysia and the Philippines but are disputed territory. This is one of the world's busiest shipping routes (30 per cent of the world's trade passes through it) and, like the Senkakus, it has large reserves of fossil fuels. China has built some structures on the islands (see Figure 3.21), but recently finished dredging the seabed to create artificial reefs and an airstrip. The Philippines government pays some of its citizens to live on the islands to reinforce its claims. Although a limited military presence surrounds the reef, it is no match for the Chinese Navy that aggressively patrols the international waters.

 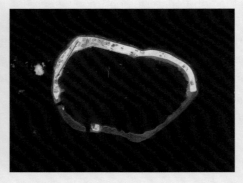

Figure 3.21: Satellite imagery of building (white areas) in the Spratly Islands, South China Sea: February 2015 (left); May 2016 (right)

In 2016 China installed two launch batteries for surface-to-air missiles on Woody Island in the Paracel archipelago. This military escalation was a possible response to the US Navy beginning military exercises in the South China Sea in 2014 and 2015. The USA is committed to defending the Philippines and Japan, and has used an aircraft carrier group to resume 'freedom of navigation' operations. The US Navy wants to maintain its capability to 'operate in an unrestricted way in the waters of our choice'. The Chinese Navy is working on quieter submarines, long-range hypersonic anti-ship missiles and so called 'carrier killer' medium-range missiles. Tensions in the East and South China Seas continue to grow (see Figure 3.22).

In July 2016 an international tribunal in The Hague gave a legal ruling in favour of the Philippines' claims in the disputed waters of the South China Sea. The tribunal also found that China had violated the Philippines sovereign rights by its fishing and oil exploration activities and construction of artificial islands. Therefore international pressure increased on China to reduce its illegal operations, although China initially did not accept the ruling.

Figure 3.22: Key tension points in the South China Sea

Yellow = Players, Orange = Attitudes and actions, Purple = Futures and uncertainties

CASE STUDY: Western Russia and eastern Europe

Tensions between Russia and other former communist states of eastern Europe have been growing ever since the Accession 8 countries joined the EU in 2004. Changing their allegiance wholeheartedly towards the West was a humiliating affront to Russia, and President Vladimir Putin is determined that Russia will regain and reassert its former glory and power.

Russia has opposed the US constructing a strategic missile shield in Poland, and planting the Russian flag on the Arctic seabed, and its military intervention in the Ukraine, demonstrate its intention to consolidate territorial claims. In 2006 and 2009 Gazprom (the Russia energy giant) raised the price of its gas and subsequently cut off the supply to Ukraine in winter. In 2006 it raised the price of gas for Belarus after it cut off the pipeline carrying Russia oil. When Georgia wanted to join NATO in 2010, Russia used the excuse of Russian ethnicity in both South Ossetia (part of Georgia) and North Ossetia (part of Russia) to begin an armed conflict between the two territories. A similar justification was given for Russia's annexation of the Crimean Peninsula in Ukraine in 2014. On all these occasions strategic control of land has been Russia's objective.

Control over the Caucasus would give Russia control over an alternative gas pipeline route from the South Caspian Sea to Turkey and into Europe; the Crimean Peninsula gives Russia complete access to the Black Sea and the Mediterranean for its navy. In the ensuing conflict between Ukraine and Russia, a Malaysian Airlines passenger plane was shot down, apparently by separatists armed by Russia. Russian military jets have also been flying very close to British airspace, and Russia's intervention in Syria has controversially used its military capability to attack both extremists and troops supported by the UK and the USA.

The environmental consequences of such conflicts are often overlooked. As well as the impact on human life, made worse through famine or drought, munitions aimed at destroying enemy targets do extensive damage. In the post-2001 Afghan conflict, US weapons are estimated to have destroyed 10,000 villages and their surrounding environments, including safe drinking water sources. The destruction of forests had a drastic impact on timber exports, and military activities made it difficult for leopards and birds to survive in the mountain environments. Pollution from explosives leaves toxic substances that can cause cancer. In the 2006 conflict between Israel and Lebanon, rocket attacks and oil spills killed fish and endangered species of turtles, as well as degrading the beaches in Beirut.

Relationships with developing nations
The challenges and opportunities of interdependence

The rising value of some metals has led to a series of 'land grabs' by emerging countries in the developing world. A European Commission report of 2012 revealed that the price of all metals, for example iron ore, rare earths and copper, had increased by an average of 59.2 per cent between 2000 and 2011 and predicted that they would increase by 13.5 per cent between 2010 and 2020. Emerging powers are rapidly increasing their extraction of raw materials across Africa, Amazonia and India. Mining technology has become more sophisticated, enabling raw materials to be extracted from areas that were previously inaccessible or uneconomic. There are associated environmental effects, such as land degradation.

Countries in Africa have experienced growing links with countries elsewhere such as Brazil and China, and 70 per cent of the world's largest mining deals in 2011 were in Africa, with countries such as Botswana, Mozambique and Namibia seen as particularly attractive for mining. However, recent falls in commodity prices have strained all these links and some large companies have sold

ACTIVITY

Explain how and why geopolitical challenges between countries arise.

off their mineral holdings. For example, the TNC Anglo-American, founded in 1917, announced in February 2016 that from now on it was going to concentrate on 'consumer-driven materials' such as diamonds, copper and platinum. During the recession the decrease in demand for products such as steel, especially from China, has lowered commodity prices, squeezing profit margins for companies such as Tata, recently forced to close or sell off expensive iron and steel plants in England and Wales.

CASE STUDY: Mozambique's resource boom

Coal, oil and natural gas are abundant in Mozambique. Mozambique's gas reserves could make it the world's fourth-largest gas producer. After a 17-year civil war that ended in 1993, the country still uses only a small amount of gas because it cannot afford the enormous capital investment required to extract and distribute the gas. In 2000 the country's GDP (PPP) was only US$8.1 billion but by 2015 it was US$33.7 billion, and its growth rate between 2013 and 2015 was 6.9 per cent (similar to India's 7.5 per cent and China's 7.3 per cent). The IMF has predicted that in 2020 the GDP (PPP) of Mozambique will be US$59.2 billion.

With the prospect of a resource boom, tension has been rising between the government and opposition groups over who is allocated building and security contracts or gains from new ports and railways. In 2010 the country was dropped from lists of 'electoral democracies'. International concerns have focused on land grabs, forced relocations, poor working conditions, bribery and the unwillingness of mining companies to stick to promises to build new infrastructure (roads, hospitals, schools and housing for workers). The World Bank is putting pressure on investors to help develop social responsibility plans. However, TNCs are already frustrated over the hugely complicated tax and concession arrangements in the country, with many waiting three years to start drilling.

As the prospects of Mozambique's economic wealth continue to grow, Portuguese companies are beginning to return to a country that was previously part of their empire. Brazilian, Australian and UAE companies have already begun mining projects, as well as the construction of rail links to move resources to the coast for export. China has also created a 'special economic zone' in Beira, a neglected part of the country. The Chinese government has gone further and bought a significant stake in an Indian company (Videocon) that is exploring natural gas supplies. Figure 3.23 shows the railway running from Beira through Tete, centre of the coal-mining region, on the main trading route linking Zimbabwe, Malawi, Zambia and Mozambique. The reopening of the railway link to Beira has brought other economic activities such as hotels, banks, an airport and car-hire companies.

Figure 3.23: Map showing recent infrastructure changes in Mozambique

Literacy tip

Make sure you understand the term PPP when it is used with economic data such as gross domestic product (GDP). PPP stands for 'purchasing power parity'. This is an adjustment, to GDP figures for example, that takes into account currency exchange rates so that the cost of products in one country is adjusted to that of identical products in another country (usually the USA, as the US dollar is used for most international comparisons). It is a more accurate way of comparing economic data as it nullifies the effect of taxes, subsidies and government currency support.

Synoptic link

Emerging powers like India and China are important players with key decision-making roles in accelerating a global shift in manufacturing and outsourcing of services towards Asia. This shift brings many benefits to people in emerging countries and their economies, but also problems for the environment and people, particularly those in rural areas and disadvantaged groups. (See Book 1, page 189.)

Yellow = Players, Orange = Attitudes and actions, Purple = Futures and uncertainties

The world's changing centre of gravity

In 2012, the McKinsey Global Institute published a report highlighting the movement eastwards of the world's **economic centre of gravity** (Figure 3.24). These were some key headline statistics in the report:

- By 2025, the 600 cities with highest GDP will generate nearly 65 per cent of world economic growth.

- Of those cities, 440 will be in emerging countries.

- One billion people will have enough income to be classified as 'significant consumers of goods and services'.

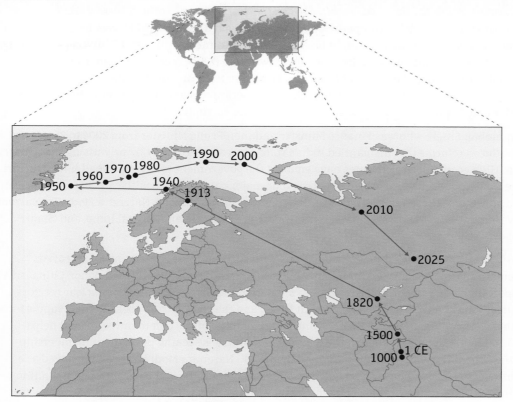

Calculated by weighting national GDP by each nation's geographic centre of gravity; a line drawn from the centre of the Earth through the economic centre of gravity locates it on the Earth's surface.

Figure 3.24: The world's shifting economic centre of gravity, 1 to 2015 CE

ACTIVITY

1. Using Figure 3.24, describe how the strength of power has shifted between the world regions from 1000 to 2025.

2. To what extent does the length of the arrows between the featured years suggest that Russia has been by-passed during the shifts in the economic centre of gravity?

The aim of the report was to help companies recognise that most of the 600 cities were not well known outside their own countries and that each city was uniquely different in its demographic and income structures. Cities play a crucial role in their home countries and regions, and most growth is projected not to occur in megacities such as Mumbai or Shanghai, but in what have been described as 'middleweight cities' such as Surat, 180 miles north of Mumbai, India's capital for synthetic textiles. Some megacities in emerging economies, such as Shanghai and Beijing, will grow to overtake developed world cities such as Los Angeles and Paris. China and South Asia, particularly India, will account for almost 90 per cent of Asia's urban population growth and this will create new markets for TNCs; for example, in recent years UK trade has increased most rapidly with China and South Korea.

China sees itself as a regional superpower. This creates tensions, especially where there are concentrations of military forces, as evidenced in disputes over the South and East China Seas as well as in the escalating hostility towards Taiwan and North Korea. China wishes to control maritime trade routes but faces the US Navy operating in international waters and those of its allies. The Philippines and Vietnam signed a strategic partnership in 2015 to increase their power in the region, to which China responded by providing loans for infrastructure projects across the ASEAN alliance (for example US$10 billion in loans and US$560 million in aid for poorer countries) and an economic takeover of a Malaysian power company, linked to the President, to increase its political leverage in the country.

CASE STUDY: The growing importance of India

In a 2015 report on global competitiveness, India was cited as the third-largest national market after China and the USA. This might be expected for a country with over 1.2 billion people, but India's GDP per capita is small; however, its growth rate (7.5 per cent) was higher than China's. Crucially, India's population is continuing to grow, free from the constraints of birth control policy. The country is democratic and, although chaotic and sometimes violent, this makes it easier to hold political leaders to account and counter corruption.

Manmohan Singh, who was Finance Minister and then Prime Minister from 2004 to 2014, introduced reforms that dismantled India's state-controlled economy by privatising many parts of it. High economic growth rates have enabled the government to invest in road infrastructure (the Golden Quadrilateral project links the four major cities), rural health care, national security and education. These reforms have continued under Prime Minister Narendra Modi. The current government is investing in improved transport infrastructure and 'Digital India' and 'Smart Villages' schemes.

During this time, India's diplomatic relations with surrounding countries have improved dramatically, particularly with Afghanistan, to whom India is the largest aid donor. China is an important trade partner, and the USA has supported Indian development of nuclear reactors. Critically, links have been developed with Israel, and the country now looks to their support rather than Russia. India is a BRICS country and member of the Commonwealth, and helps to stimulate economic change in the region. The country is seeking permanent membership of the United Nations, strongly supported by the UK. Narendra Modi has been perceived as a controversial leader, however, and some wonder whether India will be able to find a solution to the divisions that exist between ethnic and social groups within the country, or resolve to disputes with Pakistan and Bangladesh.

India's growing economy and middle-income groups make the country part of a global emerging market. Its graduates are very entrepreneurial and the country is home to many technology start-ups as well as a space programme with remote sensing. Many workers are under the age of 25, creating a large and lasting workforce, many of whom speak fluent English. The Indian diaspora stretches across the globe, helping to spread culture, technology and ideas and create links. The Bollywood film industry sold 3.6 billion tickets in 2014, in comparison to 2.6 billion for Hollywood. The Indian Ocean remains a key sea-trade route, and India is centrally located with deep natural harbours. Tourism contributes 6 per cent of India's GDP. With its low latitude and many hours of sunshine, renewable solar energy could be an affordable energy source for the country.

Although India's development is still at an early stage, as highlighted in a 2015 report by McKinsey, some geopolitical strategists recognise that its strong democratic government could be a vital counterbalance to an autocratic China. India's investment in its space programme and its ability to be the first Asian nation to put a satellite into orbit around Mars are indicative of its ambition – a form of hard power to match its status as a nuclear power with significant armed forces, such as two aircraft carriers.

ACTIVITY

Evaluate the evidence that India is an emerging superpower.

Extension

How does the recent history of China compare with that of India? Is there evidence that India, with its larger young adult population, might start to grow faster than China, and therefore have a greater global influence in the long term?

Yellow = Players, Orange = Attitudes and actions, Purple = Futures and uncertainties

Continuing tension in the Middle East

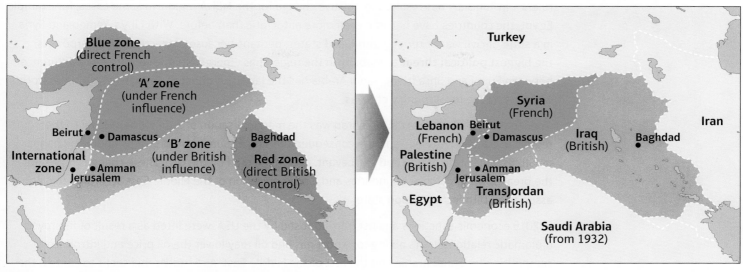

Figure 3.25: The legacy of the First World War in the Middle East

The involvement of superpowers in the Middle East dates back to the early part of the 20th century. Figure 3.25 shows how the Sykes-Picot Agreement of 1916 divided up the former Ottoman Empire into zones of influence for Britain and France, eventually supported by US President Roosevelt and British Prime Minister Winston Churchill after 1945. The West's post-colonial involvement in the Middle East continued with the establishment of the new sovereign states of Iraq, Jordan and, most controversially, Israel – with the associated eviction and landlessness of Palestinians. Tensions between Palestinian territories – represented by terrorist groups Hamas and Hezbollah – and Israel have been both supported and challenged by the establishment of the Arab League. It has been difficult to resolve this dispute, not least because of disagreement over how Jerusalem, a holy site in both Judaism and Islam (and for Christians also), should be governed. Ongoing conflict over Israeli settlement expansion, the blockade of Palestinian workers and trade, access to water in the River Jordan and the Golan Heights in Syria, have continued to sustain these tensions.

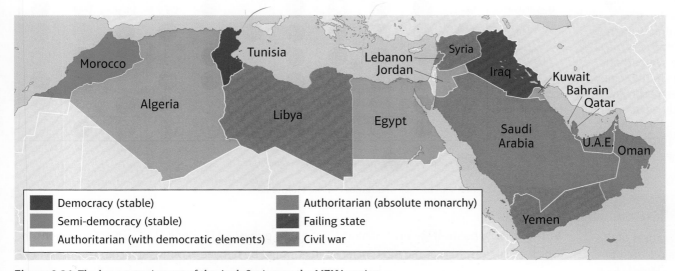

Figure 3.26: The long-term impact of the Arab Spring on the MENA region

Figure 3.26 suggests that in 2016, five years after the Arab Spring, only one uprising against former rulers – in Tunisia – has resulted in positive change for the MENA region. In some cases, for example Egypt, the countries have become even more autocratic than before. With Libya, Yemen and Syria in a state of civil war or turning into failed states, perhaps because of inaction by superpowers, the biggest **political threat** to stability in the region has emerged as the centuries-old difference between Sunni and Shi'a Islam. Saudi Arabia has the largest number of Sunni Muslims in the world, while Iran is the focal country of Shias.

One legacy of USA withdrawal from Iraq was the marginalisation of Sunni groups loyal to the regime of Saddam Hussein, and the consequent uniting of Sunni ethnic groups under the banner of ISIL (Islamic State of Iraq and the Levant, a term rejected by the UN). These changes represent the breakdown of colonial boundaries and a reorganisation of the Middle East according to cultural association rather than nation state.

In 2015 economic sanctions against Iran, imposed by the USA, were lifted as a result of improved diplomatic relations. Iran's ability to export gas and oil may lower the oil price and introduce competition for Saudi Arabia and the rest of the Middle East. As Saudi Arabia seeks to contest the USA's influence over oil prices, the web of political alliances across the region shown in Figure 3.27 seems set to become even more complex.

Synoptic link

The complexity created by contrasting cultural ideologies in the Middle East helps to explain the difficulty of justifying military aid and intervention by the USA and other European countries. Their stated wish is to tackle problems in countries with questionable human rights records, but because the Middle East has the world's largest reserves of oil, and while this remains the most important primary energy source, superpowers may wish to have sympathetic governments in the region. This has been particularly true of the UK government's decision about whether to use military force in Iraq and Syria. (See page 231.)

	Syrian Government	Syrian Rebels	ISIS	Jabhat al-Nusra	Kurds	USA and Allies	Iraq	Iran and Hezbollah	Russia	Saudi Arabia and Gulf States	Turkey
Syrian Government		Enemies	Enemies	Enemies	It's complicated	Enemies	Friends	Friends	Friends	Enemies	Enemies
Syrian Rebels	Enemies		Enemies	It's complicated	It's complicated	Friends	Enemies	Enemies	Enemies	Friends	Friends
ISIS	Enemies	Enemies		Enemies	Enemies	Enemies	Enemies	Enemies	Enemies	Enemies	Enemies
Jabhat al-Nusra	Enemies	It's complicated	Enemies		Enemies	Enemies	Enemies	Enemies	Enemies	It's complicated	It's complicated
Kurds	It's complicated	It's complicated	Enemies	Enemies		Friends	Enemies	Friends	Friends	It's complicated	Enemies
USA and Allies	Enemies	Friends	Enemies	Enemies	Friends		Friends	It's complicated	It's complicated	Friends	Friends
Iraq	Friends	Enemies	Enemies	Enemies	Enemies	Friends		Friends	Friends	It's complicated	It's complicated
Iran and Hezbollah	Friends	Enemies	Enemies	Enemies	Friends	It's complicated	Friends		Friends	Enemies	It's complicated
Russia	Friends	Enemies	Enemies	Enemies	Friends	It's complicated	Friends	Friends		Enemies	Enemies
Saudi Arabia and Gulf States	Enemies	Friends	Enemies	It's complicated	It's complicated	Friends	It's complicated	Enemies	Enemies		Friends
Turkey	Enemies	Friends	Enemies	It's complicated	Enemies	Friends	It's complicated	It's complicated	Enemies	Friends	

Friends · Enemies · It's complicated

Figure 3.27: The political mosaic of alliances in the Middle East

ACTIVITY

Evaluate the causes of ongoing tension and conflict in the Middle East.

In 1993 the US political scientist Samuel Huntington published his book *The Clash of Civilisations*, in which he suggested that the next major conflict would be between contrasting cultures (implying Islam) rather than between countries. His controversial theory gained support after the attacks by Al-Qaeda on the USA in September 2001, and the retaliatory military action by the USA in

Yellow = Players, Orange = Attitudes and actions, Purple = Futures and uncertainties

Afghanistan and Iraq. The idea has also attracted considerable criticism, not least by Edward Said, a leading post-colonial thinker famous for his book *Orientalism* (1978). Said challenged Huntington's characterisation of the relationship between the West and the Middle East, pointing out the mutual dependency of their societies and rejecting the assumption that the West was superior or different to the Arab World.

Whichever perspective is supported, the tensions in the Middle East are a real challenge for the superpowers and emerging powers. Much of the current conflict and post-colonial fallout has arisen out of the historical and current actions of past and present superpowers, particularly the USA, France and the UK. In many ways, with the fastest-growing population in the world, difficult climates (semi-arid and arid), food and water supply issues and finite oil resources, this world region seems likely to be the focal point of geopolitical instability for decades to come.

Challenges to the existing superpowers

Economic problems

The global shift in manufacturing has left some European countries and the USA in a post-industrial economic stage, with the majority of jobs in the tertiary (services) and quaternary (research and development knowledge and information) sectors. This restructuring has created a series of challenges and opportunities for the existing superpowers. The USA and the EU face four broad future economic challenges:

- **Debt:** The global recession (from 2008) created higher public debt in rich countries. In the UK a significant budget deficit is predicted to increase the national debt by US$1 billion a week by 2018; in 2015–16 the UK national debt was expected to be over £1.5 trillion in total. In some countries debt increases the need to raise taxes to pay debt interest, and this can slow an economy even further. The alternative is to introduce austerity measures, spending less and lowering costs such as wages, but this can also slow economic growth. So countries are faced with a difficult balancing act. Other countries believe that they are economically strong enough to allow economic processes to continue, so that the debt burden gradually decreases over time. A geopolitical risk is that these countries begin to rely on capital investment from emerging superpowers, which may continue to grow economically during at least part of a recession, increasing their political and economic power.

- **Unemployment:** Increased competition, with more efficient and creative manufacturers in emerging or developing countries, has led to a decline and readjustment of major Western TNCs. One of the highest-profile US companies to collapse was General Motors, headquartered in Detroit. Its business model was too rigid and depended entirely on an in-house supply chain, rather than outsourcing. With European TNCs increasingly owned by larger conglomerates from emerging countries, such as Tata, secondary-sector workers in Europe depend on decisions made abroad. This makes the workforce vulnerable to changing commodity prices and consumer spending patterns in large Asian markets with high populations. The 2008 recession caused the unemployment rate to rise from 5 to 10 per cent, and although this has recovered in many developed countries, some groups – such as factory workers and African-Americans in the USA – remain disadvantaged. For example, in 2016 Detroit had an unemployment rate of 10 per cent.

- **Economic restructuring**: The EU and the USA are being forced to shift their economies away from the secondary sector (manufacturing) towards the tertiary and quaternary sectors. A UK government report on the labour market in 2014 concluded that the country was ranked 19th out of 30 OECD countries for low skills (Key Stage 3), 24th for intermediate (GCSEs and A levels) and 11th for high skill (tertiary), concluding that the country was falling behind in developing

A level exam-style question

Explain how contrasting cultural ideologies may affect the relationships between developing and developed countries. (8 marks)

Guidance
Remember to use some examples in your answer to support your ideas and points. For this question, this might mean recalling information about the history of the Middle East and more recent ideas about the 'clash of civilisations' and orientalism. The rest of your answer might argue that relationships are also changed by global cultural and economic processes or the recent economic cultures of countries like China and India.

a skilled workforce, with other countries investing in skills more effectively. The global shift in manufacturing creates challenges for disadvantaged communities in developed countries with lower skill levels, who are not able to fully participate in the tertiary/quaternary job market, regardless of any equal opportunities policies or legislation. Some suggest that the decline of UK trade unions has reduced the voice of many groups and communities, making it more difficult for those affected to participate in making positive changes.

- **Social costs**: The consequences of unemployment in disadvantaged communities have been the deterioration of employment networks and a decaying living environment, with fewer maintained public spaces with parks and recreational areas. It has been suggested that this reduces the gross motor development of children and limits the development of the social skills they need to perform well in a services and knowledge economy. Others suggest that there is a lack of compassion for unemployed families from middle-income groups in the UK, as highlighted by Owen Jones in his 2011 book *Chavs: The Demonization of the Working Class*, which reminds us that inequality still exists in the historical superpower countries (Figure 3.28).

The global economy has a long-term cycle, and the Kondratiev Cycle (see Figure 3.11) suggests that the next wave of economic growth will focus on investment in green technology, biotechnology and nano-technology. If UK and US universities, and large and small companies, continue to invest in scientific research, generate patents, and control global trade of these products, then power and world status will be renewed in the West. The 50-year cycle of 'boom and bust' will change economic fortunes from the current recession into economic growth, and the four broad challenges may fade away. However, another possible future is that the EU and USA may be forced to rely on emerging powers to finance, and provide the resources for, the innovation projects, and therefore a multi-modal world will exist with other players such as the UAE, India and China. In addition, the relatively open nature of countries like Chile and Mexico may mean that they start to attract foreign direct investment from TNCs looking for new locations for global operations and headquarters.

ACTIVITY

1. Explain how recent economic changes, such as in the UK and the USA, have undermined the power and influence of the 20th-century superpowers.

2. Study Figure 3.28. Describe and explain the change in inequality in the existing and emerging superpowers.

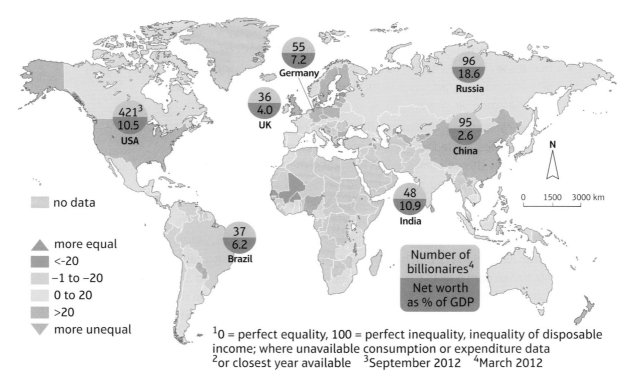

Figure 3.28: Increasing inequality in developed and emerging countries. Gini coefficient[1], % change in inequality 1980–2010[2]

[1]0 = perfect equality, 100 = perfect inequality, inequality of disposable income; where unavailable consumption or expenditure data [2]or closest year available [3]September 2012 [4]March 2012

Yellow = Players, Orange = Attitudes and actions, Purple = Futures and uncertainties

Questioning global military power

For many European countries, maintaining a policy of austerity while continuing to expand military power is a difficult and controversial policy decision. Difficult military campaigns in Afghanistan and Iraq have dented confidence in the effectiveness of traditional Western military power and strategies. The cost of military technology is extremely high, because it is often designed in a bespoke way for each country and involves the latest technologies. Overall, global defence spending is increasing, with the largest increases, proportionally, by China and India (Figure 3.29), with spending falling in Europe, the USA remains the biggest military spender overall.

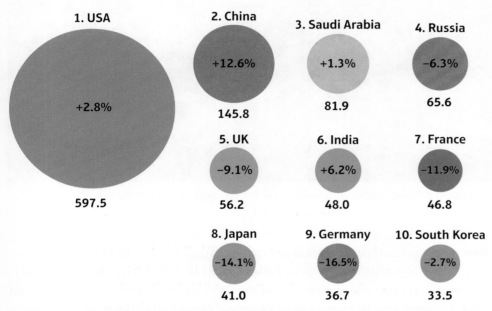

Figure 3.29: Changes in global defence spending, 2015 budgets in US dollars and percentage change from 2014

Western military power has expenditure in four key areas:

- **Navy:** Aircraft carriers are seen as vital for projecting power, but are increasingly vulnerable to land-based missile technology. While the cost of defence is theoretically higher than the cost of attack, new laser-based technology is vital for lowering costs significantly. The UK government has reduced its aircraft carrier capability to two ships (Figure 3.30) with launch dates in 2014 and 2017, but then a period of extensive fitting of military equipment will take place at a total cost of about £6 billion, including the purchase of appropriate aircraft.

Figure 3.30: HMS *Queen Elizabeth*, under construction, next to the retired HMS *Illustrious*

- **Nuclear deterrent:** This is a topic of political and cultural debate. In 2015 the UK's Labour leader Jeremy Corbyn rejected the need for air strikes against targets in Syria, and the need to maintain and deploy an expensive replacement for the UK's nuclear deterrent (Trident, delivered by submarine). The large, arguably unnecessary, expense is one aspect of the argument, since nuclear weapons were never deployed during the Cold War. But others argue that the deterrent worked and was therefore worth the investment, and that many other countries of the world, including emerging superpowers, have a nuclear capability. In July 2016 80 per cent of UK MPs voted to renew the Trident weapons system.

- **Air power:** Although fighter jets are being replaced, the introduction of autonomous drones is cheaper in the long term, more agile and able to destroy precise targets in hostile environments. Larger bombers have to be built in a way that can be upgraded as new technology is developed, rather than replaced, and this is difficult to anticipate. With a perceived increased threat from Russia and Chinese military modernisation, NATO countries are spending more money on military air power.

- **Intelligence:** The necessity and cost of human spies now contrasts with artificial intelligence, satellite technology and computer programming. There is now a strong possibility of cyber attack because of globalised computer networks and so there is debate about the kind of personnel needed by intelligence services and the need for their cooperation around the world. The theft of intelligence material by Edward Snowden in 2013 affected public confidence in the ability of the West's intelligence services to keep material secure, as well as jeopardising undercover UK M16 agents. However, an independent report commissioned by the UK government strongly justified the case for intelligence to counter terrorism and criminal threats.

Yellow = Players, Orange = Attitudes and actions, Purple = Futures and uncertainties

One of the most notable symbols of global power has been the Russian and US space programmes. Russia was the first country to put a man into space (in 1961), while the USA was the first to land a human on the moon (in 1969). Figure 3.31 shows changes in government spending on these programmes. The USA did not replace the Space Shuttle programme in 2011, after 30 years of operation, and reduced funding for NASA (although funding in 2016 was still US$19.3 billion). Instead, the USA is developing a new launch rocket and favours the commercialisation of space travel. The first commercial spacecraft docked with the International Space Station in 2012. NASA plans to send manned missions to Mars at some point after 2030, but commercial operators such as Mars One may get there first. India and China have their own space race today, with India launching its first low-cost spacecraft (US$74 million) to Mars in 2014. Although a clear statement of its ambitions as an emerging superpower, this action, together with the US$1.2 billion allocated for its space programmes, has been questioned by some, as the country still has many poor people and it continues to receive economic aid from other countries.

> **ACTIVITY**
>
> Explain why advanced technologies are important to helping existing superpowers maintain their influence or emerging superpowers to increase their influence.

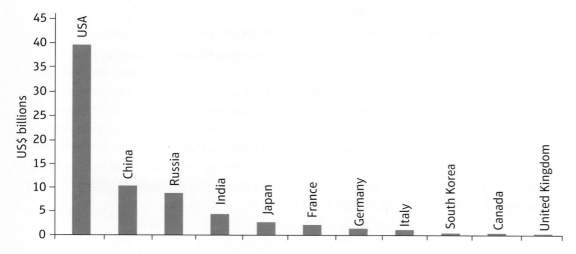

Figure 3.31: Total space programme spending (US$ billion PPP) 2013

Future global power structures

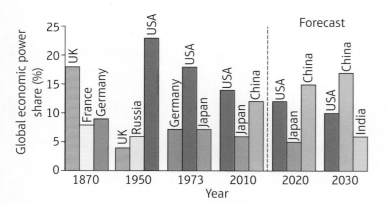

Figure 3.32: The changing share of global economic power

ACTIVITY

GRAPHICAL SKILLS

1. Study Figure 3.32. Explain how the information predicts future challenges to superpower influence.

2. Study the potential scenarios in Table 3.6 and research recent world events and issues. Given recent and current patterns, suggest which of these scenarios is most likely and explain your choice.

There are many ways to define power, such as cultural, political and economic. Figure 3.32 shows what many would consider to be the most significant form of power: economic. This measure of power shows that there has been a shift from Europe to the USA and, most recently, to China. Recent GDP data supports this idea, with China's economy now larger than that of the USA. *The Economist*, analysing superpower status in 2015, suggested that GDP data on its own does not tell the whole story, however:

- In 1931, the Waldorf Astoria Hotel in New York was seen as a symbol of cultural confidence by the USA during the Great Depression. Today it is owned by a Chinese company but still run by a US company, Hilton, which owns 'intellectual-property' around the world.

- The USA's superpower capacity continues in the form of huge aircraft carriers, the power of e-commerce and social media and the global popularity of Hollywood stars.

- The USA replaced the UK as the leading global superpower only ten years after the Second World War. The USA today has more power than the UK ever had, while China is still less powerful than the USA at its peak.

- Global networks and organisations support the status quo, discouraging change.

Every four years, after a presidential election, the US National Intelligence Council produces a global trends review. In 2004 it predicted that the USA would continue to dominate globally, but by 2008 the report predicted that China, India and Brazil would grow at the expense of the USA and the EU, with increased fragmentation of power. In 2012 the report focused on the key changes, summarised in Table 3.6. The end of the table shows four possible global scenarios, but perhaps the future will combine elements of all of them. What is clear is that the future is unlikely to be simply unipolar or bipolar. The most likely future is an emerging multipolar world where players on all levels (supranational organisations, IGOs, countries, individuals) must work together to tackle the global challenges of climate change, trade and access to resources.

Yellow = Players, Orange = Attitudes and actions, Purple = Futures and uncertainties

Table 3.6: Key changes from the USA's *Global Trends Review*, 2012

Megatrends	• Individual empowerment following poverty reduction, a move to urban areas, and consequent increased demand for food, water and energy. • Multipolar world governed by networks – so tackling problems in one country creates supply and demand consequences for others. • An ageing population in some countries that limits economic growth.
Game changers	• Increasing uncertainty about the resilience of the global economy. • Internal and international governance or conflict has potential to spill over into regional conflicts (for example, the Middle East and South Asia). • An important role for the USA in the increased development of technology.
Potential worlds (scenarios)	• **Worst case:** The USA and Europe start to protect their own economies and withdraw from free-trade arrangements; globalisation stalls, reducing the flow of goods, people and money around the world. • **Best case:** China and the USA collaborate, leading to global cooperation on global challenges, with enough food, water and energy for everyone while limiting the effects of climate change, resource shortages and associated conflicts. • **Inequalities explored:** Some inequalities between and within countries remain, so organisations like the EU fail to be effective; political and religious extremists start to dominate and lack of clear governance causes some states to fail. • **Restructure:** Non-government players take the lead in solving problems, as increasing numbers of middle-income groups reject the authoritarian and manipulative structures introduced by the global power elite and take advantage of the internet to share values and ideas, challenging human injustice and promoting women's rights.

Synoptic link

The uncertainty over the future of global power suggests that the role of governments in countries such as the UK (and other European countries) is increasingly variable. In the UK, the 2016 referendum to leave the EU reflects the kind of decisions individual countries must make, and will shape the future path of Europe and the UK as major world players. Nationalist movements within countries (for example political parties like UKIP in the UK) could create significant political tensions in the future, perhaps as a stage of cultural rejection of globalisation forces. (See page 300)

Summary: Knowledge check

Through reading this chapter and by completing the tasks and activities, as well as your wider reading, you should have learned the following and be able to demonstrate your knowledge and understanding of superpowers (Topic 7).

a. Assess the characteristics used to identify superpowers, emerging powers and regional powers.

b. Explain the differences between hard and soft power, and how the use of these has changed over time.

c. Explain how neocolonial mechanisms have become important and how the different patterns of power may bring stability or risk.

d. Explain the strengths and weaknesses of the ways in which emerging countries exercise their power help or restrict their future geopolitical and economic roles.

e. Explain how world systems theory, dependency theory and modernisation theory explain changing patterns of power.

f. Assess how IGOs, TNCs and Western culture influence power over the global economy.

g. Describe how superpowers exercise their power over global issues through interdependence and alliances.

h. Explain how the UN provides geopolitical stability.

i. Assess how increasing consumption and attitudes in superpower countries influence the state of the global environment.

j. Explain why tensions arise over ownership of resources, intellectual property rights and territory.

k. Explain how the world's economic centre of gravity has shifted and created new economic links and geopolitical influences.

l. Understand the complex economic and cultural situation in the Middle East.

m. Explain how the need for economic restructuring could challenge existing superpowers.

n. Assess how superpowers could continue to use their military capability.

o. Evaluate the different geopolitical future scenarios for the uni-, bi- or multipolar world.

As well as checking your knowledge and understanding through these questions, you also need to be able to make links and explain the following ideas:

- Soft power is linked to global cultural influence in various ways, some clear and others more subtle, and the source of power may not always be clear. Hard power may be more direct but the geopolitical motives are often complex.

- Neocolonialism is leading to the exploitation and extraction of rare-earth resources in developing countries. There are underlying economic motives as cultures seek to improve their quality of life, although sometimes elites may cause inequalities.

- The principles of Mackinder's geo-strategic location theory might be linked to world systems theory and tensions over the acquisition of physical resources (for example Arctic oil and gas).

- Modernisation theory partly explains the formation of IGOs, as well as the patents system and intellectual property rights. These organisations and systems increase linkages and create interdependence in the contemporary world. Counterfeiting undermines the principles behind the patent system and jeopardises the global economy and economic power bases.

- All countries, including superpowers, have key roles to play in the 21st century regarding influence and action on concerns for the physical environment. These roles may be indirect through economic systems or direct through crisis management.

- The UN is responsible for arbitrating disputes over land and resources, as well as facilitating the use of hard and soft power. It has many IGOs which play an important role in economic areas, socio-cultural areas, and the physical environment of the world.

- Historical situations have an influence on situations found in the modern world, legacies persist, not least through language and culture, but also influence is still contested in several world regions.

- The BRICS countries decision to set up rival institutions to the IMF and World Bank illustrates the principles of dependency theory. Alliances between groups of neighbouring countries may change relationships between developing countries and the superpowers (a multipolar world).

Yellow = Players, Orange = Attitudes and actions, Purple = Futures and uncertainties

- Emerging countries are increasingly important in the world economy, shifting the economic centre of gravity towards the East. However, economic cycles may favour certain world regions and shift the world economic power base at regular intervals.

- Economic restructuring in existing superpowers creates the kind of opportunities illustrated in world systems theory for emerging powers to exert new influence. Investment in new technologies such as space exploration, information, and bio-technology may be future ways of exercising soft and hard power.

Preparing for your A level exams

Sample answer with comments

Assess the extent to which the USA's superpower status is threatened by emerging economies such as the BRICS. (12 marks)

The world depends on oil supplies. In the USA, oil reserves are diminishing, and it imports over a million barrels per day as it only produces 700,000 so it has to rely on other countries to supply more. The emerging countries such as China also want the oil as they are expanding quickly, so quickly that it has been estimated there will be more cars on the road in China than there are in America by 2020. If the price of oil were to reach $140 as it has done previously, countries such as Brazil and Russia which have their own supplies of fuels would pose a threat to the US, due to the fact they are less vulnerable to the stability of the market.

Prior to 1991 Russia had been the traditional threat to the USA, dating back to the Cold War era. The USA is said to have won, due to the fall of communism as Perestroika was brought in and the Warsaw Pact fell to pieces. Even to this day, the USA and Russia are the largest holders of nuclear weapons. So Russia still has immense military and economic power, through control of natural resources, on which Europe so heavily relies. In the region of 80% of Russia's natural gas is sent through the Trans-Siberian pipeline.

Potentially the USA's greatest threat is China. It has been described as 'the workhorse of the world', manufacturing goods cheaply for foreign markets. The Chinese have found economic success by setting up export processing zones (EPZs) such as Hong Kong. However, it could also be argued that this makes them 'the periphery' and the Western world 'the core', meaning they will never rise above them. This may be the case as many Chinese still live in relative poverty, although the workforce in China is substantially larger than the USA's, with a population of 1.3 billion people, compared to the 300 million people living in the USA. The actions of China in the South China Sea, over the Senkaku Islands, suggest that China is willing to stand up to the USA with military force. However, institutional arrangements (e.g. international maritime law) may prevent China acting with any legal grounds and it might still lose when the United Nations and other countries get involved.

So to conclude, the BRICS states do pose some threat to the USA, as they possess all the attributes needed, such as large workforces, natural resources and, to an extent, military power. However, they are still only emerging, so the USA may still prevail, but only time will tell.

> An attempt to get straight to the point of the question and so avoid the problem of a meaningless introduction. The focus is on control of energy supplies as a factor important to superpower status and a suggestion of how the USA's superpower status could be threatened. But perhaps clarity is lacking; a plan might have helped – for example what about the role of OPEC (or Saudi Arabia) and new resources?

> A lost opportunity here to point to the role of emerging economies. It would have been useful to question the political role of emerging countries when dealing with the USA or Russia, or show that the roles of Brazil, India and South Africa are limited to economic factors.

> Core/periphery ideas are included, showing synopticity, but referring to the Kondratiev cycles would have allowed the cyclical nature of superpower status to be assessed.

> A final assessment, but it perhaps does not recognise the full range of points or complexities involved. Some emerging countries, including some BRICS, may overtake the USA, but others will not. With 'assess' questions, a fully developed conclusion is important.

Verdict

This is an average answer. Clear ideas, such as energy and military power and their significance, are assessed, but economic power is not directly assessed and the role of emerging economies in international organisations is not explored. Knowledge and understanding are applied logically, with relevant connections and some factual evidence. However, each paragraph needed to show more clearly the significance of the factors, and the conclusion needed to consider all the factors and identify which may be the most important.

Preparing for your A level exams

Sample answer with comments

Assess the extent to which the balance between direct and indirect superpower influence has changed over time. (12 marks)

In the times of a unipolar world dominated by the British Empire, political power was exercised to gain direct influence over colonies. After 1945 the world became bipolar, as seen by the Cold War, which produced an increase in the use of economic power for indirect influence. The USA still uses neocolonialism in order to indirectly influence other nations today.

> This introduction sets the scene and integrates mentions of direct and indirect power, hinting that both may still exist today. Terminology is used well but the theme of the answer could have been made clearer.

Before 1945 the British Empire used colonised nations, such as India, for trade and cheap labour. As they were under rule, India received direct influence in a number of ways: The English language was taught primarily in schools, churches were built to impose Christianity on the people and British sports such as cricket were played. These influences can still be seen today, with English remaining a main language of India and cricket becoming the national sport.

> Examples of direct power through colonialism are included, but what about change over time? The paragraph should have considered the extent and type of influences on India today, including globalisation links.

After the fall of the British Empire, in which former colonies became independent, Britain struggled with economic debts and the USA and USSR emerged as new superpowers. This resulted in a proxy war in a battle between capitalism and communism. The economically wealthy USA directly influenced much of the world with a global military presence and satellite surveillance technology, but also began to indirectly influence other nations through both trade and the spread of capitalism after winning the Cold War. The USSR mainly directly influenced through a large military population and in neighbouring countries.

> This is a historical narrative and needed to be made relevant to the question, perhaps by using examples of different countries and how they were influenced – such as Somalia and Somaliland.

In the latter part of the 20th century the USA dominated as a single superpower. With globalisation and the almost universal acceptance of economic capitalism, the USA and its IGOs (World Bank, IMF) have been able to indirectly influence through terms of trade, perhaps matching dependency theory, benefiting economically by importing cheap raw materials from peripheral countries and exporting expensive manufactured goods.

The USA has also been able to indirectly influence the world by the spread of media and culture via globalisation and the 'American Dream'. As the world is 'shrinking' due to globalisation, the international presence of US TNCs, such as McDonald's and Universal, has created a common world culture that caters to the capitalist ideals of the USA. This has been described as McDonaldisation. The US film industry (Hollywood) also projects this ideology through movies by the likes of Disney, watched throughout the world, although Bollywood is also influential.

> This presents ways in which indirect influence is spread and the importance of globalisation processes, but where is the assessing? There was also an opportunity to show how some cultures have reacted radically to 'Western' influences.

> This paragraph shows some of the ways indirect influence has come to dominate, but needs more examples of how the USA has indirectly influenced other countries, and does not mention how the USA continues to exercise direct hard power – e.g. in the Middle East. Doing so would have allowed the candidate to start to assess.

Verdict

This is an average answer. It shows an understanding of direct and indirect superpower influence and uses knowledge to place it in an historical framework. However, it needed to show evidence of whether or not there has been a change in the significance of direct and indirect power. A better answer would have referred to examples of influence over countries to show change, such as direct hard power through military actions in Iraq and Syria and the potential conflict between the USA, China and Philippines over the jurisdiction of islands in the South and East China Seas. Even though indirect power is perhaps more common today, direct power is still used in some situations.

THINKING SYNOPTICALLY

Read the following extract carefully and study the figure. Think about how all the geographical ideas link together or overlap. Answer the questions posed at the end of the article. This extract is part of an article that first appeared in the December 2015 edition of the *National Geographic* magazine (pp. 99 to 118) by Alexandra Fuller.

HAITI ON ITS OWN TERMS

Haiti is the country in the Western Hemisphere most vulnerable to the effects of natural disasters. Hurricanes and floods are common. The first recorded earthquake hit in 1562. Quakes aren't nearly as frequent as hurricanes and floods, but since the early 1900s concrete block and reinforced-concrete construction – which hold up better than wood against wind, fire, and rushing water – has been used for houses, hospitals, and schools. Yet when the ground shakes, concrete buildings crack and collapse easily. Haiti's latest and most catastrophic earthquake – a magnitude 7 – struck just west of Port-au-Prince on January 12, 2010. Untold thousands perished in the disaster. The Haitian government eventually put the figure at 316,000. A team funded by the U.S. Agency for International Development (USAID) estimated that the number could not have exceeded 85,000. A group of American academics calculated fatalities at 158,000. ...

With each disaster, in an effort to help, foreign nongovernmental organizations (NGO) and missionaries flood the country with such predictability that some locals call the period in the aftermath of hurricanes 'missionary season'. ... Though many foreigners stay for only a few days, in what amounts to a mercy vacation, others remain for years of gruelling, often vital, work in a country that lacks basic services. Haiti has more than 4000 registered NGOs, but there is no effective oversight of foreign aid institutions, no formal impartial measure of the efficacy of the aid, not even a tally of how many missionaries are in the country. ... Nixon Boumba, a Haitian human rights activist, told me. 'They change the parts, but they don't fix the car. And of course, things got worse after the earthquake. People were so desperate for relief. They put out their hands for help.' He stretched out his hands in an impression of the walking dead. 'But after too long like that, you can become a zombie.'

Of the more than six billion dollars in international aid donated to the country for humanitarian and recovery work following the disaster, only 9.1 percent was channelled directly to the government and less than 0.6 percent went directly to Haitian NGOs and businesses. ... What is not in dispute is that more than a million Haitians were displaced ... the International Monetary Fund loaned Haiti US$24.6 million. In return the Haitian government was required to reduce tariffs on imported rice and other agricultural products. A trade liberalization push in the mid-1990s – championed by President Bill Clinton, a longtime visitor to Haiti and a self-proclaimed supporter of its people – pried open Haiti's markets even more, and rice tariffs were lowered from 50 percent to 3 percent. Heavily subsidized U.S. rice flooded the Haitian markets, much of it from Arkansas, Clinton's home state. Haitian farmers' rice couldn't compete with the cheap and donated imports. Many farmers, after chopping down the last of their trees to sell for charcoal, gave up and flooded the cities, crowding into slums. ...

... [The Spanish] mined every ounce of gold they could easily find, enslaving the native Taino to do so. As a result, almost all the Taino subsequently died, either from overwork or introduced European diseases, especially smallpox. Then came the French colonists, who took over the western third of the island for 140 years and made themselves among the wealthiest people on Earth at that time. They brought up to a million African slaves to the colony they called Saint-Domingue to raze the land's legendary forests – 'tall trees of different kinds which seem to reach the sky', Columbus had written – for hardwood to furnish their mansions in Europe and to make room for lucrative sugarcane and coffee plantations. The incipient environmental disaster – Haiti is now one of the most deforested nations on Earth, with less than 2 percent of its land covered by forest – paled in comparison with the human rights catastrophe that was under way.

French masters in Saint-Domingue treated their slaves so brutally that they died in the thousands. To replace their dead slaves, the French imported more. By the night of August 22, 1791, when a Vodou priest called Boukman gave the signal to begin the uprising that would become the most successful slave revolt in history, slaves – two-thirds of them African born – outnumbered masters by ten to one. In 1804, after 13 years of bloody insurrection, Haiti emerged as the world's first independent black state. The impression of Africa in Haiti remains indelible.

... In the view of some of her citizens, Haiti had become less democratic than anarchic. Wittily desperate graffiti was splashed across the capital city. 'Occupation = Martelly', a reference to the ongoing presence of the UN Stabilization Mission in Haiti (MINUSTAH) and President Michel Martelly, which since 2004 has kept thousands of troops in the country. 'Martelly =

cholera', a reference to an ongoing cholera epidemic that first hit Haiti in late 2010, presumed to have been brought by a Nepali contingent of MINUSTAH.

… The value of the gold and other minerals – copper, silver, iridium – under Haiti's ground isn't known, but exploratory drilling suggests that they may be worth US$20 billion. In December 2012 the Office of Energy and Mines issued the first three permits to mine gold and copper. A member of parliament later complained that he'd learned about the permits from the radio. Two months later the senate passed a nonbinding resolution calling for a moratorium on mining. To get around the deadlock, Haitian government officials invited the World Bank to redraft the mining law, which it did, in close consultation with mining-company officials. In January 2015, with the assistance of the New York University School of Law's Global Justice Clinic and the California-based Accountability Counsel, the Haiti Mining Justice Collective lodged a complaint with the World Bank. It alleged that Haitians had been left out of World Bank–funded efforts to draft new legislation intended to attract foreign investors to finance extraction of Haiti's gold and other minerals. …

Some Haitian activists see the World Bank's cosy relationship with foreign mining companies and disregard for the concerns of Haitian civil society groups as an exhausting repetition of the disastrous arrival of cheap U.S. rice. 'Recolonization comes in two forms', Boumba warned. 'Either the foreign entities use your space to invade your markets with their own products, or they simply steal what you have. But there are a group of us prepared to fight the extractive habit.' … Chansolme, a community on the Trois Rivières, but it feels like a country apart from the city … a place of refuge and nurturing, as home is supposed to be. Mango and palm trees fringed the rough dirt road. There was also the occasional stand of ceiba trees, giants up to 200 feet tall with buttressed trunks like pylons. Sacred to Vodou's Loko – spirit of vegetation and guardian of sanctuaries – the trees haven't been chopped down. The wide river flowed clear and strong, coming in and out of view as we drove along its bucolic banks. Small herds of sleek cattle grazed its shores; villagers and their children bathed and swam.

Figure A: Rice-threshing time in Haiti

ACADEMIC SKILLS

1. In terms of literacy, what do you need to be aware of in this article?
2. **a.** Why are there different estimates of the death toll for the 2010 earthquake?
   b. What may be the consequences of data with such wide estimates for the players involved with the development of Haiti?
   c. How can more accurate data be ensured for Haiti in the future?
   d. Why is accurate data important to helping solve Haiti's natural and human problems?
3. This article has a style (written and photographic) that is sometimes described as a 'travelogue'. How useful is this style in supplying qualitative geographical evidence?

ACADEMIC QUESTIONS

4. **a.** Explain why the coastal areas of Haiti experience multiple hazards.
   b. Which factors have increased the vulnerability of the Haitian people?
5. **a.** Outline how human rights issues in Haiti have changed over time.
   b. Assess the role of global superpowers in creating and solving these issues.
6. **a.** Evaluate the role of globalisation processes in shaping modern Haiti and its future (such as international politics, aid and migration).
   b. Examine the views of the Haitian people towards their island, and the external influences on their country.

Human systems and geopolitics

Global development and connections: Health, human rights and intervention

Introduction

Levels of development in different countries are still measured in economic terms but, as many countries have become wealthier, attention has turned towards broader measures such as people's wellbeing and the condition of the natural environment. Objective data on economic progress is readily available, but data on social and political progress is often more subjective, or is incomplete. The measurement of development now includes analysis of governance and rights to freedom and a decent life. National and international organisations are increasingly influential in situations with negative effects on human welfare, such as disease or armed civil conflict. Even countries that are wary of foreign influence have willingly opened their borders to allow in humanitarian aid after a disaster such as an earthquake, as in Pakistan (2005) and China (2008).

In development terms, human rights measures were previously considered too controversial because of the civil and political overtones, and because they diverted attention from the links between human rights abuses and poverty. The UN's Millennium Development Goals (MDGs), 2000–15 helped change perceptions of human rights.

Human rights are guaranteed through international law; they apply to every person and these rights cannot be taken away. All aspects of human rights are linked, and countries are obliged to ensure that individuals and groups have these rights. Globalisation has raised awareness of human rights and there has been a move towards geopolitical intervention in the form of development aid and military action. While sovereignty remains important, if a state fails to protect its citizens from war crimes or crimes against humanity, the international community now feels it has a responsibility to step in to protect them.

In this topic

After studying this chapter, you will be able to discuss and explain the ideas and concepts contained within the following enquiry questions, and provide information on relevant located examples:

- What is human development and why do levels vary from place to place?
- Why do human rights vary from place to place?
- How are human rights used as arguments for political and military intervention?
- What are the outcomes of geopolitical interventions in terms of human development and human rights?

Figure 4.1: Local community and Rwandan UN peacekeepers building a new school as part of a UN mission in South Sudan (September 2013). What is the role of international intervention in countries? How important is intervention in the 21st century? Can intervention be justified?

Synoptic links

There are many players at different levels, from the United Nations and all its affiliated intergovernmental organisations (IGOs) at a world scale, to regional organisations such as NATO and the African Union, to national governments responsible for the sovereignty of their territory and the wellbeing of people within it. Many non-governmental organisations (NGOs), such as Amnesty International and Médecins Sans Frontières/Doctors Without Borders (MSF), are also involved with humanitarian issues. In addition, a huge range of organisations produces databases and indices to show patterns in health, freedom, human rights and aid. The combined actions of these players help determine the overall wellbeing of the human population, as measured in many ways. This topic links with population geography, especially forced migrations (refugees and internally displaced people), as well as health and welfare. The international links explored are part of globalisation processes, which have created stronger connections between countries through organisations and large companies. The role of sovereignty in the 21st century is brought into question, with the increasing potential for interventions for socio-political reasons and as the result of the deterioration of the natural environment linked to climate change.

Useful knowledge and understanding

During your previous studies of Geography (GCSE and AS level), you may have learned about some of the ideas and concepts covered in this chapter, such as:

- Theories of development such as Rostow.
- Measures of development such as GDP per capita and HDI.
- Development schemes and projects.
- Population geography, including migration, life expectancy and infant and child mortality.
- Patterns of health and education.
- The role of international organisations such as the UN and the EU.
- The role of NGOs such as charities supporting development schemes.
- Top-down and bottom-up development approaches.
- The role of TNCs in development and globalisation.
- Trade and the role of the WTO, the World Bank and the IMF.
- Food and water supplies.
- The causes and patterns of deprivation and inequality.

This chapter will reinforce this learning and also modify and extend your knowledge and understanding of health, human rights and intervention.

Skills covered within this topic

- Comparison of different measures of development, including using ranked data.
- Use of scattergraphs and correlation techniques to show links between development indicators.
- Use of proportional circles to compare government spending on welfare, health and education between countries at different stages of development.
- Use of qualitative and quantitative data to derive an index of corruption and compare world patterns on global maps.
- Use of flow-lines to show aid sources and destinations for world regions.
- Evaluation of a range of source material to assess the impact of development aid.
- Interpret visual images to evaluate environmental impacts of economic activity in ethnic area.
- Critical analysis of sources to show errors in judging success of intervention with European or Asian boat people.
- Use of the Gini coefficient and income proportion for deciles of the population to identify inequalities between and within countries.
- Critical analysis of source materials to identify the possible misuse of data in a qualitative assessment of a named military intervention.

CHAPTER

4

What is human development and why do levels vary from place to place?

Learning objectives

8A.1 To understand that concepts of human development are complex and contested.

8A.2 To understand why there are notable variations in human health and life expectancy.

8A.3 To understand the significant role played by governments and international government organisations (IGOs) in defining development targets and policies.

Concepts of human development

Traditional measures of development

There are various ways of measuring the **development** of countries, and these have usually been based on economic criteria linked to the amount of wealth a country generates through its businesses and trade (Table 4.1). However, more recent measures are based on socio-economic or socio-political criteria such as happiness or corruption. Some measures also consider the state of the natural environment, because ecosystem services are important in providing essential resources for humans. Economists believe that traditional measures are best because they are based on objective measurable data, but others believe that such measures do not accurately assess the full range of human wellbeing.

Table 4.1: Traditional economic measures of development

Measure	Definition
Gross domestic product (GDP)	The total value of goods and services a country produces in a year (or a quarter); it reflects the country's economic activity and broadly represents the standard of living in a country.
GDP per capita	GDP divided by the number of people in the country, giving a measure of mean wealth per person. However, this disguises disparities between the very rich and the very poor.
GDP per capita (PPP based)	GDP per capita adjusted according to Purchasing Power Parity (PPP); it considers the difference in costs of living between countries (usually compared with the USA).
Gross national income (GNI) per capita	The total wealth created by a country, including income from exports (minus taxes and debts). Since currency exchange rates vary, this measure can change considerably over time.

Many measures of development cannot cover all the countries of the world because data for those countries is either unavailable or unreliable: The Better Life Index covers only Organisation for Economic Co-operation and Development (OECD) countries. Some emerging or developing

Yellow = Players, Orange = Attitudes and actions, Purple = Futures and uncertainties

countries use proxy criteria; this is where a development indicator is based on an indirect measure by assuming that they are linked.

The Human Development Index (HDI) – produced by the United Nations Development Programme (UNDP) since 1990 – is a socio-economic measure of development based on GDP, adult literacy levels and life expectancy. Many regard this as a better measure of development because it considers wealth, education and health (but not the state of the natural environment). The links between the wealth of a country or the people within it and contentment (or happiness) are complex: a country may be wealthy, but **inequalities** between its people may still exist, and freedoms may not be guaranteed. Development itself is not without problems: economic development may be unsustainable for some countries if there are obstacles such as high levels of pollution affecting human and environmental health, overuse of resources such as water and forests, disparity between ethnic groups producing tension and conflict, or corruption in political and economic systems. Measures of development based on these issues show sustainable development over a longer timescale, judging whether future generations will have the same opportunities as current generations. Development approaches also vary between countries: although the 'Western' economic approach is the most common due to globalisation forces, although some countries try to follow a more socialist or environmentally sustainable path.

Table 4.2: Measures of development: the top and bottom ten ranked countries (2014/15)

GDP	GDP per capita	GNI per capita	HDI
USA	Luxembourg	Monaco	Norway
China	Switzerland	Liechtenstein	Australia
Japan	Qatar	Bermuda	Switzerland
Germany	Norway	Norway	Denmark
United Kingdom	USA	Qatar	Netherlands
France	Singapore	Switzerland	Germany
Brazil	Denmark	Macao SAR (China)	Ireland
Italy	Ireland	Luxembourg	USA
India	Australia	Isle of Man	Canada
Russia	Iceland	Australia	New Zealand
St Vincent & the Grenadines	Mozambique	Guinea-Bissau	Mali
Comoros	DR Congo	Guinea	Mozambique
Dominica	Liberia	Gambia	Sierra Leone
Tonga	Gambia	Madagascar	Guinea
São Tomé & Príncipe	Niger	Niger	Burkina Faso
Micronesia	Madagascar	DR Congo	Burundi
Palau	Malawi	Liberia	Chad
Marshall Islands	Central African Rep.	Central African Rep.	Eritrea
Kiribati	Burundi	Burundi	Central African Rep.
Tuvalu	South Sudan	Malawi	Niger

ACTIVITY

CARTOGRAPHIC SKILLS

1. Locate the countries listed in Table 4.2, and on a world outline map colour the top ten countries from all measures in one colour, and the bottom ten countries in another colour. Describe and explain the pattern shown by your completed map.

2. Considering the countries listed, suggest the advantages and disadvantages of each measure shown in Table 4.2. You may wish to investigate these measures further.

ACTIVITY

1. In 2014 Bolivia had a HDI score of 0.662 (rank: 119) in the 'medium human development' category, in 2015 a GDP per capita of US$2,886 (rank: 121), and a happiness score (2013–15) of 5.822 (rank: 59). Investigate the latest data for Bolivia and evaluate the success of its development strategy.

2. Describe and explain how Islamic countries such as Iran and Tunisia have adopted a different approach to development, and how successful these might be considered to be.

Extension

Is the Gaia hypothesis, first suggested by James Lovelock, real or not real? To what extent is the Bolivian approach to development based on the Gaia hypothesis?

CASE STUDY: Bolivia's development strategy since 2006

Bolivia is a presidential republic where 68 per cent of the population are of mixed white and Amerindian ancestry (mestizo). President Evo Morales (2006–) represents the Movement Toward Socialism political party and the philosophy of indigenous peoples. In 2009 a new constitution was drawn up and in 2012 the 'Law of Mother Earth' came into force, which recognised that 'Mother Earth is a living dynamic system made up of the undivided community of all living beings, who are all interconnected, interdependent and complementary, sharing a common destiny'. This was a new development approach that put nature first, in response to climate change, the excesses of mining operations and the Andean spiritual world. Bolivia has experienced a recent decline in agricultural production and increasing rural-to-urban migration, pests and diseases, water shortages, mudslides, glaciers melting (for example Chacaltaya) and lakes drying up (for example Lake Poopo). Average temperature is predicted to be 4°C higher by the end of the century.

Bolivia's resources have been reclassified as 'blessings from nature', and a new emphasis has been placed on conservation, with controls on industry and pollution – including the rights of nature and indigenous groups not to be affected by mega-infrastructure or development projects. However, such an approach requires major economic change – reorientating Bolivia's economy (which has been based on mining exports) and attracting foreign direct investment. In addition, 20 per cent of the population lacks clean water and 40 per cent do not have sanitation. The country faces a difficult transition to *Vivir Bien* (living well in harmony with nature). It requires renewable and efficient energy, an ecological audit of economic activities, reduction of greenhouse gas emissions, food and energy **sovereignty**, investment in ecological practices and organic farming, and accountability for pollution and degradation of the environment.

CASE STUDY: Sharia law - contesting the 'western' development model

Most 'western' countries have separated national government (state and law) from religion (church and morality), this is known as secularisation. An assumption has been that this is necessary for a country to become democratic with full respect for universal human rights which allows development. However, the Muslim world does not see secularisation as necessary to development and Islamic law (Sharia) is not restricted to religious matters but also covers, for example, inheritance, marriage, contracts and criminal punishments. Some countries have embedded Sharia law into their constitutions, for example, Iraq has included the clause 'no law can be passed that contradicts the undisputed laws of Islam'. Some Islamic countries have used this to justify an authoritarian approach but many Muslims believe that Sharia is not so rigid and immutable but is actually flexible and fully compatible with contemporary human rights and some elements of 'western' development. For example, Sharia has been used in some countries to restrict the rights of women, but during the life of the prophet Muhammad women were involved in every aspect of society; in Egypt, the Supreme Constitutional Court asserts jurisdiction over any questions of Islamic law and reconciles Sharia with international human rights and economic liberalisation; Islamic financial businesses are growing with products similar to the 'western' world (but without interest payments which is not allowed by Sharia), there are banks, such as in Malaysia, and international banks have created Islamic departments and deal with Islamic bonds (sukuk).

Yellow = Players, Orange = Attitudes and actions, Purple = Futures and uncertainties

Alternative measures of development

- **The Happy Planet Index**, devised by the New Economic Foundation (NEF), combines impacts on the natural environment (the ecological footprint) with the wellbeing of people (life expectancy), and considers the efficiency of resource use in improving people's lives without damaging the environment. However, it does not include a strong economic component.

1 Pollution
2 Resources
3 Resources for power base
4 Environmental protection laws
5 Culture and refugees
6 Human rights laws, democracy and military intervention
7 Wealth and spending on health care and education
8 Population structure/workforce
9 Ecosystems services
10 Population pressures
11 Views on development, sanctions and other interventions
12 Development aid

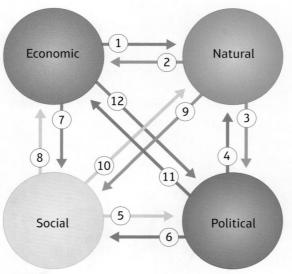

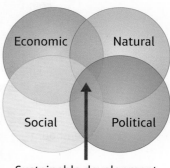

Figure 4.2: Development dimensions and links

- **The World Happiness Index** considers dystopia as a benchmark against which to measure a country's levels of social support, generosity, life expectancy, corruption, GDP per capita and freedom to make choices.

- **The KOF Index of Globalisation** measures the strength of links (Figure 4.2) between countries, using economic, social and political criteria. It indirectly measures development because the countries with the strongest links are likely to have developed in terms of trade, investment and socio-political power (Table 4.3). Overall, the 2015 KOF Index showed that while economic and social globalisation has slowed, political globalisation has slowly increased.

- **The Freedom Index** considers political rights, civil liberties and freedom status; in 2016 the Middle East and North Africa (MENA) region had the least freedom, with 72 per cent of countries 'not free', followed by Eurasia (58 per cent) and sub-Saharan Africa (41 per cent). The 'best' area was Europe (86 per cent free), and so it is not surprising that many asylum-seekers from Africa and Asia migrate to Europe (Figure 4.3) for sanctuary.

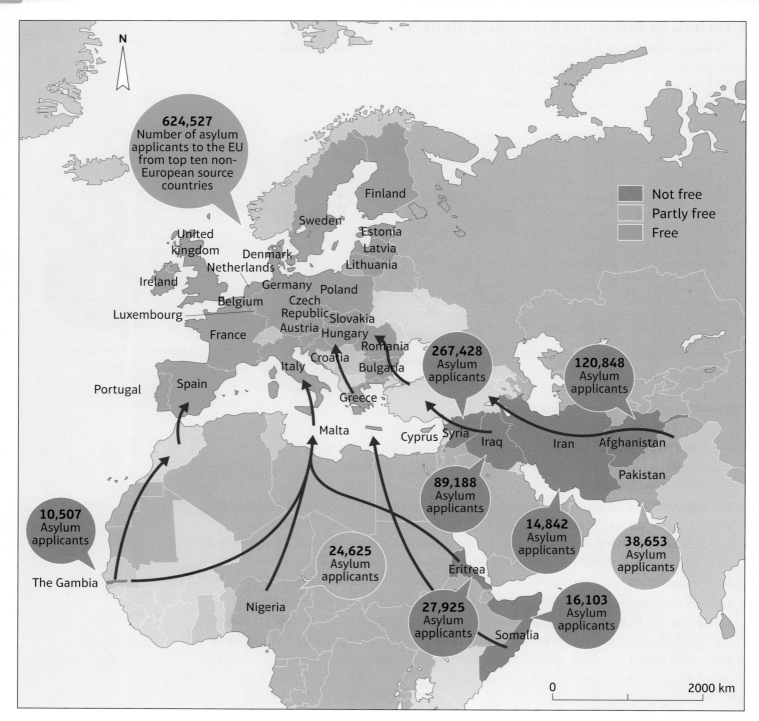

Figure 4.3: Freedom and asylum-seekers, 2015

Yellow = Players, Orange = Attitudes and actions, Purple = Futures and uncertainties

Table 4.3: Alternative measures of development – the top and bottom ten countries

Happy Planet Index (2012)	World Happiness Index (2015)	KOF Index of Globalisation (2015)	Freedom Index (2016)
Costa Rica	Norway	Ireland	Finland
Vietnam	Switzerland	Netherlands	Iceland
Colombia	Denmark	Belgium	Norway
Belize	Finland	Austria	San Marino
El Salvador	New Zealand	Singapore	Sweden
Jamaica	Canada	Sweden	Canada
Panama	Netherlands	Denmark	Netherlands
Nicaragua	Australia	Portugal	Australia
Venezuela	Sweden	Switzerland	Barbados
Guatemala	Israel	Finland	Denmark
South Africa	Niger	Sudan	Central African Rep.
Kuwait	Tanzania	Comoros	Sudan
Niger	Benin	Afghanistan	Turkmenistan
Mongolia	Madagascar	Bhutan	Western Sahara
Bahrain	Haiti	Equatorial Guinea	Eritrea
Mali	Guinea	Eritrea	North Korea
Central African Rep.	Rwanda	Laos	Uzbekistan
Qatar	Syria	Kiribati	Somalia
Chad	Yemen	Somalia	Tibet
Botswana	Liberia	Solomon Is.	Syria

Literacy tip

Dystopia is the term used to describe a culture or society that consists of human misery and unhappiness, characterised by poverty, squalor, lack of freedom, high crime rates, poor health and overcrowding. In terms of migration, an *asylum-seeker* is a person who has left their home country and made a formal application to a host country for protection because they have a real and great fear of being persecuted and not protected in their home country. This is different from a refugee, who would intend to return to their home country once it was safe to do so.

Health and human rights goals

The wellbeing of people is an important goal, but many factors affect it – such as access to fresh clean water, food and energy security, environmental quality, health care provision, life expectancy and human rights. These factors were reflected in the UN Millennium Development Goals (MDGs), 2000–15, and the UN Sustainable Development Goals (SDGs), 2015–30 (see page 208).

- **Environmental quality** is the quality of the air, water, land and natural environments in which people live. Pollution and environmental degradation have a negative influence on human wellbeing. The NEF assessment of the global footprint showed that 1987 was the first year in which humans used more resources than the Earth provided. The overuse of annual resources occurs earlier each year, and in 2016 it was estimated that this happened on 8 August.

Literacy tip

Malnutrition occurs when people do not eat enough nourishing food to meet their daily needs. The UN's Standing Committee on Nutrition judges that malnutrition is the biggest contributor to disease in the world. *Hunger* or *famine* is lack of sufficient food (calories).

ACTIVITY

1. Using Table 4.3, compare the listed countries and their locations.
   a. Which alternative measure produced the result most different from the others?
   b. Suggest why this measure produced such a different result.
2. Describe and explain the patterns of asylum migration shown in Figure 4.3.

Extension

Investigate the OHCHR website and explore the 'human rights by country' for Bangladesh, Thailand, Albania, Brazil and Mauritania. Expand your research to include organisations that monitor human rights. Make notes to add to your knowledge and understanding.

Synoptic link

The natural environment has an important influence on the wellbeing of people. It includes weather and climate, tectonic hazards, the hydrological cycle and the carbon cycle. Climate change is creating more difficult living conditions in some areas, especially for the poor and minority groups, who have a lower capacity to adapt. (See Book 1, page 46.)

ACTIVITY

1. Compare the correlations shown in Figures 4.4 and 4.5. Which has the strongest correlation, and why?

2. Assess the extent to which economic growth can deliver improvements to human development.

- **Health** can be assessed through mortality rates, which can be age-specific – such as infant or child mortality – or cause-specific – for example deaths from diseases or natural hazards. Food scarcity and malnutrition may exaggerate mortality rates. In addition, conflicts such as civil wars clearly bring the threat of death or injury, reducing wellbeing.

- **Life expectancy** is the number of years a newborn baby is expected to live assuming living conditions in the area of birth do not change. Usually given separately for males and females, life expectancy is an indicator of health and it reflects the living conditions and health care system of a place.

- **Human rights** are enshrined in the UN's Universal Declaration of Human Rights (UDHR) 1948 (see page 210). There are international codes (supported by the UN's Office of the High Commissioner for Human Rights (OHCHR)) but, as the Freedom Index shows (Table 4.3), this does not guarantee that people's rights are recognised.

Development through economic growth

Many think that economic growth is the way to improve health, life expectancy, human rights and environmental quality. Figure 4.4 shows that life expectancy increases with wealth, since wealthier countries are able to spend more on health care systems and water and sanitation systems.

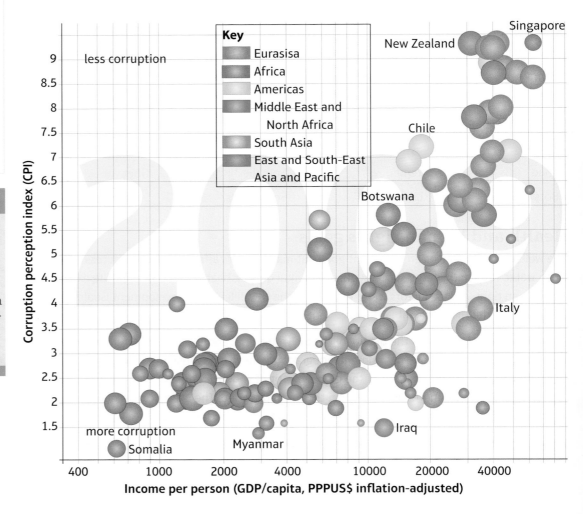

Circle size represents democracy score (Polity IV Project) with larger circle equal to greater democracy

Figure 4.4: The correlation between corruption and economic wealth, 2009

Yellow = Players, Orange = Attitudes and actions, Purple = Futures and uncertainties

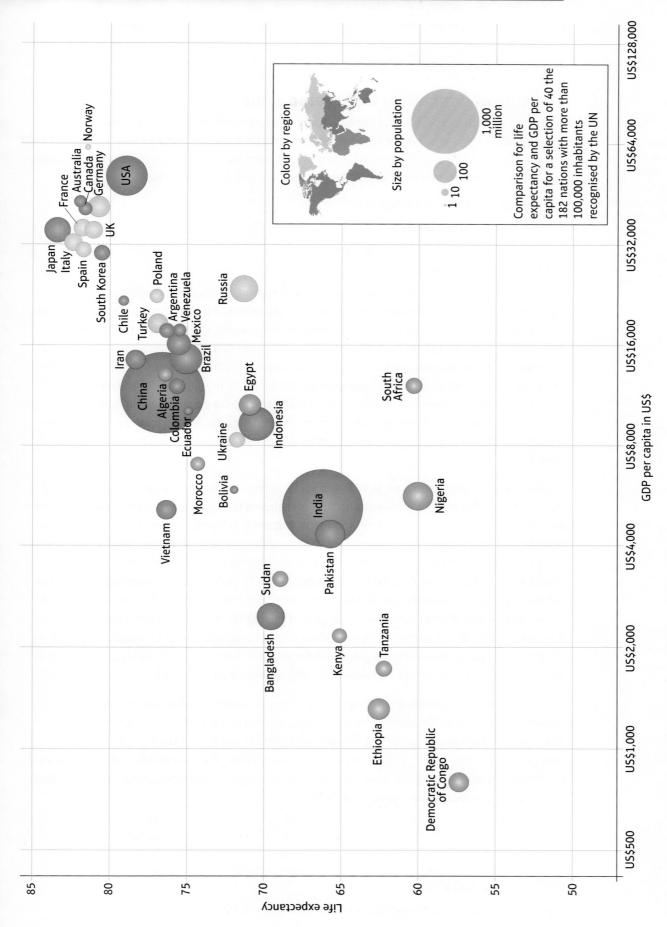

Figure 4.5: The correlation between life expectancy and GDP per capita, 2013

Wealthy countries are also able to spend more on education, which increases people's ability to improve their lives and obtain higher-paid jobs, which in turn improves their access to services and decision-making pathways.

The role of education in development

Education is central to developing human capital – because people need knowledge, understanding and skills in order to be able to improve their lives, and because better workers help the development of a country. Through education, literacy levels improve and this enables people to learn and communicate more widely. People are then able to understand:

- the need for basic hygiene and health care
- ways to control their family size
- how to become involved in decision-making
- their rights in the 21st-century world.

Children in developed countries attend primary and secondary school, and many progress to higher education. However, in developing countries many children aged 7 to 14 are working, rather than attending school – to help on family farms, as in Indonesia, or in manufacturing industries, as in Bangladesh. The UN estimated that, in 2013, 59 million children of primary school age and 65 million of lower secondary school age were not attending school; most of them were girls. In 2015 only 69 per cent of countries had equal gender access to primary school and 48 per cent to secondary education (Figure 4.6). The problem was worse where poverty was high and in areas with conflicts, epidemics or natural disasters, with the countries of the Sahel having nearly 30 million children (6–11 years of age) out of school, with twice as many girls as boys never receiving any education.

Improvements in education have not come equally to females, and historically across the world the role of females has not been equal to that of males. In modern democracies, equality laws now ensure that women have equal status and rights, but some cultures in developing countries still restrict female freedoms. Investment in female education has been shown to improve health and child mortality between 15 and 20 years after female literacy rates have increased, and the UN Educational, Scientific and Cultural Organisation (UNESCO) believes that **gender equality** can be achieved through education. UNESCO sees education as the main 'driver' of development and a fundamental right of all people, and its suggested targets for countries include spending at least 4–6 per cent of GDP on education. Many SDGs are linked to education and equality (for example Target 4.1: 'By 2030, ensure that all girls and boys complete free, equitable and quality primary and secondary education leading to relevant and effective learning outcomes').

Access to education is restricted during times of internal conflict and because of poverty. In culturally conservative Islamic countries there may also be restrictions affecting females, for example on mixed gender schooling and the roles of male and female teachers, and the belief that girls should be learning how to run a home and preparing for married life. But there is a wide difference between Islamic countries and, rather than religious influences, conservative traditions and poverty may be the most important factors.

Extension

Development may reflect different cultural beliefs. Investigate a country or region that practises Sharia Law and explain how this is affecting (or may affect) the development strategy.

Yellow = Players, Orange = Attitudes and actions, Purple = Futures and uncertainties

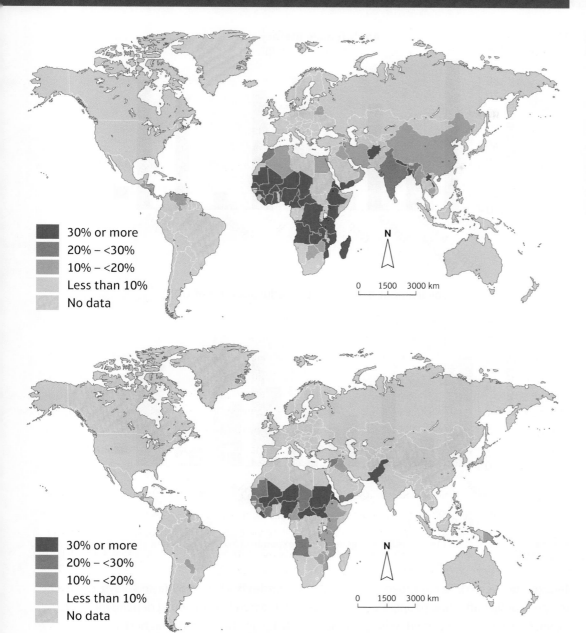

ACTIVITY

CARTOGRAPHIC SKILLS

Study Figure 4.6. Describe the changes that took place between 2000 and 2015 in the share of primary-school-age girls out of school. Refer to world regions and countries within your answer.

Figure 4.6: Share of primary-school-age girls out of school, 2000 and 2015

Patterns of human health and life expectancy
Health and life expectancy in the developing world

The world population explosion (it grew from 3 billion in 1960 to 7 billion in 2011) has mainly been due to large increases in developing countries, especially Africa and Asia, with 80 per cent of the world's population now living in developing countries. By 2050 the world's population is expected to have grown by a further 2.5 billion, with much of this increase in Africa. Population growth increases pressures on resources such as food and water, on living space and on infrastructure such as sanitation. Widespread poverty and overcrowding, especially in rapidly growing cities, ensure that health issues persist in many developing countries, despite success in eradicating some diseases (see page 200).

ACTIVITY

1. Study Figure 4.7.

   a. Describe and explain the change in adult female literacy between 2000 and 2015 in the selected countries.

   b. Is there a correlation between government expenditure on education and the changes in female literacy rates? Suggest reasons for your answer.

   c. Do the featured countries meet the UNESCO suggested expenditure on education target? What are the implications for the development of these countries?

A level exam-style question

Explain why many view education as central to enabling the economic development of countries. (8 marks)

Guidance

Do not just describe links, but make sure you offer a genuine explanation of why education may bring about greater economic development. It may be useful to include data and named countries or world regions to support your points.

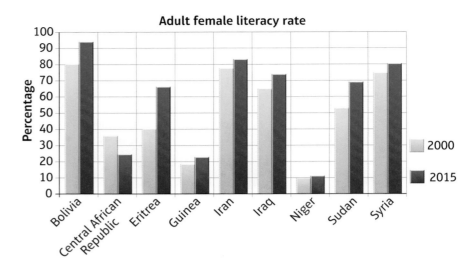

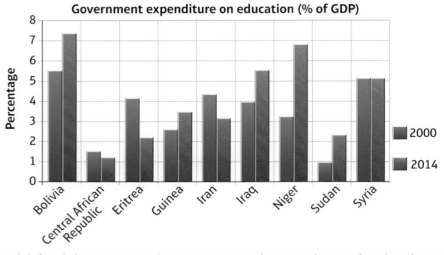

Figure 4.7: Adult female literacy rates and government expenditure on education for selected countries, 2000 and 2014/15

Tensions between different groups in developing countries, such as in Syria and Libya, can lead to open conflict affecting people's wellbeing, as do large numbers of young dependants. For example, Niger had a young dependency ratio of 107:100 in 2014, and the highest 13 nations were all in Africa. In some developing countries the numbers of elderly dependants are increasing; for example Argentina's ratio was 17:100 in 2014, with China's predicted to rise to 39:100 by 2050. Consequently, health and life expectancy vary considerably between the developing regions of the world, as reflected in a wide variety of indicators including infant and **maternal mortality** rates.

Maternal mortality has improved globally, with an average decline of 45 per cent between 1990 and 2013. However, some developing countries performed well, such as Cambodia (86 per cent) and Laos (80 per cent), while others did poorly, such as Côte d'Ivoire (3 per cent) and Kenya (17 per cent). Similarly, there has been a global average reduction of 47 per cent in child mortality, but with large variations. For example, Peru (77 per cent) and Egypt (75 per cent) do well, but DR Congo (15 per cent) and Somalia (17 per cent) do poorly.

Yellow = Players, Orange = Attitudes and actions, Purple = Futures and uncertainties

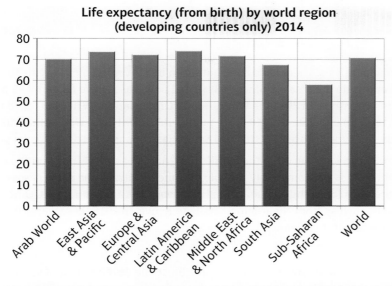

Life expectancy (from birth) by world region (developing countries only) 2014

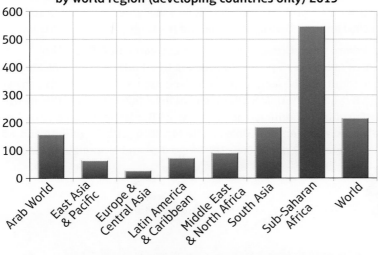

Maternal mortality ratio (modelled estimate per 100,000 live births) by world region (developing countries only) 2015

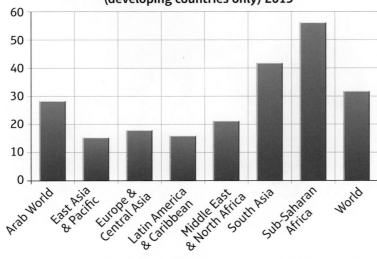

Infant mortality ratio (per 1,000 live births) by world region (developing countries only) 2015

Figure 4.8: Life expectancy, maternal mortality and infant mortality rates in developing countries (by world region), 2014/15

CASE STUDY: Polio and tuberculosis

Polio

Polio is a communicable disease that causes lifelong paralysis. There is no cure, but it can be prevented by a vaccine that developed countries have used since the 1950s to eradicate the disease. Since 1988 global polio cases have fallen by 99 per cent, but the disease is still prevalent in some developing countries. In 2013 the World Health Organisation (WHO) launched the Polio Eradication and Endgame Strategic Plan (PEESP), with an end date of 2018. By September 2015 only one wild poliovirus remained, with type 2 eradicated before 2000 and no new cases of type 3 since 2012. In 2014 South East Asia was certified free of polio.

Obstacles to complete eradication include the conflicts and insecurity in the Middle East, north-eastern Africa and Pakistan, which disrupted immunisation programmes. Monitoring is not completely reliable in some countries, and some cases escape detection; health workers are stretched where there are a lot of children and supplies of vaccines are insufficient. Outbreaks occur when polio is spread from source countries to others, such as from Nigeria to Somalia to Kenya. In 2015 there were 106 worldwide cases of polio, mostly in Afghanistan and Pakistan where it is endemic.

Tuberculosis

Tuberculosis (TB) is a contagious infection associated with overcrowding and poverty. Since the development of vaccines in the 1950s and as nutrition and living conditions have improved, TB mortality rates have declined, halving since 1990. However, in 1993 the World Health Organisation (WHO) declared TB a 'global emergency' and in 2000 the tackling of TB was one of the MDGs. In 2014 9.6 million people worldwide were ill with TB, with 1.5 million dying, and it was the largest cause of death among those with HIV/AIDS. MDG6 did ensure progress between 2000 and 2015, with SDG3.3 taking the target forward to 2030.

Between 2000 and 2014 an estimated 43 million lives were saved through diagnosis and treatment (TB is curable within six months, with the correct drugs), but this still amounted to only a 1.5 per cent reduction in TB infections a year. Some strains of TB have become resistant to medical drugs, and in some countries the management of drug use has been poor and an estimated 3.3 million cases have not been diagnosed. In 2016 StopTB called for a US$65 billion global investment to solve this health issue, and a 2015 WHO report stated that there was a funding gap of US$2.7 billion. The majority of TB infections and deaths are in 'lower middle-income countries'; China and India had the largest numbers of people affected in 2014, with 1.4 million and 1.3 million respectively, while TB mortality rates were highest in Nigeria (9.7 per 10,000), DR Congo (6.9) and Mozambique (6.7) (Figure 4.9).

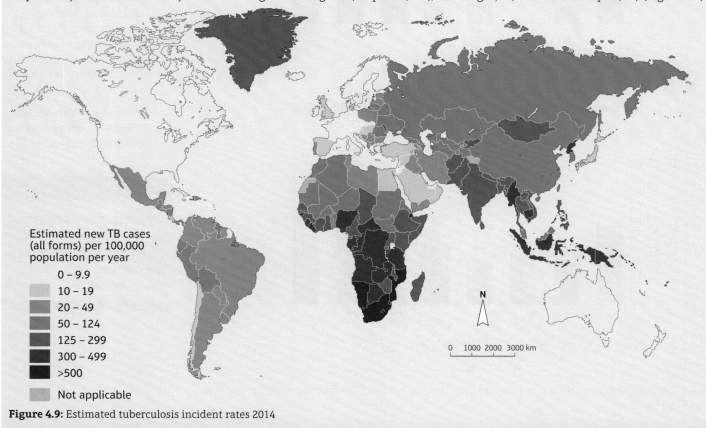

Estimated new TB cases (all forms) per 100,000 population per year

- 0 – 9.9
- 10 – 19
- 20 – 49
- 50 – 124
- 125 – 299
- 300 – 499
- >500
- Not applicable

Figure 4.9: Estimated tuberculosis incident rates 2014

Yellow = Players, Orange = Attitudes and actions, Purple = Futures and uncertainties

Health and life expectancy in the developed world

While developed countries in North America, Europe, Oceania and Japan have much better health levels and life expectancy than developing countries, there are variations in the developed world of between 70.5 years (Russia) and 83.7 years (Japan), with life expectancies also over 83 years in Switzerland and Singapore. These variations are linked to diet, lifestyle, relative **deprivation** and access to medical care. There is a positive correlation between spending on health care, linked to the wealth of a country, and life expectancy (Figure 4.10). For example, Switzerland had the highest expenditure, at US$9,673 per capita in 2014, while Russia spent only US$893. Life expectancy is highest in western and northern European developed countries and East Asian developed countries, and lowest in Middle East countries, despite their oil wealth, and Eastern European countries. It is clear that countries with shorter life expectancies are not spending enough on health care to raise standards. Some countries, such as South Korea and Japan, are doing well on slightly lower expenditures, while the USA has a relatively low life expectancy (79.3 years) despite having the third-highest expenditure on health care per person (US$9,403).

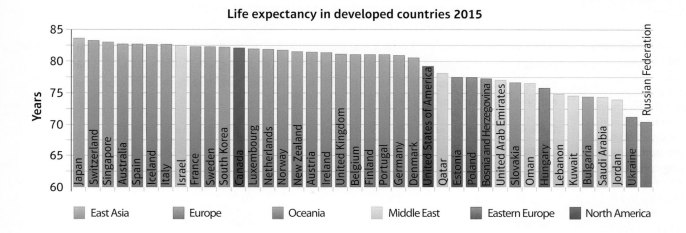

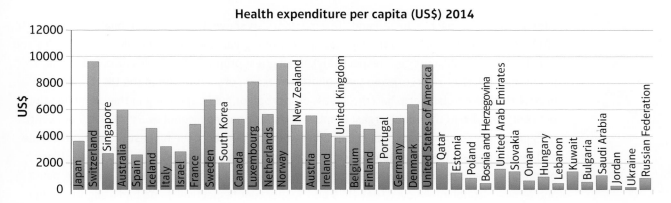

Figure 4.10: Correlation between income and life expectancy

The Japanese National Centre for Global Health and Medicine found, in a 15-year study of nearly 80,000 people (2016), that one of the main reasons for the country's long life expectancy was a healthy balanced diet. Other reasons include advances in medical treatment, staying active and elderly people being involved in the local community. In the USA, 2016 research by the National Centre for Health Statistics showed that 48 per cent of the gap between male life expectancy in the USA and Europe could be accounted for by injuries resulting from firearms, drugs and vehicle accidents. In 2012 the National Research Council and Institute of Medicine highlighted the higher poverty levels in the USA compared with other developed countries, which limited access to health care for poorer people, who tend not to have health insurance. Other common reasons for this lower life expectancy include obesity, heart and lung disease and pollution levels in cities.

Health and life expectancy variations within countries

Southern parts of the UK have a longer average life expectancy than the north. The longest life expectancies are in Dorset (2010–12) – 82.9 years for males in East Dorset, and 86.6 years for females in Purbeck – while the shortest were in Glasgow (72.6 years for men and 78.5 years for women). Life expectancy in the UK continues to increase, at a slightly faster rate for males than females, perhaps because of changing types of employment, with males now involved in less physically demanding work. The living environment also has a part to play: the two highest-ranked places to live in England (2015) – Bebington (Wirral) and Kesgrave (Suffolk Coastal) – had life expectancies that were significantly higher than Liverpool and Middlesbrough, cities that have experienced economic decline and deprivation (Table 4.4).

Table 4.4: Average life expectancies (years from birth) in selected English places (2010–12)

Location	Male life expectancy	Female life expectancy
Liverpool	76.1	80.2
Middlesbrough	76.3	80.2
Bebington (CH63)	79.2	83.0
Kesgrave (IP5)	80.6	83.9

Lifestyle has an impact on health and life expectancy, as shown by a European health survey in 2013–14: Northern Ireland had the highest number of regular smokers in the UK for both males and females (23.5 and 22.4 per cent respectively), the highest proportion of males involved in heavy labouring work (10.6 per cent), and the highest perception of poor health in the UK (11 per cent). In 2014 Northern Ireland also had the lowest proportion of public spending in the UK on health (19 per cent). Wales had the equal largest number of males drinking alcohol every day (10.9 per cent), a high proportion of men doing heavy labouring work (9.9 per cent), the highest proportion of male and female obesity (26.6 and 18.4 per cent respectively) and the highest male and female high blood pressure (20.5 and 18.7 per cent respectively). England was only highest in terms of male and female alcohol intake every day (10.9 and 5.8 per cent respectively) and had a slightly higher proportion of public spending on health care (23 per cent). Scotland did not feature as the highest in any category but still had the shortest life expectancies of any UK country – 76.2 years for males and 80.6 years for females – showing that lifestyle is not the only factor.

The north-south health divide in the UK (Figure 4.11) has been partly attributed to smoking and alcohol patterns in 2015. Easington (Co. Durham) had the highest proportion of smokers (37 per cent) and Liverpool the highest number of hospital admissions for alcohol-related problems (652/100,000), while Chiltern (Buckinghamshire) had the lowest number of smoking-related deaths (147/100,000) and Kensington and Chelsea (London) – home to many wealthy people, and with life expectancies over 80 years for both males and females – the fewest cancer deaths (81/100,000). London nevertheless has the highest rates of infectious diseases, probably because of its high population density and the large number of migrants and tourists moving through the capital.

There is not much difference in health and life expectancy within the same socio-economic group across the UK, but significant differences between socio-economic groups. An important factor would appear to be occupation type: jobs that are physically demanding or that expose workers to chemicals and particulates (dust) are riskier. Another factor in inequalities is income level: people on lower incomes tend to have lower educational attainment, which affects their attitude to diet, exercise, smoking and alcohol consumption, and may also limit their access to health care, despite having a free, universal health care system.

ACTIVITY

Suggest the extent to which Figure 4.11 shows that there is a north-south health divide in England and Wales.

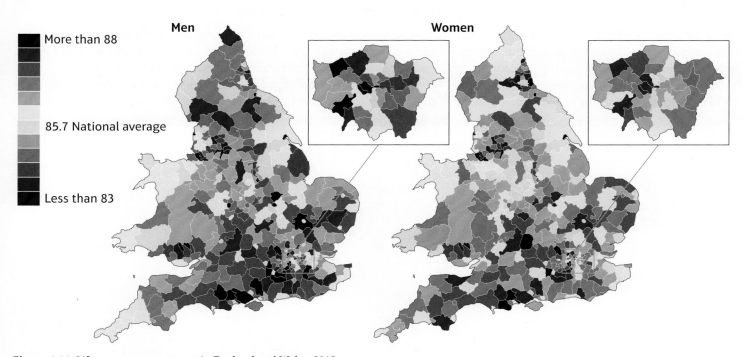

Figure 4.11: Life expectancy patterns in England and Wales, 2015

CASE STUDY: The health of indigenous people in Australia

Indigenous people (Aborigines) make up 3 per cent (713,600) of Australia's population, with a younger average age than the non-indigenous population. A 2012/13 survey by the Australian government found that the health of Aborigines was impaired by social factors such as losing connectivity with the land and their family. Aborigines often live in remote areas with limited access to health services, which may also be expensive or culturally inappropriate.

There are some major differences between the health of Aborigines and non-indigenous people:

- Life expectancy is about ten years shorter for both males and females (69.1 years and 73.7 years respectively), with the main causes of death cardiovascular disease – twice as high as for non-indigenous people – and cancer.

- Infant mortality rates are twice as high.

- Type 2 diabetes is three times more prevalent, especially in indigenous women, and responsible for seven times more deaths.

- Indigenous adults, especially women, are three times more likely to suffer psychological distress as a result of greater exposure to unemployment, alcohol and drugs issues, gambling and 'getting into trouble with the police', and suicide rates are higher.

- Diseases such as kidney, respiratory and TB are all higher, the latter over 12 times more prevalent. Cancer rates are higher, with more fatal types.

- Obesity is 30 per cent higher in indigenous children and 66 per cent higher in indigenous adults.

There are some lifestyle causes of these differences: indigenous people eat a poorer diet and 27 per cent had a vitamin D deficiency. Exercise rates were similar and average alcohol consumption was lower (although for a minority it was much higher), but smoking rates were twice as high, accounting for 20 per cent of all deaths. Educational achievement levels were lower, although retention has been increasing, and incomes were 38 per cent lower. Solutions are improving education levels so that socio-economic status can be raised, and improving health services by adapting them to the needs of the indigenous population.

Figure 4.12: Map of Brazil states

Yellow = Players, Orange = Attitudes and actions, Purple = Futures and uncertainties

Table 4.5: Health and life expectancy data for Brazil regions

State	Life expectancy 2015 (years)	Average monthly household income per capita (RUS$) 2015	Gini index (monthly income) 2015	Adults feeling discrimination in health service (%) 2013
Acre	73.3	752	0.501	11.4
Alagoas	70.8	598	0.459	9.3
Amapá	73.4	849	0.449	14.8
Amazonas	71.4	752	0.490	13.3
Bahia	73.0	736	0.506	9.7
Ceará	73.4	680	0.475	10.3
Distrito Federal	77.6	2252	0.565	10.7
Espírito Santo	77.5	1074	0.480	5.8
Goiás	73.8	1077	0.447	14.1
Maranhão	70.0	509	0.493	11.5
Mato Grosso	73.7	1055	0.458	14.6
Mato Grosso do Sul	75.0	1045	0.491	12.8
Minas Gerais	76.7	1128	0.473	11.1
Pará	71.7	672	0.468	13.3
Paraíba	72.6	776	0.501	9.5
Paraná	76.5	1241	0.450	14.8
Pernambuco	73.1	822	0.473	11.9
Piauí	70.7	729	0.494	8.3
Rio de Janeiro	75.6	1285	0.502	8.9
Rio Grande do Norte	75.2	818	0.486	10.9
Rio Grande do Sul	77.2	1435	0.469	9.6
Rondônia	70.9	822	0.451	11.7
Roraima	70.9	1008	0.495	14.3
Santa Catarina	78.4	1368	0.429	11.5
São Paulo	77.5	1482	0.476	8.9
Sergipe	72.1	782	0.484	9.8
Tocantins	72.8	818	0.501	18.4
n = 27 Means:	73.88	983.89	0.480	11.53

ACTIVITY

CARTOGRAPHIC SKILLS

1. Using a copy of the outline map of Brazil and its regions (Figure 4.12) and Table 4.5, produce a choropleth map to show the pattern of life expectancy within the country.

2. Using the data in Table 4.5, carry out statistical analyses, such as central tendency and dispersion indicators, or correlations, to help understand the patterns.

3. Describe and explain the pattern with reference to your map and statistical analyses, and data on climate, ethnic variations, income levels, inequalities and health care for the regions.

Maths tip

Remember that the *Gini index* is a measure of income distribution among the people of an area. It has a range from 0 to 1 – where 0 is perfect equality and 1 is perfect inequality. In Brazil, Santa Caterina state is the most equal and the Distrito Federal (Brasilia – the capital) the most unequal. The *Gini coefficient* converts the Gini score to be out of 100 instead of 1. Measures of central tendency are the *mean, median* and *mode*, and dispersions are the *range, standard deviation* and *interquartile range*. Correlation can be judged through *Spearman's rank correlation* analysis.

The role of governments and IGOs
Links between economic and social development

The links between economic development, as measured by wealth, such as GDP, and social development, as measured by improvements in health and freedom, such as life expectancy, are complex and vary greatly between countries of different types. Some countries are committed to a **welfare state**, within which the wellbeing of all citizens is regarded as a priority and spending on health care and education systems is high, such as in Norway and Sweden. Other countries are committed to economic development in the short term and this may mean that budgets are arranged in favour of industrial and business infrastructures, such as in Brazil and Ghana. A few countries have prioritised other areas, such as military spending, as in North Korea, and others have **totalitarian regimes** where the political or ruling elite may allocate only small budgets to health and education as a means of controlling people, as in China and Uzbekistan.

Public spending in the UK for 2016 was £759.5 billion (US$1,102.8 billion), with 18 per cent on health care and 12 per cent on education (Figure 4.13). The National Health Service (NHS) offers free medical care for all, with some means testing for prescriptions: Spending on health amounts to US$3,102 per person for 2016. Brazil has a Unified Health System that entitles every Brazilian to free health care, which amounted to US$1,471 per capita in 2013, or 4.4 per cent of GDP. However, a quarter of Brazil's population has private medical plans, which suggests that there may be unequal access (see Tables 4.5 and 4.6). Information from some developing countries and especially from totalitarian countries is often not available, for example North Korea, or may be unreliable, for example Niger.

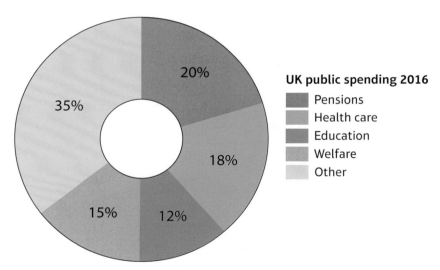

UK public spending 2016
- Pensions
- Health care
- Education
- Welfare
- Other

Figure 4.13: UK public spending 2016

Yellow = Players, Orange = Attitudes and actions, Purple = Futures and uncertainties

Table 4.6: Government expenditure on health and education for selected countries, 2012–14 (% of total expenditure)

Democratic rank (Economist 2015) (/167)	Country	Health expenditure	Education expenditure
9 (democracy)	Australia	17.3	13.2
16 (democracy)	UK	16.5	13.0
35 (flawed democracy)	India	5.0	14.1
51 (flawed democracy)	Brazil	6.8	15.6
85 (hybrid)	Bolivia	11.8	16.9
121 (authoritarian)	Niger	7.6	21.7
136 (authori/totalitarian)	China	10.4	12.6 (1999)
142 (authori/totalitarian)	Oman	4.77	10.9
158 (totalitarian)	Uzbekistan	10.7	n/a

The views and programmes of IGOs

Since 1945 the main IGOs have emphasised the importance of economic development through free trade and self-regulating markets, allowing economic processes to follow their course without interference or regulation. These views were based on **capitalist** economic theories, as reflected in Kondratiev's cycles or Rostow's stages (see pages 148 and 149).

The IMF was established to provide loans to countries in financial difficulties, so that they could continue to participate in international trade and transactions. Similarly, the World Bank offers loans to countries that need money during an economic recession or for spending on expensive infrastructure and industry to bring about the development of the country. However, many of these loans did not lead to economic success and countries became indebted, leading to the Highly Indebted Poor Countries Initiative (HIPC) in 1996, which provides debt relief from the IMF and World Bank cancellation of debts or sustainable repayment schedules. Most loans or debt relief agreements also have conditions attached, which can lead to austerity within the country and divert government spending away from health, education and welfare.

The WTO has operated since 1995, regulating world trade by overseeing negotiations between countries and promoting the reduction or removal of tariffs and other trade barriers, on the premise that all countries benefit. However, this is not always the case, as demonstrated by the Fair Trade movement.

IGOs have emphasised the importance of social and environmental themes, such as environmental quality, health, education and human rights. The United Nations Development Programme (UNDP) helps countries make policies and national development plans, and provides expertise for working towards the SDGs. The UNDP's Human Development Index (HDI) reflects the desire to combine economic and social progress. The Food and Agriculture Organisation (FAO) promotes food security and nutrition, the World Health Organisation (WHO) promotes health and disease eradication (see page 200), and the UN International Children's Emergency Fund (UNICEF) has helped look after the rights of children since 1946 (Figure 4.14). The UN Educational, Scientific and Cultural Organisation (UNESCO) identified 31 natural and cultural world heritage sites in May 2016 that are at risk from climate change, including Yellowstone National Park (USA), the Galapagos Islands (Ecuador), the Great Barrier Reef (Australia) and Komodo National Park (Indonesia). The UN Environment

---

ACTIVITY

GRAPHICAL SKILLS

1. Study Table 4.6.

a. Create proportional circles to represent the expenditure by governments on health care in the selected countries.

b. What is the correlation between government expenditure on health care and level of democracy?

c. Suggest why 'flawed democracies' may have the lowest expenditure on health care.

Extension

Visit the World Bank database, especially the World Development Indicators, or the UNESCO database. Find the latest information on government expenditure. Compare the different ways that the data on expenditure is presented, such as proportion of total expenditure or proportion of GDP. To what extent is data available for totalitarian states? Explain your answer.

Extension

Investigate and analyse how UNICEF protects the human rights of children around the world.

Figure 4.14: UNICEF Goodwill Ambassador David Beckham promoting children's rights in the Philippines, 2014

Assembly (UNEA) promotes improvements in air pollution levels and the eradication of the illegal wildlife trade, assisted by other IGOs.

Millennium Development Goals and Sustainable Development Goals

The eight Millennium Development Goals (MDGs) ran from 2000 to 2015, when the more extensive Sustainable Development Goals (SDGs) replaced them. Developed countries had mostly already fulfilled all the MDG goals, but had an important role in assisting emerging and developing countries to meet them, and this role will continue with the SDGs.

MDG1: Eradicate extreme poverty and hunger. By 2015 the proportion of people in developing countries living on less than US$1.25 a day had fallen to 14 per cent, but this still left 836 million people living below this level. The proportion of undernourished had fallen by 12.9 per cent by 2015. While all world regions improved their poverty and hunger levels, the specific targets for poverty were not met in sub-Saharan Africa or Western Asia, and those for hunger were not met in sub-Saharan Africa, the Caribbean, South Asia, West Asia or Oceania.

MDG2: Achieve universal primary education. By 2015 primary school enrolment rates had reached 91 per cent, but 57 million primary-aged children were still not attending school. All world regions improved, with sub-Saharan countries on average reaching 80 per cent enrolment. In the whole of Africa, the gender balance was still unequal and conflicts were also responsible for children not going to school.

ACTIVITY

ICT SKILLS

Investigate the most recent MDG progress reports on the UNDP website. Compare the progress between 2000 and 2015 for one country from each of the following world regions: Africa, Arab states, Asia and Pacific, Europe and CIS, and Latin America and Caribbean. Using the information and data you find, evaluate the factors that created mixed progress towards the eight MDGs in your selected countries.

Yellow = Players, Orange = Attitudes and actions, Purple = Futures and uncertainties

MDG3: Promote gender equality and empower women. Many developing countries eliminated gender disparity in education, with South East Asia having more girls than boys in primary school by 2015. More women were in paid employment and in parliaments but parity has still not been reached. Oceania and sub-Saharan Africa were furthest from the target. In Latin America and the Caribbean the number of women in poor households increased.

MDG4: Reduce child mortality. All regions improved significantly, with child mortality halving between 1990 and 2015. Between 2000 and 2013, 84 per cent of children received at least one measles vaccination, preventing an estimated 15.6 million deaths. Sub-Saharan Africa improved rapidly from a low base but missed the target, as did most world regions except North Africa, Latin America and the Caribbean and East Asia.

MDG5: Improve maternal health. Maternal mortality declined by 36 per cent between 2000 and 2013, with notable decreases in South East Asia and sub-Saharan Africa. More health personnel were present at births (71 per cent in 2014) and antenatal care improved, as did the use of contraception. However, no world region met the target.

MDG6: Combat HIV/AIDS, malaria and other diseases. New HIV infections fell by 40 per cent between 2000 and 2013, and the distribution of treatments increased six times. Between 2000 and 2015, 6.2 million malaria deaths were prevented and the incidence of the disease reduced by 37 per cent. TB prevention, diagnosis and treatment saved 37 million lives (2000–13), with mortality rates decreasing by 45 per cent. Despite these improvements in all world regions, new diseases, such as ebola, brought new threats.

MDG7: Ensure environmental sustainability. Protection of marine and terrestrial areas increased; for example, protected land areas in Latin America and the Caribbean increased from 8.8 per cent in 1990 to 23.4 per cent in 2014. Some forms of pollution greatly decreased, such as stratospheric ozone, but greenhouse gas emissions increased. Forest and marine resources continued to be over-exploited, causing environmental degradation, which affected the poorest people the most. Many people had improved access to water and sanitation, but in 2015 42 per cent of the world population still did not have access to water in their homes. Fewer countries met the sanitation target despite 2.1 billion people benefiting from improved sanitation. The drinking water and sanitation targets were not met in Oceania, sub-Saharan Africa and parts of Asia.

MDG8: Develop a global partnership for development: There were improvements, with official **development aid** from developed countries increasing by 66 per cent (2000–14), reaching US$135.2 billion, with several European countries such as the UK exceeding the UN aid target (0.7 per cent of GNI). Trade balances became more favourable for developing countries. Mobile phone networks spread to cover 95 per cent of the world's population, and internet cover for 43 per cent by 2015.

The 17 SDGs adopted in September 2015, with a target date of 2030, are broader than the MDGs and each has many specific targets.

Extension

Using the internet, find and read the most recent UN SDG Report. You will find that much of the information links to geography topics; make notes to add to your breadth of understanding and synopticity.

ACTIVITY

ICT SKILLS

1. Investigate the UN Sustainable Development Goals website.

a. Investigate each SDG and make brief notes for each on (i) why it matters, (ii) facts and figures that show its importance and (iii) the IGOs that will help.

b. The MDGs had difficulties collecting data and accurately measuring progress. How will the SDGs be measured? Will their monitoring be more accurate than that of the MDGs? Explain your answer.

Why do human rights vary from place to place?

Learning objectives

8A.4 To understand why human rights have become important aspects of both international law and international agreements.

8A.5 To understand the significant differences between countries in both their definition and protection of human rights.

8A.6 To understand the significant variations in human rights within countries, which are reflected in different levels of social development.

Human rights laws
The Universal Declaration of Human Rights

The UN's Universal Declaration of Human Rights (UDHR) of 1948 set out the fundamentals of human rights that everyone is entitled to. It contains 30 Articles specifying these rights, such as freedom, justice, peace and no persecution, to be applied within countries and internationally. The UDHR and its Covenants on Civil and Political Rights and Economic, Social and Cultural Rights (1976) are collectively known as the International Bill of Human Rights, which has led to a range of treaties. These form the basis of **international law**, national constitutions and laws, and are reflected in cultures and government policies. As all aspects of human rights are linked, countries are expected to adhere to the whole 'package'. However, some countries are selective for economic or political reasons, and sometimes human rights may be more important than sovereignty. Some rights are contradictory or difficult to define, which leads to different interpretations and contested points of view, so rights are inconsistently applied.

Since the start of the 21st century there has been an increase in the number of 'authoritarian' countries, which has further limited human rights. For example, there are countries where executions still take place (Pakistan), slavery exists (Mauritania), and discrimination against women continues (Saudi Arabia). Not all countries have signed and ratified the UDHR; for example, Saudi Arabia with its conservative Islamic beliefs does not agree with freedom of religious choice or equal rights of women in marriage. Some Islamic countries believe that the UNDR is too 'Westernised', and in 1990 they produced their own similar version – the Cairo Declaration of Human Rights in Islam.

Developed 'Western' countries have often tied development aid to human rights, or have occasionally undertaken military operations to intervene where there have been human rights violations, as international law can be regarded as more important than sovereignty. Some nations, especially developing countries, dispute the balance of civil and political rights and economic, social and cultural rights, arguing that economic development must have priority for them. The USA and the EU have condemned human rights violations in a number of countries, such as Syria, Russia and China.

ACTIVITY

ICT SKILLS

1. Find and read the UK Government's latest *Human Rights and Democracy Report.*

   a. Describe the government policies.

   b. Describe and explain the actions taken for three of the 'human rights priority' countries.

Yellow = Players, Orange = Attitudes and actions, Purple = Futures and uncertainties

Russia cited protection of ethnic minority groups as the reason for its military action in eastern Ukraine. Similarly, the USA cited the suppression of human rights in Iraq to gain support for military action. However, failure to act in Libya and Sudan in the recent past shows the lack of consistency in the global governance of human rights. In 2015 the UK identified 30 human rights priority countries (from Afghanistan to Zimbabwe), and the Department for International Development (DFID) based their overseas development assistance or aid (ODA) on economic and social rights arising from the SDGs. In 2015 59.6 per cent of the UK's ODA went to African countries and 38.6 per cent to Asian countries; and the UK supported sanctions on Burundi and South Sudan as well as Yemen and Syria.

The European Convention on Human Rights

The European Convention on Human Rights (ECHR) was formulated by the Council of Europe, based on the UDHR, in 1950, and ratified by the UK in 1951. As political Europe expanded and unified, more countries ratified the **convention,** and there are now 47 signatory countries. The purpose is to achieve 'greater unity' and a 'realisation of human rights and fundamental freedoms'. There are 59 Articles and various **protocols**. The European Court of Human Rights was set up in 1959 to interpret and ensure compliance with the ECHR, acting as a checking mechanism on national laws; for example, the UK Human Rights Act (1998) was based on the ECHR to make it more effective in UK law.

The ECHR has been controversial – some see it as undermining national sovereignty when its rulings override national courts' decisions. For example, the European Court has blocked the deportation of people to countries where they may be tortured, allowed prisoners to vote, overturned abortion laws and judged surveillance to infringe privacy. Criticisms of the Court are the length of time it takes to make rulings and the backlog of cases that built up, although this was streamlined in 2010. In 2015/16 these factors led to the UK Government proposing to replace the 1998 Act with something 'more British', but this is also controversial because many believe that the ECHR has helped the police and juries reach clearer verdicts. Also, only a minority (1.4 per cent) of judgements made by the European Court have concerned the UK, and 99 per cent of these were 'thrown out'. The vast majority of judgements and cases have involved eastern European countries, including Russia and Turkey (Figure 4.15).

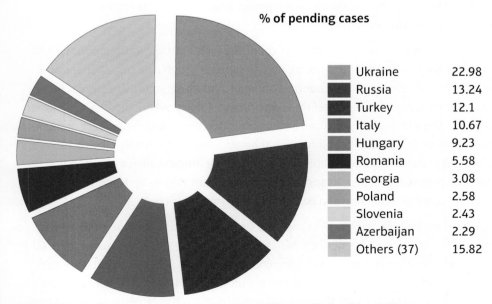

% of pending cases

	Ukraine	22.98
	Russia	13.24
	Turkey	12.1
	Italy	10.67
	Hungary	9.23
	Romania	5.58
	Georgia	3.08
	Poland	2.58
	Slovenia	2.43
	Azerbaijan	2.29
	Others (37)	15.82

Figure 4.15: Pending cases at the European Court of Human Rights (May 2016)

ACTIVITY

Study Figure 4.15 and note the countries shown. Suggest why these ten countries featured more than others in early 2016.

Literacy tip

There are several legal terms in this section. Make sure you know what the following terms mean: act, article, convention, covenant, international law, protocol, ratified, treaty, tribunal.

The Geneva Convention

The four 1949 Geneva Conventions (GC) have been agreed by 196 countries, or 'observers'. However, not all have agreed to the 1977 and 2005 protocols. Collectively, these are international treaties that create rules for war situations, especially offering protection to civilians, medical workers, hospital ships, aid workers, those unable to fight such as the wounded, and prisoners of war. Since in the 21st century most armed conflicts are internal, many regard Article 3 as the most important because it prohibits torture, hostage-taking and degrading treatment. There is also recognition of the neutral emblems of the Red Cross, Red Crescent and Red Crystal in the 2005 protocol. The GC applies to both international and internal armed conflicts. The GC is strongly interlinked with international humanitarian laws (UDHR and ECHR), and breaches of the GC are investigated, perpetrators caught and judged by an international court. However, the process is complex and many breaches are dealt with by regional or internal legal systems, although ultimately the UN Security Council is the final international tribunal. The UN also can call upon its peacekeeping forces to monitor humanitarian situations and ensure that the GC is being applied. War crimes have been tried at International Criminal Tribunals for the civil wars in Rwanda (1994) and the former Yugoslavia (1992–95) (see page 244), but other situations such as Sudan (2003–04), Ukraine (2014>) and Syria (2011>) were not dealt with in the same way.

NGOs such as Amnesty International (AI) and Human Rights Watch (HRW) have reported that many countries breach the GC by torturing or mistreating their citizens. This occurred in 82 per cent of the 160 countries surveyed, including the Central African Republic, Syria, Libya, Iraq, Nigeria and Ukraine. AI described the international response as 'shameful and ineffective' and called for greater action. One consequence of this situation is the increase in the total number of refugees, asylum-seekers and internally displaced people to its highest level since the Second World War, reaching 59.5 million in 2014. These migrations have significant international repercussions, such as the EU debate on admitting refugees, and fears over international terrorism and political extremism.

Definitions and protection of human rights
Approaches to human rights

A country's priorities dictate the emphasis of its policies and plans, and are linked to its existing levels of economic and social development, including its political philosophy. Some countries put human rights first while others put economic development first, and they may move back and forwards along this continuum over time.

A logical premise would be that democratisation of countries such as Brazil, India, Indonesia and South Africa would create a culture for promoting human rights through new constitutions and laws. However, many newly democratised countries appear to prioritise security of the country, energy security, economic development, and trade and financial flows over human rights. Former colonies may wish to show their ability to make autonomous decisions, regarding direct or indirect **superpower** influence on human rights as infringing their sovereignty, as in Iraq, Libya and Syria. International NGOs and world regional organisations such as the African Union may intervene, but some influential organisations such as the Association of South East Asian Nations (ASEAN) prioritise economic growth in emerging and developing countries.

Sometimes countries may ignore the UDHR or the UN's Universal Periodic Review (UPR), which assesses human rights issues of all member countries and provides help and assistance (over 40 countries are reviewed each year). Some developing countries that do not wish to publicise their own human rights record, such as India with Kashmir, are unwilling to implicate other countries too quickly. The people in a country may themselves be reluctant to change systems, even if they

Extension

Investigate and evaluate the role of the African Union in protecting human rights in the continent of Africa.

Yellow = Players, Orange = Attitudes and actions, Purple = Futures and uncertainties

are authoritarian, if the systems appear to be working economically, or they may reject democratic or semi-democratic systems when these do not appear to be improving jobs and wealth. These differences have led to the share of 'free' countries declining by 2 per cent between 2005 and 2015, according to the 2016 Freedom Index.

Globalised communication means that reports of human rights abuses are not easy for democratic nations to ignore. This has led to crowdsourcing and pressure on the UN to take action, for example over North Korea. The UN, ASEAN and bilateral diplomacy imposed sanctions on Myanmar to confront and encourage the military dictatorship to respect human rights within that country; this pressure eventually led to democratic elections in 2015. The international community often uses the Responsibility to Protect (R2P) principle to gain the cooperation of countries in sanctions, as in the Côte d'Ivoire (2011) and the Central African Republic, working with the African Union.

CASE STUDY: Human rights in Canada and Indonesia

Canada: Human rights are promoted to a high level, although there are problems with the rights of indigenous peoples and gender inequality; in 2008 there was still concern about discrimination against women. All suspected abuses of human rights are investigated, the government has its own human rights bodies and it recognises and uses Canadian and international human rights groups. The law protects indigenous people, but there have been ongoing issues and protests over their land and living conditions: dams and pipelines are still built across their territory. In 2014 the government paid compensation to the Nunavut Inuit for failing to provide enough resources for education. In 2016 a public inquiry investigated reports of violence towards indigenous women and girls. The Canadian constitution allows freedom of speech and a free press, which is supported by a democratic political system. The Supreme Court has ruled, though, that the government may limit free speech if it helps stop discrimination, ensures social harmony or promotes gender equality.

Indonesia: Indonesia increased its level of democracy between 2006 and 2015 (Table 4.7), with democratic elections in 2014, but its corruption remained worse than the world average. A 2015 report by the US government suggested that the government has not conducted transparent public investigations into all alleged abuses of human rights, used torture 84 times in 2014/15, and applied treason, blasphemy, defamation and decency laws to restrict freedom of expression. The police or the army often break up protest gatherings, such as in 2014 when at least four civilians were shot dead at a peaceful protest in Papua Province, and there was no investigation. The separatist movement in Papua and West Papua provinces continues to be a source of human rights violations, with frequent arrests by the authorities. This includes the rights of indigenous peoples; in 2013 the Constitutional Court ruled in favour of their rights to own the forests, but due to corruption the forest is still used for commercial purposes. There is an independent national media and in 2014 restrictions on foreign journalists visiting Papua or West Papua were partially lifted. Aceh province now operates Sharia law and carries out public canings of those accused of gambling, alcohol consumption or adultery.

Variations in human rights and freedom of speech

Emerging economies of the world need political leaders and systems able to support economic growth. However, the economic recession (from 2008 onwards) led many to internal public and political unrest, revealing that they were not well governed. The transition of emerging powers towards democracy has slowed (Figure 4.16), even though reforms and pledges in some – such as India, Indonesia and Malaysia – promise progress. Some of the established superpowers also struggled, according to *The Economist*'s Democracy Index, with authoritarian Russia and democratic USA moving backwards (from 5.02 to 3.31 and from 8.22 to 8.05 respectively) between 2006 and 2015, totalitarian China making a small improvement, and the democratic UK taking a significant step forwards (from 8.08 to 8.31) (see Table 4.7).

The Democracy Index looks at a country's electoral process and pluralism, civil liberties, the functioning of government and the political culture. The year 2015 was reported as a 'fearful' one, with civil wars, terrorism and migration issues, which appeared to decrease democracy in some

ACTIVITY

1. Study Figure 4.16.

   a. Describe and explain the world changes in democracy between 2006 and 2015.

   b. Suggest how these changes might have affected human rights and freedom of speech.

2. Study Table 4.7 (overleaf). Assess the extent to which economically emerging nations have made the transition to democracy and freedom.

countries because of fears of insecurity and extremism, resulting in only 20 full democracies. Europe experienced 'populist' or 'nationalistic' political movements and the USA experienced a political 'mood' based on some extreme rhetoric. Latin America, Asia and Eastern Europe were characterised by flawed democracies and political instability, while autocratic regimes were concentrated in Africa, the Middle East and CIS countries in Eastern Europe. The largest annual decline in democracy was in the Middle East and North Africa (MENA) region, as the Arab Spring (2010) failed to improve freedom and human rights (Figure 3.26). However, some countries moved forward with democratic elections, such as Myanmar, Nigeria, Madagascar and Burkina Faso, and the public outcry against **political corruption** in South American countries sowed the seeds of greater democracy. An analysis of the link between GDP growth rates and democracy in 18 economically emerging countries (Table 4.7) showed no correlation between the two variables.

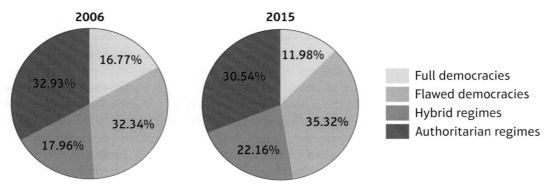

Figure 4.16: Proportion of countries in each democracy category, 2006 and 2015

Levels of political corruption

The Corruption Perceptions Index (produced by NGO Transparency International) measures the perceived level of public sector corruption and so it is a qualitative measure. In 2015, 68 per cent of the 168 countries surveyed – including developed, emerging and developing countries – had serious corruption problems (score under 50) (Table 4.7, Figures 4.17 and 4.18), causing an estimated financial loss of over US$1 trillion a year. Between 2012 and 2015 Greece, Senegal and the UK (74/100 in 2012 to 81/100 in 2015) reduced their levels of corruption, but in Australia, Brazil, Libya, Spain and Turkey corruption increased, as did conflict areas such as Sudan, Somalia and Afghanistan. An analysis of the link between GDP growth rates and corruption in 18 economically emerging countries (Table 4.7) shows a significant negative correlation between the two variables, so the greater the democracy the lower the corruption. There are few quantitative measures of corruption, because data is difficult to obtain (Figure 4.4). The World Bank assesses 'policies for social inclusion and equity' and 'public sector management and institutions', but for only 95 countries.

Corruption reduces levels of trust and threatens human rights, as systems become unfair and support the groups with power, or even persecute the poor and other disadvantaged groups. Corruption may affect the economic system through favouritism of certain businesses, and may also work against health and safety laws, such as in Bangladesh textile factories, or affect the political system through ignoring ethnic minority groups such as in Myanmar (Rohingya Muslim minority), or affect the judicial system by retaining people in detention without a fair trial, such as in Russia, China and even the USA (Guantanamo Bay) (see page 232).

The UN Convention against Corruption was established in 2003, and the anti-corruption summit meeting held in London in 2016 recognised that corruption was at the core of many world problems. The meeting concluded that corruption should be exposed, the corrupters pursued and punished, victims of corruption supported and corruption eradicated (linked to SDG16):

Yellow = Players, Orange = Attitudes and actions, Purple = Futures and uncertainties

Table 4.7: Emerging countries (BRICS, MINT and others, indicated by economic media in 2015): economic, democratic and corruption data

Emerging countries	Av. annual GDP growth rate 2000–14 (%)	Type of government	Democracy Index 2015 (2006) (higher is more democratic)	Corruption Perceptions Index 2015 (higher is less corrupt)
Brazil	3.7	Federal presidential republic	6.96 (7.38) (Flawed democracy)	38
Russia	4.3	Semi-presidential federation	3.31 (5.02) (Authoritarian regime)	29
India	7.6	Federal parliamentary republic	7.74 (7.68) (Flawed democracy)	38
China	10.3	Communist state	3.14 (2.97) (Authoritarian regime)	37
South Africa	3.3	Parliamentary republic	7.56 (7.91) (Flawed democracy)	44
Mexico	2.3	Federal presidential republic	6.55 (6.67) (Flawed democracy)	35
Indonesia	5.5	Presidential republic	7.03 (6.41) (Flawed democracy)	36
Nigeria	8.2	Federal presidential republic	4.62 (3.52) (Hybrid regime)	26
Turkey	4.5	Parliamentary republic	5.12 (5.70) (Hybrid regime)	42
Chile	4.1	Presidential republic	7.84 (7.89) (Flawed democracy)	70
Colombia	4.6	Presidential republic	6.62 (6.40) (Flawed democracy)	37
Czech Republic	2.6	Parliamentary republic	7.94 (8.17) (Flawed democracy)	56
Hungary	1.4	Parliamentary republic	6.84 (7.53) (Flawed democracy)	51
Kenya	4.8	Presidential republic	5.33 (5.08) (Hybrid regime)	25
Malaysia	4.9	Federal constitutional monarchy	6.43 (5.98) (Flawed democracy)	50
Peru	6.0	Presidential republic	6.58 (6.11) (Flawed democracy)	36
Poland	4.0	Parliamentary republic	7.09 (7.30) (Flawed democracy)	62
Romania	3.6	Semi-presidential republic	6.68 (7.06) (Flawed democracy)	46
World av.	**2.6**		**5.55**	**43**

ACTIVITY

CARTOGRAPHIC SKILLS

Study Figures 4.17 and 4.18. Describe and explain how the patterns of corruption vary according to the level of democracy.

Extension

Choose one authoritarian, one hybrid and one flawed democracy country from Table 4.7. Investigate the degree of democratic freedom in your selected countries and compare the variation in human rights and freedom of speech between them. Useful sources include the US Department of State Country Reports on Human Rights, Amnesty International country profiles and Human Rights Watch countries.

A level exam-style question

Explain why the protection of human rights varies between countries. (8 marks)

Guidance

There are several interlinked reasons to be identified; think carefully before writing, so that your structure allows you to make key points, such as priorities, freedom, corruption, stage of development and recognition of global conventions.

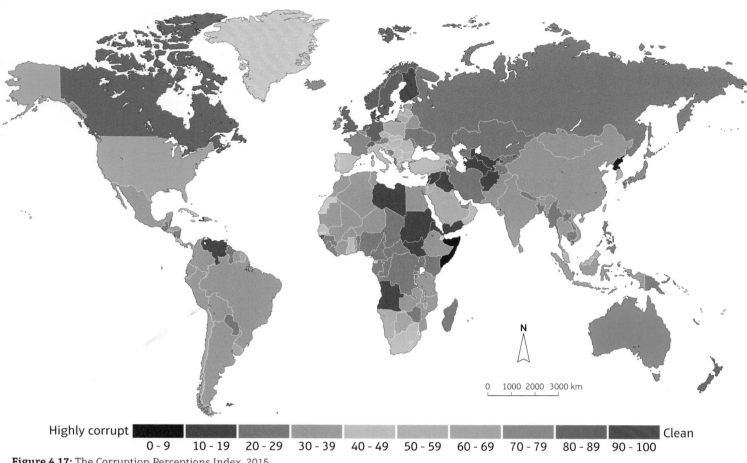

Highly corrupt | 0 - 9 | 10 - 19 | 20 - 29 | 30 - 39 | 40 - 49 | 50 - 59 | 60 - 69 | 70 - 79 | 80 - 89 | 90 - 100 | Clean

Figure 4.17: The Corruption Perceptions Index, 2015

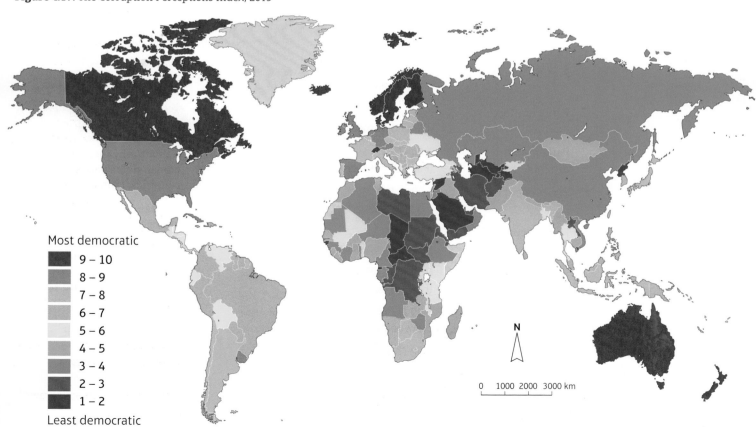

Most democratic

- 9 – 10
- 8 – 9
- 7 – 8
- 6 – 7
- 5 – 6
- 4 – 5
- 3 – 4
- 2 – 3
- 1 – 2

Least democratic

Figure 4.18: The Democracy Index, 2015

Yellow = Players, Orange = Attitudes and actions, Purple = Futures and uncertainties

[Corruption] erodes public trust in government, undermines the rule of law, and may give rise to political and economic grievances that may, in conjunction with other factors, fuel violent extremism. Tackling corruption is vital for sustaining economic stability and growth, maintaining security of societies, protecting human rights, reducing poverty, protecting the environment for future generations and addressing serious and organised crime.

Communiqué from the Anti-Corruption Summit, London, May 2016

Social development and human rights
Gender and ethnicity rights

As a result of their colonial history, many modern countries have borders that were drawn without reference to traditional boundaries between ethnic or cultural groups. Some boundaries are simply straight lines, drawn by the colonists on a map without any consideration for physical or social geography. This has created ethnic tension and group conflicts within newly independent states. For example, Iraq was largely created in 1920 after secret negotiations between European countries, but after its independence in 1932 the country experienced conflict between Sunni, Shiite and Kurdish ethnic groups, which led to civil war, authoritarian rule, modern wars and debate about dividing the country into three autonomous regions. After its independence in 1947, India immediately experienced widespread conflict between Hindu and Muslim groups, especially in the Kashmir region, which led to the creation of Pakistan and Bangladesh and continuing tensions between these three countries. In West Africa there have been at least seven post-colonial conflicts, with civil wars in Nigeria (1967–70), Liberia (1989 and 1999), Sierra Leone (1991–2002), Guinea-Bissau (1998–99) and Côte d'Ivoire (2002–07 and 2010–11). These conflicts were costly in terms of damage to the environment, economy and human life, and in many countries minority groups continue to have fewer rights than dominant groups, such as the Roma population in Eastern Europe and the Dalits in India.

The 2014 Gender Inequality Index, produced by UNDP, shows that while European countries have the greatest gender equality, with Slovenia (0.016) and Switzerland (0.028) at the top, most of the countries with the highest levels of inequality are in developing countries, especially Africa (Figure 4.19). Women in countries with high inequality have fewer rights than men; these are often entrenched in the culture as traditional or religious beliefs and then reflected in the country's laws.

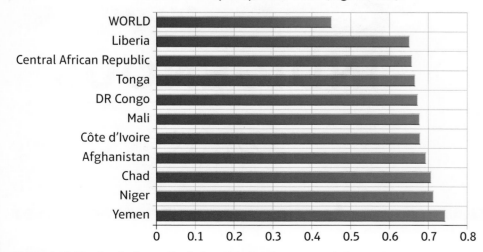

Figure 4.19: The Gender Inequality Index, 2014 (worst ten countries)

Health and education differences

Since 1995 the UN has recognised the rights of indigenous peoples, with awareness campaigns, a Permanent Forum on Indigenous Issues (2000), a Declaration on the Rights of Indigenous Peoples (2007), and a World Conference on Indigenous Peoples (2014). However, as with the Aborigines in Australia (see page 204), they still suffer inequalities when compared to the non-indigenous population, and in many parts of the world are only moderately satisfied with progress. The UN Special Rapporteur on the Rights of Indigenous Peoples stated that 'indigenous women experience a huge complex range of human rights abuses arising from discrimination and marginalisation, with lack of access to land, livelihood, decision-making bodies and physical abuse and atrocities.'

CASE STUDY: North American Indians

American Indians and Alaskan natives comprise about 2 per cent of the US population, with 78 per cent of these living outside tribal territories. Indian territories are theoretically sovereign, but subject to treaties and federal laws through the Bureau of Indian Affairs, and only tribes officially recognised are entitled to assistance (Tribal Recognition Act of 2015), although state laws may provide for this also. The health and education of recognised tribes is provided through the Indian Health Service (IHS), which provides care for about 2 million. Since 1972 the IHS has tried to make provisions off reservation areas, but is finding it difficult due to the spread and isolation of the indigenous population. The IHS has been found to be underfunded, with problems with sterilisation equipment, errors in medical records and unqualified medical staff. Less than half of the indigenous population has health insurance, and thus their access to the US health care system is restricted, but their needs are greater than the US white population, with a high prevalence of infant mortality, diabetes, injuries, suicide and TB.

In terms of educational attainment, in 2012 performance at all levels was about 10 per cent below the US white population and, while 5 per cent of indigenous children attend schools with a curriculum emphasising 'native ways of knowing', these schools, run by the Bureau of Indian Education (BIE), have poor-quality buildings and poor governance.

Barack and Michelle Obama called for better support for American Indian youth and launched the Generation Indigenous initiative in 2015 (Figure 4.20). This initiative focuses on programmes in education, health and nutrition, juvenile justice, housing and youth engagement. The 2016 US budget proposed US$20.8 billion for indigenous programmes, a US$1.5 billion (8 per cent) increase from 2015, including US$5.1 billion for the Indian Health Service (IHS), a US$461 million increase over the 2015 level. This is to improve federal, tribal and urban programmes in over 650 facilities in 35 states. In 2016 funding for BIE schools was US$904 million, an increase of US$94 million (12 per cent) from 2015, directed at increasing opportunities and improving outcomes in the classroom. The money included a US$59 million increase for buildings and infrastructure.

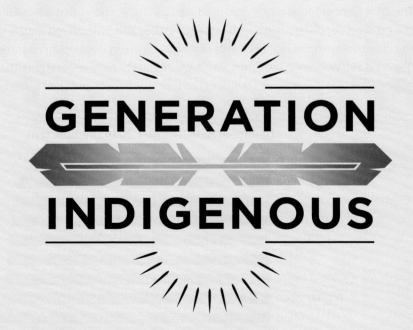

Figure 4.20: The Generation Indigenous logo, 2015

Yellow = Players, Orange = Attitudes and actions, Purple = Futures and uncertainties

Equality movements within countries

Women in all countries have struggled to attain equal rights with men. In developed countries these rights are now entrenched in laws, although remuneration levels often remain significantly below those of men. In developing countries much progress is still required, especially where traditional or distorted religious beliefs place severe restrictions on girls and women, limiting their access to education, freedom of choice and freedom of movement. Campaigns continue in many countries to protect and improve the lives of females.

CASE STUDY: Women's rights in Afghanistan

Afghanistan has experienced a long period of political transition after decades of internal conflict and external intervention. This continues to threaten ethnic and women's rights, and civilians are targets of indiscriminate terrorist attacks. Only 2.4 per cent of full-time workers are women, and only 14 per cent enrol for secondary education. Legal protection for women and girls in the country remains fragile. Changing the culture is difficult, and female activists face threats and attacks. A 2009 law 'Elimination of Violence Against Women' (EVAW) has not been enforced, even though government leaders have indicated their commitment to support women's rights. In 2014 a code passed by the Afghan parliament stated that relatives of those accused of domestic violence could not be made to testify in court, which makes it very difficult for prosecutors to gather evidence. The government also rejected calls by the international community for the abolition of the ability to charge women with 'moral crimes'.

Humanitarian Assistance for Women and Children of Afghanistan is one partnership working in the country with local communities and women's groups such as the Afghan Women's Network (AWN) (established in 1995) and the Afghan Women's Resource Centre. These bodies aim to raise awareness of the EVAW law, and they provide legal aid, encourage women to speak out and use the courts, and protect females within their communities. The AWN liaises with law enforcement agencies to monitor the use of the EVAW law and collect evidence, uses the media to publicise the rights issues and abuse of women, and trains public service personnel in how to support victims of abuse.

UNDP's Gender Inequality Index shows that Afghanistan has made a small improvement, from 0.743 in 2005 to 0.693 in 2014 (developed countries score around 0.1). In 2014 the HDI for females was 0.328, lower than the averages for all sub-Saharan African countries, while male HDI was 0.546. Only 5.9 per cent of women over 25 had attended secondary school, compared to 29.8 per cent of men, although there was a slowly growing proportion of female MPs (27.6 per cent).

The International Work Group for Indigenous Affairs (IWGIA) 2016 report notes at least 11 countries with human rights violations and conflicts within indigenous territories, with the loss of many lives. Many world regions have organisations representing indigenous peoples, such as in the Americas and Pacific (International Indian Treaty Council) and the Arctic (Arctic Council).

Extension

- Investigate the progress of Bolivia towards gender equality, from its 1939 General Labour Act to recent laws such as the Domestic Violence law of 2013. Consider evidence of change being put into practice.

- Investigate the progress of Colombia towards improving the rights of its 1.5 million indigenous peoples in the face of 'land grabbing' and deforestation.

A level exam-style question

Explain how corruption can threaten human rights in different types of country. (8 marks)

Guidance

Rather than defining what corruption is, the question is asking you to make links between types of corruption and how each of these may affect the human rights of different groups of people. While full case study details are not required, do use examples of the situation in various countries to illustrate your points.

How are human rights used as arguments for political and military intervention?

Learning objectives

8A.7 To understand the different forms of geopolitical intervention in defence of human rights.

8A.8 To understand how some development improves human rights and welfare but that other development has negative environmental and cultural impacts.

8A.9 To understand why military aid and direct and indirect military intervention are frequently justified in terms of human rights.

Geopolitical interventions
Aid, embargoes and military action

Indicators of human rights (see page 237) show that there is not always a positive correlation between economic development and improvements in human rights. However, there is a wide range of possible development and human rights **interventions** that can be used by superpowers such as the USA, developed countries such as Australia, or regional organisations such as NATO or the African Union to bring about improvements.

Development aid

During the Cold War era, much aid was given for political reasons, for example to friendly regimes, former colonies, or to persuade countries to support the 'West' or the 'East'. Following the end of the Cold War, the purpose of aid was realigned to help poor people anywhere in the world. However, in the 21st century aid has again become politicised and is given to support countries combating radical extremism, such as Afghanistan and Syria, or to help control migration into Europe, such as Turkey and Nigeria. Aid is often in the form of money allocated for specific purposes; in 2014, 17.52 per cent of EU aid was for 'government and civil society', which includes human rights matters, and 17.63 per cent for 'humanitarian aid' in response to disasters (Figure 4.21).

An analysis of development aid by *The Economist* in 2016 suggested that the quality of governance and level of poverty in a country did not significantly affect the amount of aid received. The countries that received the most aid per capita were those with smaller populations, for example Samoa and Bhutan, perhaps because it is easier to direct aid to where it is needed and publicise the positive results.

Extension

Investigate specific projects supported by development aid from the EU or one European Union country, for example the European Commission's EU aid explorer or the UK's Department for International Development case studies. Note how these have helped development or improved human rights.

Yellow = Players, Orange = Attitudes and actions, Purple = Futures and uncertainties

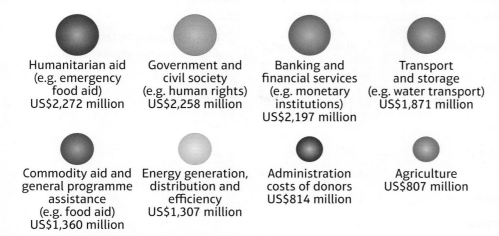

Figure 4.21: Distribution of EU development aid by purpose, 2014

Trade embargoes

These are **foreign policies** or laws that ban exports to and imports from a country in protest against actions by that country. Trade embargoes can be effective in bringing about change, because without exports national income is reduced. Strategic goods and technology, especially military arms and equipment, are banned from export when there are concerns about human rights violations, internal instability and repression, or about international security threats – including the development of weapons of mass destruction. However, exports of medicines and foods can be allowed to continue for humanitarian reasons.

The US **trade embargo** against Cuba (condemned by the UN), imposed in the 1960 after Fidel Castro's takeover and support of communism, costs the country an estimated US$685 million a year and restricts its development. However, the USA also loses as a result of the embargo – an estimated US$1.2 billion a year. In 2016 President Obama called for the embargo to be lifted.

Embargoes are usually imposed by the UN or EU, and effectively become international law, such as with Iran (concerns over enriched uranium – lifted in January 2016), DR Congo (internal repression and instability), and North Korea (a totalitarian state with nuclear weapons development).

Military aid

This may be in the form of technology and equipment, or peacekeeping forces, such as NATO or UN. The purpose of military aid is to help a country defend itself, to counter insurgents or to support pro-democracy factions within a country against extreme authoritarian regimes (although these are not always easy to identify). For example, in 2015 the USA provided military aid to 63 countries through its foreign military funding programme. By far the biggest recipient was Israel, with over 55 per cent of the budget (Table 4.8). The USA has traditionally supported Israel in the troubled Middle East, despite its often aggressive policies. In 2016 a new agreement was reached to provide up to US$40 billion to update Israel's air force and missile defences, although a condition may be that they provide 'dignity and self-determination' for Palestinians.

ACTIVITY

1. Study Table 4.8.

   a. Suggest why the USA provided military aid to these countries in 2015.

   b. To what extent can the provision of military aid to these countries be justified? (You may wish to research development and human rights information.)

Table 4.8: US foreign military spending, 2015 (top 15)

Recipient country	% of total budget 2015	Freedom Index 2016 (higher value = greater freedom)
Israel	55.59	80
Egypt	23.31	27
Jordan	5.38	36
Pakistan	5.02	41
Iraq	4.48	27
Lebanon	1.43	43
Philippines	0.72	65
Colombia	0.45	63
Tunisia	0.45	79
Yemen	0.45	17
Indonesia	0.25	65
Georgia	0.18	64
Vietnam	0.18	20
Poland	0.16	93
Bahrain	0.13	14

In May 2016 there were 16 UN peacekeeping operations, with 89,098 troops deployed, 12,611 police and 1,801 military observers as well as civilian personnel. These forces were drawn from 123 countries. Perhaps unsurprisingly, nine peacekeeping operations were in Africa and four in the Middle East, with significant forces assigned to Sudan (Darfur and Abyei) (20,513 and 4,788 peacekeepers respectively), South Sudan (16,014) and the DR Congo (DRC) (22,721). The operation in the DRC protects civilians, humanitarian personnel and human rights defenders under threat, and supports the DRC government in its efforts to stabilise the country and bring peace.

Military action

Direct military action has been used to defend human rights, sometimes controversially. For example, no-fly zones were enforced over Iraq (1991) and Libya (2011) by military aircraft from other countries (NATO) – the latter authorised by the UN. Indirect action may take the form of covert support for factions, such as Russia's involvement in eastern Ukraine, or providing weapons to one side in a conflict. The UN often relies on the willingness of militarily powerful members such as the USA and the UK to provide the force necessary.

Literacy tip

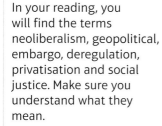

In your reading, you will find the terms neoliberalism, geopolitical, embargo, deregulation, privatisation and social justice. Make sure you understand what they mean.

Interventions and consensus

In 2016 Amnesty International promoted campaigns to bring attention to:

- Brazil, for unlawful killings in Rio de Janeiro
- Saudi Arabia, for ill-treatment and lack of protection for human rights defenders and activists
- Venezuela, for attacks on human rights workers
- Iran, for imprisonment of human rights campaigners
- Germany, for failing the victims of racial violence.

Yellow = Players, Orange = Attitudes and actions, Purple = Futures and uncertainties

To put collective international pressure on a country, aid may be withdrawn if corruption is involved. For example, Malawi received US$1.17 billion in 2012, about 28 per cent of the country's GNI, but in 2013 around US$30 million was stolen through corruption, and investigators suffered physical violence. Consequently aid the following year was reduced by 20 per cent and donors tried to avoid using government systems, taking the aid directly to those needing it.

In the UN, Russia frequently uses its veto powers to block support for action against its perceived allies, countries or individuals, such as supporting Assad in Syria because this regime was Russia's only influence in the Middle East, and withdrawing support for NATO action in Libya when it believed that the military action had gone beyond the UN mandate and it appeared that a 'Western-friendly' government would be installed. There is great concern among humanitarian groups that powerful developed countries intervene only when it matters to them politically, but from 2014, EU development aid directly to emerging economies was phased out to be replaced by a concentration of aid on the poorest places in the world (75 per cent of EU aid 2014–20 to be of the latter type).

In recent decades, the intervention approach has stressed 'trade not aid', based on neoliberal ideas that social justice and human development would follow economic development when corrupt governments, which slow social development, are removed. The IMF adopted this approach through its structural adjustment policies, which attach conditions to loans such as **deregulation** and **privatisation**, as seen in Chile and Mexico. There has been debate over the development aid needed to meet the Sustainable Development Goals (SDGs) by 2030: It has been estimated that US$17 trillion will be required and the UN target is 0.7 per cent of GNI to be allocated by developed countries to overseas development assistance (ODA), with at least US$100 billion a year donated by 2020. However, some say that this will merely prolong structural inequalities and the dependence of developing countries on aid from developed countries, although the G77 group has called for more financial assistance. The World Bank and other international banks believe that there needs to be 'intelligent development finance', providing more development aid but targeted at countries with poverty, vulnerability and limited fiscal capacity, and used to encourage public and private sources of finance. The UN Environment Programme (UNEP) said that there should be a proper valuation of natural capital and ecosystems services when considering aid projects so that damage is limited.

Justifications for intervention

In 2009 the UN General-Secretary said that there may be times when countries should respond collectively 'in a timely and decisive manner' when a country has failed to protect the welfare or allow human suffering of its citizens – including genocide, war crimes, ethnic cleansing or crimes against humanity. In these situations, some argue that a country sacrifices its sovereignty.

Synoptic link

Sovereignty issues are often raised at times when intervention is suggested. With globalisation and the plethora of international organisations (IGOs and NGOs), considerable pressure can be applied to countries, which may be powerless to resist. While some international pressures may be direct, such as military intervention or global trade, some are subtler, such as the spread of 'Western' culture or diplomacy. Climate change is another threat to sovereignty, as some island states may physically disappear. (See Book 1, pages 147 and 155.)

ACTIVITY

Compare the characteristics of the different types of geopolitical intervention.

A level exam-style question

Explain why geopolitical interventions are often contested. (8 marks)

Guidance

This question is not asking for definitions of geopolitical interventions but the reasons why some countries or organisations support these types of intervention while others do not. Brief mentions of examples with a few facts would be useful.

CASE STUDY: Interventions in Yemen

Yemen is a small country on the Arabian Peninsula next to the Red Sea and the Gulf of Aden. In 2015 a civil war broke out when the Houthi (Shia Islam), who make up about a third of the population, mounted an armed rebellion against the Sunni Islam President Hadi. The situation was further complicated by the presence of Al-Qaeda and IS within the country. After the first year of conflict the UN estimated that 6,500 people had been killed, half of them civilians, and most of the population needed humanitarian assistance. Over 2.5 million people were internally displaced from their homes.

The UN took a passive role, making no resolutions to support foreign interference beyond diplomatic dialogue to find peace. However, President Hadi requested international help against insurgents and to prevent civilians being killed by Houthi rebels. Several countries – the USA, the UK, Turkey, Egypt, Kuwait, UAE, Qatar and Bahrain – supported Saudi Arabia's belief that military intervention was required on the grounds of 'collective self-defence' (the reasoning also used with Syria and Iraq). The situation was complex because:

- there were doubts over the legitimacy of President Hadi's position

- some countries wished to suppress the rebellion because of Yemen's potential to train and support terrorists

- the rebels were directly or indirectly supported by Iran and Eritrea.

The situation raised questions about how international law related to the 'responsibility to protect' principle, perhaps suggesting the need for a political solution, rather than a military solution that would increase risks for civilians.

The Saudi Arabia (Sunni Islam)-led coalition carried out air strikes on rebel positions and the US supported this coalition with weapons, intelligence-sharing, coordination and planning, including drone strikes against known terrorists. 'War crimes' accusations have been made by both sides, with both Houthi rebels and Saudi Arabian air strikes killing civilians (Figure 4.22). In 2016 Médecins Sans Frontières (MSF) had eight medical projects in Yemen and the Red Cross has delivered aid supplies. Iran mobilised naval vessels to 'safeguard shipping routes' and support the rebels, while Pakistan helped to enforce the UN arms embargo against the rebels. These foreign interventions challenge the present and future sovereignty of Yemen.

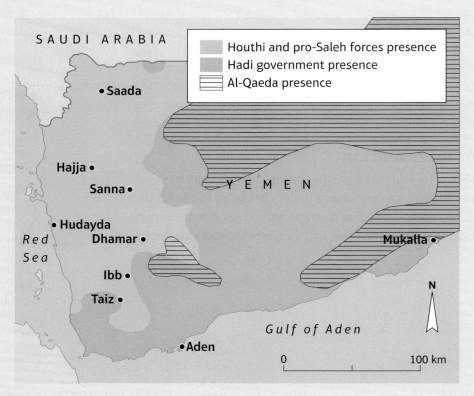

Figure 4.22: Civil war in Yemen: the situation in January 2016

Yellow = Players, Orange = Attitudes and actions, Purple = Futures and uncertainties

The impact of development aid
Forms of development aid

Some suggest that foreign direct investment (FDI) and trade have become more important than aid, as developing countries prioritise economic development. The amount of aid given has been in decline and the plethora of IGOs and NGOs such as Oxfam and Christian Aid, and smaller charities, sometimes creates coordination issues so that aid falls below the required standard. In 2015 US$130 billion was donated, mainly from Germany, the UK, France, Japan and the USA, although 9 per cent of this amount was spent in donor countries to help refugees. The main recipients of aid were India (US$4.8 billion), and Vietnam (US$4.8 billion), although per capita this meant that Vietnam received more than India (US$1,658, compared to just US$17). A trend with aid is that it is given in smaller amounts, with the average project loan in 2014 only one-third the size it was in 2000.

Humanitarian aid is Official Development Assistance (ODA), given mostly by NGOs in times of emergency, such as a natural disaster like an earthquake, or human-made disaster such as a civil war. This aid usually concentrates on the basics of life, including food, clean water, shelter and medical care. Community ODA is given to help poorer people improve their lives through projects based on 'bottom-up' schemes decided in partnership with local communities, who are best placed to identify their own needs. The small, low-cost and manageable projects are usually work- or infrastructure-related, such as micro-HEP (Practical Action) or Playpumps (One Water). Welfare ODA is to provide access to education and health care and improve the status of women. Bilateral aid is where ODA is agreed between the governments of two countries (Table 4.9), sometimes for specific purposes and sometimes with conditions attached. Multilateral aid organised by IGOs such as UNICEF, UNDP, the World Bank or the IMF usually aims to reduce poverty and raise economic performance, but the MDGs and SDGs have broadened the focus to include social and environmental matters since 2000.

CASE STUDY: Christian Aid in Haiti

In Haiti, 77 per cent of the population live on less than US$2 a day. The country is in desperate need of development aid. It has a history of political turmoil based around disparate political groups, a power elite and strong external influence; the tax system is unfair and ineffective and the country is vulnerable to economic shocks and natural hazards (hurricanes and earthquakes). Recovery since the major earthquake of 2010 has been slow: despite new homes, better roads and schools, living standards declined after 2010 and in 2015 85,000 people were still living in emergency tents, facing a desperate situation every day.

Christian Aid (an NGO) has worked in Haiti since the 1980s, with the aims of empowering citizens and supporting communities. After the earthquake, rural areas needed help as internally displaced persons (IDPs) moved away from the capital Port-au-Prince. While many NGOs and IGOs concentrated on the urban area, Christian Aid (CA) worked with its partners at the rural community level, for example the farmers' group GRAMIR (Figure 4.23). CA reached 180,000 people through its £14 million appeal fund, initially providing 237,000 hot meals, 10,000 hygiene kits, helping 2,500 families with emergency shelter, 5,600 families with cash, and 7,000 with clean water. In a second phase it helped construct 550 earthquake-proof homes, provided 32,000 farmers with seeds, and trained 35,000 people in how to prepare and respond to disasters. Health spending in 2010 amounted to US$46 per person, with just 21 per cent of funds from the government; 40 per cent of the burden falls on households and foreign aid contributes 38 per cent. Haiti continues to face problems, but many aid providers are now reducing their operations in Haiti; even CA only committed 0.97 per cent of its funds to the country in 2015.

ACTIVITY

ICT SKILLS

1. Explore the geospatial dashboard on the AidData website. Study the development aid project information for China in the most recent year available. Explain the distribution and types of project.

2. Investigate the Oxfam website and select a contemporary example of their development aid (not Haiti). Make notes to go with the Christian Aid case study included on this page.

Synoptic link

Tectonic hazards create situations where countries at all stages of development may need humanitarian or emergency aid. However, the most vulnerable people are those in developing countries, because they are often excluded from decision-making processes before, during and after a hazardous event. Rates of recovery are represented by Park's model. Haiti is perhaps an example of a location where the situation has not returned to normal and remains in the deterioration category. The risk of disease, such as cholera in October 2010, remains a real possibility, given the poor health care. (See Book 1, page 38 and this book pages 184–185.)

Figure 4.23: Recovery in Haiti – house built by CA after the 2010 earthquake

The strengths and weaknesses of development aid

Historically, development aid has had successes, such as the USA's support for South Korea and Taiwan, which enabled those nations to become 'tiger economies' and then established economically developed countries. In terms of world health, aid helped eradicate smallpox in the 1970s and has nearly eliminated polio (see page 200).

Countries in strategic positions have tended to receive more aid – for example, Turkey received US$3.4 billion in 2014 – even if they are more authoritarian or have internal human rights issues. Countries that received aid to help them to become more democratic, such as Peru, may find their aid reduced even though there may still be many needy people in the country. China is becoming a larger global donor, often to countries where aid from other sources has been reduced or withdrawn, but is less concerned with human rights and democracy and may overlook abuses.

Development aid may encourage corruption because of the large sums of money involved; it may also support authoritarian regimes at times and postpone democracy. The administration systems of developing countries may be overwhelmed or take a substantial proportion of aid, so that not

Extension

- Investigate the links between development aid and Haiti's recovery rate after the 2010 earthquake.

- To what extent was Haiti's cholera outbreak later in 2010 a result of ineffective interventions after the earthquake?

Yellow = Players, Orange = Attitudes and actions, Purple = Futures and uncertainties

all the aid reaches the people who need it. Aid may distort the actual financial and economic status of a country by artificially adding to its GDP, and when aid is reduced or withdrawn it may cause economic uncertainty and collapse. Sometimes the ruling elite of a recipient country diverts the aid budget to military spending, which may prolong unrest and civil war. Sometimes it has not been possible to direct humanitarian aid to people in developing countries because the government has banned foreign aid workers from the country, as happened in Myanmar.

Some research suggests that there is a trend of aid going to poorly governed 'middle-income' countries rather than to those in poverty, but other research suggests that countries attempting to become more democratic receive more aid. This illustrates the complexity of this type of intervention, and why there is uncertainty and controversy.

Table 4.9: Countries receiving development aid (2014) from Australia, the UK and Russia

Australian aid	% of total Australian aid	UK aid	% of total UK aid	Russian aid	% of total Russian aid
Indonesia	18.3	Ethiopia	7.5	Kyrgyzstan	34.0
Papua New Guinea	16.9	India	6.6	Cuba	29.7
Afghanistan	6.0	Pakistan	6.2	North Korea	11.5
Solomon Islands	5.6	Nigeria	5.7	Tajikistan	3.3
Philippines	5.3	Sierra Leone	5.5	Nicaragua	2.9
Vietnam	5.1	Bangladesh	4.8	Guinea	2.8
East Timor	3.6	Afghanistan	4.7	Serbia	2.7
Myanmar	3.6	South Sudan	3.9	Mozambique	1.3
Bangladesh	3.5	DR Congo	3.9	Syria	1.2
Cambodia	3.2	Tanzania	3.5	Armenia	1.1

CASE STUDY: Development aid to eradicate malaria

The Global Fund (established in 2002) is a partnership between governments, private organisations and the general public with the aim of eradicating epidemics, especially AIDS, TB and malaria. It invests US$4 billion a year in projects run by local experts and communities. Between 2000 and 2014 there was a 48 per cent decline in deaths from malaria, with children and pregnant women being targeted. Malaria has been combated through the distribution of 548 million insecticidal nets in 100 countries, indoor spraying and medicines (artemisinin-based) treating more people – an increase of 19 per cent from 2013 to 2014.

Ethiopia has 3 million deaths a year from malaria, but in 2015 US$611 million was allocated for action, with 41.6 million nets distributed. In Haiti the number of mosquito nets increased from 380,000 in 2010 to 3.37 million in 2015 when the Global Fund allocated US$35.88 million to a grant 'Towards the Elimination of Malaria in Haiti' (2004>). However, mismanagement of funds made it necessary to transfer the grant to another provider in 2013.

ACTIVITY

CARTOGRAPHIC SKILLS

1. Using the data in Table 4.9, use an outline map of the world to draw proportional arrows to show the flow and level of aid given in 2014 to each recipient country. Colour these according to the donor country.

2. Study the patterns in Table 4.9 and your completed map regarding countries receiving aid, the donor countries, and the proportion of development aid given. Suggest reasons for the patterns observed.

Extension

Using a variety of internet sites for a range of news sources, government aid and marketing sites of charitable organisations in the UK, Australia and Russia, (a) assess the impact of development aid on one country from each list in Table 4.9, and (b) evaluate the source material (for example, the reliability of the data and judgements made).

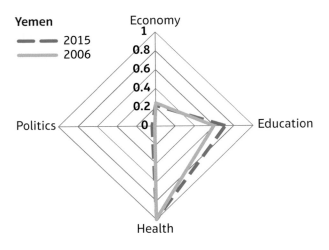

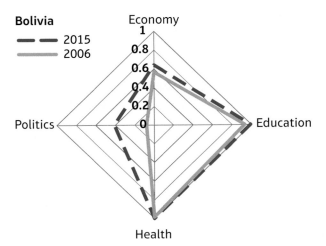

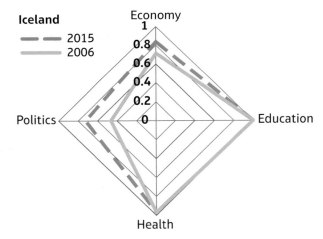

Figure 4.24: Radar graphs showing the Gender Gap Index for Iceland, Yemen and Bolivia

Yellow = Players, Orange = Attitudes and actions, Purple = Futures and uncertainties

ACTIVITY

GRAPHICAL SKILLS

Figure 4.24 shows radar graphs for the Gender Gap Index for three countries.

1. Using the data in Table 4.10, create two more radar graphs, using Excel or similar software.

Table 4.10: Gender Gap Index data for Sri Lanka and Mali

		Economy	Education	Health	Politics
Sri Lanka	2015	0.577	0.995	0.980	0.193
	2006	0.545	0.990	0.980	0.365
Mali	2015	0.605	0.755	0.949	0.086
	2006	0.665	0.674	0.968	0.091

2. Using the five radar graphs and case study information, suggest why the impact of development aid is contested.

CASE STUDY: Gender equality and the 'gender gap'

The Global Gender Gap Report 2015 by the World Economic Forum showed that between 2006 and 2015, gender parity in education and health was almost achieved, but not at all in economic and political terms. Between 2006 and 2015, 0.25 billion more women were earning, but the earnings gap grew from US$5,000 to US$10,000. Parliamentary representation by women was only 19 per cent, although half of the 145 countries surveyed had a female head of state. Overall, the global gender gap closed by 4 per cent, with the least progress being made in the MENA region. Progress was made by Iceland, Nicaragua, Bolivia and Nepal, with some low-ranked countries such as Yemen, Chad and Saudi Arabia improving, but Iran and Pakistan stalled and gaps widened in Sri Lanka, Jordan, Mali, the Slovak Republic and Croatia. Data on gender-related aid (ODA grants) from the OECD show that Yemen received an aid increase of 150 per cent between 2006 and 2014, to US$285.9 million, and Bolivia an increase of 160 per cent, to US$269.5 million. However, despite an increase of over 104 per cent, to US$523.5 million, Mali slightly declined on the Gender Gap Index; Sri Lanka had a decrease in gender aid of –32.5 per cent, which may partly account for its decline in the index.

Economic development, the environment and human rights

Economic development has been encouraged by colonial superpowers and the dominant superpower, the USA, following the theoretical economic principles espoused by **neoliberal views** such as Rostow (see page 148). However, these economic aims often overlooked the effect of development on the natural environment and the human rights of some groups of people, especially minority groups such as indigenous peoples living in peripheral areas.

Synoptic link

Economic models such as modernisation theory developed by Rostow (1960s) can be used to explain intervention by superpowers on the basis of stimulating economic 'take-off' and reducing the spread of communism. This is a neoliberal approach with investment by TNCs from economic core regions. Dependency theory, developed by Frank (1970s), highlighted interventions by TNCs that led to exploitation and an unequal relationship between developed and developing countries. World systems theory, developed by Wallerstein (1970s), reveals interventions as part of political and economic competition, which creates tensions. (See page 149.)

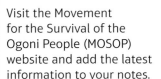
CASE STUDY: The impact of the oil industry in the Niger Delta

The Niger Delta is the third-largest wetland and fourth-largest mangrove forest in the world. Ogoniland was one of the first areas of the delta to be exploited for oil after commercial production started in 1958. The oil industry forms the core of the Nigerian economy, with 55 per cent of the joint venture with TNCs being owned by the Nigerian National Petroleum Corporation (other owners are Shell Petroleum Development Company of Nigeria Ltd (SPDC) with 30 per cent, ELF Petroleum Nigeria Ltd (TotalFina) with 10 per cent and Nigeria Agip Oil Company with 5 per cent). There is a long history of oil spills in the area, with 550 spills in 2014 alone, according to the oil companies. These spills are the result of an extensive old pipeline system, sabotage and thefts on different scales. Corruption and weak regulations have allowed the situation to arise and continue; it is estimated that as much as 20 per cent of production is being stolen, amounting to US$20 billion in losses a year.

In the 1990s the Ogoni people started peaceful protests about the damage being caused by oil pollution, targeting the TNCs involved and expelling SPDC from their lands. Ken Saro-Wiwa led the protests, but he and eight tribal leaders were executed by the Nigerian government in 1995 on charges of murder. Friends of the Earth (FoE) believe that oil has been a disaster for the country and that oil exploitation should stop immediately. The lives of the Ogoni have been badly affected by the damage to their farmland, natural resources and fishing areas (Figure 4.25) but there has been no definitive action to clean up the oil spills. The Ogoni continue to protest, and their May 2016 'Break Free from Fossil Fuels' event at Bori in Ogoniland was supported by FoE and Health of Mother Earth NGOs.

In 2010 SPDC made an out-of-court compensation settlement with some of the Ogoni, with 15,500 people receiving US$3,000 each, and in 2014 contributed US$14.8 million in scholarships, education programmes and health care. A UNEP report in 2011 concluded that environmental restoration would take 25 to 30 years, and recommended eight emergency measures to improve the health and livelihoods of the Ogani people. In 2015 the Nigerian government finally agreed to act on the UNEP report.

Figure 4.25: Images of oil pollution in Ogoniland in the Niger Delta, Nigeria

Military aid and intervention
Motives for military intervention

Countries' global strategic interests have long influenced their intervention in other countries. Examples are Russia's support for Cuba and Syria, China's support for North Korea and the USA's support for Saudi Arabia. In this way the superpowers are able to maintain their influence in world regions from which they might otherwise be excluded. However, they give humanitarian reasons as public justification of their military intervention, using international human rights conventions to explain their involvement in situations where people are at risk of persecution.

Yellow = Players, Orange = Attitudes and actions, Purple = Futures and uncertainties

CASE STUDY: Intervention in Iraq in the 21st century

By 2003 Saddam Hussein had been in power in Iraq for 23 years. During that time he had been responsible for terrorising his own people and conducting wars against Iran (1980–88) and Kuwait (1990). According to some estimates, he caused as many as 635,000 deaths; evidence of mass graves in Iraq suggests that he may have killed 250,000 Iraqis. His was a regime where disappearance and murder were commonplace, minority groups in the country were persecuted, a secret police operated, people were imprisoned for political crimes, and enemies of the state and their families were tortured and executed.

It was known from at least 1990 that the regime violated human rights conventions, but it took a long time to get a UN resolution for action. It was feared that Iraq was capable of using weapons of mass destruction against Iraqis and within the Middle East region, so threatening stability. Given the important oil resources of the region, there was international concern that disruption to this energy pathway could cause economic problems in many countries. Iraq was also seen as a supporter of the terrorist group Al-Qaeda, increasingly a global threat to democratic countries. A further reason for action was the credibility of the UN, given its responsibility for protecting people once a state fails to do so.

Critics of the invasion by US forces and its allies pointed out that there was no recent evidence of genocide in Iraq, and that weapons of mass destruction were never found. Also, the UN Security Council's resolutions did not justify or legally authorise direct military intervention; innocent civilians would be killed in a military conflict, and it was unlikely that Iraq would ever become a liberal democracy, even with a new government.

Between 2003 and 2013 184,512 people were killed in Iraq, 97 per cent of them Iraqis and 5,272 foreigners (including military personnel). Of these deaths, it is estimated that 72 per cent were civilians, with 11 per cent due to friendly fire from US coalition forces. An analysis of development data from before (2002) and after the intervention (2014/15) shows that health spending dramatically improved, from US$8.4 per capita to US$292, maternal mortality and infant mortality rates improved (by 9 and 8 per thousand respectively), but life expectancy and control of corruption remained the same, and 'voice and accountability' improved but remained negative. However, the country faced further internal conflict, especially when IS started its insurgency in 2013.

Military aid

Military aid may sometimes be given to countries whose human rights record over time is poor, occasionally at a time of regime change so that there is support for a change in human rights, such as in Bolivia, Egypt and Zimbabwe. Table 4.8, which lists US foreign military spending in 2015, shows that three of its top contributions were to some of the worst human rights offenders in 2014 (Pakistan, Iraq and Yemen). However, a considerable length of time is usually required for there to be consolidated political and cultural change, and some countries may return to 'old ways'. Military aid may also be given for strategic or political reasons, to gain influence in a country or key region of the world, such as Russia supporting the Assad regime in Syria to keep a foothold in the Middle East, and the USA supporting Jordan for similar reasons.

Extension

What is the latest situation in Iraq in terms of development data, unrest and security, and international interventions?

ACTIVITY

Evaluate the motives behind the intervention in Iraq between 2003 and 2011.

A level exam-style question

Explain why military interventions are often contested. (8 marks)

Guidance
This question does not require lengthy definitions of the types of military intervention; instead, make sure that your answer concentrates on the reasons why some countries and organisations support military intervention and others do not. Use some examples and show that you recognise that each situation has significant differences.

CASE STUDY: Human rights in Jordan

According to Human Rights Watch, Jordan has some significant human rights issues. For example, it accepted many refugees from Syria, but refused entry to all Palestinian refugees escaping from Syria and even detained and deported Palestinians who had entered Jordan illegally. The government narrowed the scope of its judicial courts and broadened anti-terrorism laws in such a way as to threaten freedom of speech: it has criminalised criticism of the king, the government or Islam. These restrictions extended to peaceful protests and the media, with nine news websites shut down in 2014, and journalists and students arrested and faced with possible charges under the Terrorism Act. Torture exists and is not recognised by a corrupt justice system, and people are detained without charge. There are still restrictions on normal lives, such as the use of child labour, abuse of migrant domestic workers, bans on marriages between Muslim women and non-Muslim men, and child marriages.

However, in 2014 the USA renewed a five-year aid package, which will provide US$360 million in economic assistance, US$300 million in foreign military financing and US$340 million towards costs caused by regional instability (for example looking after refugees and border security). The EU also pledged US$298 million to help Jordan with the refugee crisis, and Saudi Arabia provided US$232 million for development projects. In 2015 Jordan received over 5 per cent of US foreign military finances (third-highest country), despite having a Freedom Index score of only 36/100 in 2016 (see Table 4.8).

ACTIVITY

Explain why the USA, the EU and Saudi Arabia have provided development aid and military aid to Jordan.

Direct military intervention

Since the terrorist attacks on civilians in the USA in 2001, some countries have become involved in direct military intervention in places where terrorist groups are known to operate, such as Afghanistan and Pakistan. The US Department of State currently lists 60 terrorist organisations, with ISIL-Khorasan and ISIL-Libya added in 2016. Islamic State (IS) in Iraq and Syria has been the most organised, with 31,000 fighters, and the wealthiest, with oil revenue. In 2014 it managed to gain territory in Iraq and Syria, while committing atrocities and mass executions including those of minority groups such as the Albu Nimr tribe in Anbar Province in Iraq in October 2014. The US used air strikes to halt and weaken IS.

The Institute for Economics and Peace Global Terrorism Index of 2015 showed that in 2014 terrorism was concentrated mostly in five countries: Iraq, Afghanistan, Nigeria, Pakistan and Syria, which together accounted for 78 per cent of the lives lost globally. While terrorism has spread widely in the world, nearly all attacks (92 per cent) were in countries that supported political violence, with a long history of armed conflict, corruption and a lack of respect for human rights. However, there have been accusations of torture of prisoners by US and UK forces as well, especially of suspected terrorists detained in the infamous camp at Guantanamo Bay.

CASE STUDY: The Guantanamo Bay detention camp

The Guantanamo Bay prison was constructed in the US naval base in Cuba in 2002 to hold extremists and terrorists viewed as posing a threat to the US. Hundreds of prisoners, mostly captured in Afghanistan and Iraq (Taliban and Al-Qaeda), were held in detention without charge or access to the US legal system. President Bush justified this because the people were considered war criminals, and the facility was not on US soil and therefore beyond US jurisdiction. However, the prison was increasingly criticised and condemned by organisations such as Amnesty International, Human Rights Watch, the EU and the Organisation of American States (OAS) because of the abusive treatment of prisoners. In 2006 the US Supreme Court said that there were violations of the Geneva Convention and the military code of justice, and that detainees did have the right to challenge their detention in US courts.

By 2009 it was clear that torture had been used on terrorist suspects, and President Obama instructed that the facility be closed and a way found to bring the prisoners to trial in the US. However, the US Congress blocked these moves, arguing that the detainees would be a greater threat once on the US mainland. In 2013 the prison had 166 detainees reduced to 80 by 2016, with ten transferred to Oman in January and nine to Saudi Arabia in April, where they will go through a rehabilitation programme to reintegrate former jihadists. In May 2016 it was reported that at least a further 22 prisoners will be transferred to different countries within two months. However, at least 40 detainees – regarded by the US as too dangerous ever to release – remain, and so the facility may be kept open indefinitely.

Yellow = Players, Orange = Attitudes and actions, Purple = Futures and uncertainties

CASE STUDY: The Yazidi minority community

In northern Iraq in 2014, IS forced Yazidi people to flee their homes and take refuge in the Sinjar Mountains after hundreds of people had been abducted, forced to convert to Islam or killed. In August 2014 the US stated that they considered the systematic attacks on the Yazidi people to be genocide, which broke UN human rights conventions. The US and others regarded direct military action against IS as justified and began air strikes. Initial action was to clear a safe corridor for 50,000 Yazidis to escape from IS and make their way to Syria where they could be protected by Kurdish forces (Figure 4.26). This was successful and military personnel from the US and UK visited the Yazidis left in the mountains and arranged humanitarian aid drops of food and water – this was considered sufficient, as the military threat had greatly reduced. The allied air strikes on IS continued and were used to support an advance by the Iraqi army to retake territory from IS.

Figure 4.26: Yazidi flee genocide in 2015

Extension

Investigate the extent to which 'crowdsourcing' is an objective way of informing people about human rights and military interventions.

4 What are the outcomes of geopolitical interventions in terms of human development and human rights?

Learning objectives

8A.10 To understand the various ways of measuring the success of geopolitical interventions.

8A.11 To understand why development aid has a mixed record of success.

8A.12 To understand why military interventions, both direct and indirect, have a mixed record of success.

Measuring the success of interventions
Measurements of success

Ways of measuring the development of a country are diverse (see page 188) (Table 4.11) and the success of **geopolitical interventions** is difficult to judge, especially if the accuracy of data is in doubt (Table 4.12). It may be years or decades before it is possible to make judgements on interventions in countries currently in conflict situations, especially if government structures, economic systems, health and education systems and infrastructure have been damaged and need investment.

Table 4.11: Changes showing measures of intervention success, 2000–14

Country	Change in % annual GDP per capita growth	Change in life expectancy at birth (years)	Change in primary school completion rate (% both genders)	Change in child mortality rate (/000)	Change in net ODA received per capita (US$)
Cuba	− 3.1	+ 2.8	+ 1.2	− 2.8	+ 19
Egypt	− 3.5	+ 2.5	+ 14.3	− 21.7	+ 19.3
Ethiopia	+ 4.4	+ 12.1	+ 31.4	− 85.9	+ 26.7
Haiti	+ 2.2	+ 5.0	+ 59.9	− 33.8	+ 78.2
India	+ 4.0	+ 5.1	+ 24.4	− 39.1	+ 0.6
Indonesia	+ 0.2	+ 2.7	+ 6.8	− 24.1	− 9.3
Israel	− 5.5	+ 3.1	− 2.9	− 2.7	− 14.0
Kyrgyzstan	− 2.7	+ 1.8	+ 11.5	− 26.1	+ 63.1
Papua New Guinea	+ 11.3	+ 3.8	+ 23.5	− 19.4	+ 26.2
Yemen	− 1.8	+ 3.3	+ 11.3	− 53.5	+ 27.0

ACTIVITY

Study Tables 4.11 and 4.12 carefully. All the countries listed have received overseas development assistance (ODA), and some have had military interventions. Analyse the data provided and suggest whether the interventions have been successful in helping these countries.

Yellow = Players, Orange = Attitudes and actions, Purple = Futures and uncertainties

Table 4.12: Changes in gender inequality and freedom of the press, 2000–14

Country	Change in Gender Inequality Index (UNDP Human Development Report data: lower score = more equal, 0–1 scale)	Change in Freedom of the Press Index (Freedom House data: lower score = more freedom, 0–100 scale)
Cuba	− 0.063	− 3
Egypt	− 0.087	+ 8
Ethiopia	+ 0.011	+ 19
Haiti	+ 0.029	− 7
India	− 0.058	− 1
Indonesia	− 0.057	+ 2
Israel	− 0.091	+ 2
Kyrgyzstan	− 0.107	+ 6
Papua New Guinea	− 0.053	0
Yemen	− 0.070	+ 14

It is difficult to correlate data measuring success with data on interventions, and other intangible factors that are difficult to measure may also affect the results. The data for the 21st century appears to suggest an improvement in gender equality, life expectancy, health and education in countries receiving intervention. However, freedom of speech does not appear to have improved, and changes in wealth have been very mixed.

An analysis of population migration shows that movements of refugees and internally displaced people have increased, particularly in the MENA region, where there is a high level of conflict as well as the usual push factors of poverty and natural hazards. In 2016 there were a reported 4 million Syrian refugees, and also significant displacement of people in Iraq and Yemen. People smuggling and trafficking has increased, and these illegal activities have cost refugees not only their money but also their lives. In 2015 the International Organisation for Migration (IOM) estimated 5,411 refugee deaths. Around the world in 2016 the main reason for refugee fatalities was drowning, as people tried to cross open water in craft not designed for such journeys. In Syria an estimated 10.8 million people were in need of humanitarian aid, including 6.5 million IDPs, but because of the conflict it was difficult to get aid to people. UNHCR did reach over 3 million IDPs in 2014 and delivered key relief items, but more was needed. Similarly, Iraq was declared a level-3 emergency situation by UNHCR in August 2014, which mobilised a huge logistics operation to get aid to people before winter; this was their largest operation in over a decade. Working with IGOs and NGOs, in 2015 UNHCR launched the Regional Refugee and Resilience Plan (3RP), which brought together humanitarian and development interventions to produce a 'sustainable resilience-based response' to the refugee crisis in each country.

In 2015 Reuters news agency reported the plight of Rohingya Muslims fleeing persecution and poverty in Myanmar by boat, risking sickness and starvation at sea. Political complexities in the region made it difficult to assess the success of interventions: Thailand, Indonesia and Malaysia all turned away refugee boats while also accepting some people, while the UN expressed 'alarm' and ASEAN was 'silent'. The Guardian reported that the UNHCR, USA, other governments and international organisations held emergency meetings in 2015 but no clear action plan for the region emerged. Assessing intervention success is therefore difficult, especially as the numbers involved in illegal human trafficking are uncertain. A reason for the lack of reported intervention may be the uncertainty over the legal status of the 'boat people', as Myanmar considers the Rohingya to be illegal Bangladeshi immigrants in their country.

Democratic institutions and freedom of expression

Democratic institutions are based on the concepts of equality and freedom, using political voting systems based on 'one person, one vote', with majority rule enforcing human rights. Economic development has often been seen as a promoter of democratic institutions in capitalist societies, with a good standard of living highly prized. Once established, democratic institutions help to develop socio-political values that promote freedom and human rights. Education is important in this process as it makes people aware of their rights and gives them the means of expressing them. Wealth also facilitates greater independence from the state and the ability to resolve conflicts through local and national democratic institutions such as councils and courts. Economic growth creates global links that include a flow of information that may weaken authoritarian states, a fact illustrated by China's policy of limiting access to the global internet. Research suggests that democratic countries do make greater economic progress than autocratic ones but that this is not continuous over time – probably because of economic cycles, as predicted by Kondratiev (see page 149).

However, there is debate about how democratic institutions are created initially. It has been found that economic development may prolong democracy and allow it to mature, and when per capita income (PPP) rises above US$10,000 (2015 values), democracy should last forever (Przeworski and Limongi, 1993). Global democracy advanced after the shock of the Second World War, which led to the establishment of the UN and its IGOs, and waves of independence of former colonial countries. Other research has shown that a rapid increase in per capita income may change moderately authoritarian countries into democracies, but not strongly authoritarian countries (Londregan and Poole, 1990). What is clear is that wealth, democracy and economic freedom all contribute to better governance.

The UDHR states that everyone has the right to freedom of expression (free speech), to hold an opinion without fear of interference with the right to have that opinion, and to be able to receive and give information and ideas through any form of communication and across any boundaries. Assessing this is usually done by looking at press freedom, censorship and of arrests of journalists. Countries where freedom of expression was curtailed in 2016 included areas of unrest such as Syria and Ukraine, areas with political change such as Turkey, authoritarian states such as Russia and China, and communist countries such as Vietnam and Cuba (see pages 191–193).

Economic growth versus wellbeing and rights

ACTIVITY

Assess the factors that are important in ensuring the development of human rights and human development.

For some countries, especially those with flawed economies or those that are authoritarian-based, economic development has been a priority and systems and institutions are organised to support this aim. This means that there may be less public money to invest in human wellbeing, and regulations and laws may create human rights issues. In recent decades China has experienced economic growth despite being an authoritarian regime, raising millions of people out of poverty and linking in with the world economy. Between 2000 and 2014 China increased its GDP per capita (PPP) by 4.5 times, while South Korea's PPP only increased by 1.9 times. However, at the same time China's economic freedom and human rights decreased while South Korea's increased (Table 4.13).

Yellow = Players, Orange = Attitudes and actions, Purple = Futures and uncertainties

CASE STUDY: Economic growth vs human rights in China

Since the late 1970s China has been a relatively stable country. This has allowed the ruling Chinese Communist Party (CCP) to make changes without yielding control. In the late 1980s and early 1990s, reform of the rural commune system allowed for village elections and for petitioning of officials and local courts to improve local governance and farm efficiency, and enabled workers to move to industries in eastern cities. However, the main purpose of increasing local democracy appears to have been to reduce unrest in rural areas; reports in 2006 suggested that the petition system had been supressed.

At the turn of the century economic liberalisation began in China, with a market-orientated approach but no democratisation, which many Chinese people appear to have accepted because it has brought increased wealth. The country used its strengths – a young, educated workforce and a political structure that allowed regional and local initiatives – to rapidly develop its economy (Table 4.13). This rapid development was aided by foreign direct investment from the Chinese diaspora, in addition to global trade links.

Some democratic institutions and processes helped support the authoritarian regime by diffusing opposition, using legal systems or co-opting key players such as entrepreneurs and allowing intra-party democracy. For example, in 2000 public input was allowed on draft legislation and in 2010 the country's twelfth five-year plan (2011–15) emphasised public participation. Some strict policies have been eased or removed, but restrictions on political freedom and civil liberties continue, with harassment of human rights activists and the detention of journalists. The 2015 National Security Law had very broad definitions (Table 4.13). The CCP continues to control economic reforms, ethnic relations, domestic security and the internet, as well as many judicial and financial institutions. However, since 2012 there has been a strong anti-corruption campaign, with over 400,000 officials disciplined, some through the courts, although this was mainly to re-establish Party authority.

Table 4.13: Indicators of change in China and South Korea

Indicator	Year	China	South Korea
Economic Freedom Index (The Heritage Foundation: higher is freer)	2000	56.4	69.7
	2016	52.0	71.7
Freedom Index (Freedom House: higher is freer)	2000	7	71
	2016	16	83
GDP per capita (PPP) (World Bank)	2000	US$2,915	US$18,083
	2014	US$13,206	US$33,395
Fragile States Index (The Fund for Peace: lower is more stable)	2005	72.3	39.9
	2015	76.4	36.3

The 2008 world economic recession affected China later than other countries, but this perhaps shows that to continue the rate of economic progress, its institutions need to be more democratic. Hong Kong still remembers democracy, and there have been pro-democracy demonstrations (Figure 4.27) and demands for freedom of speech, transparent government and rule of law which may influence China in the future. At present, Chinese leaders remain fearful of democratic change and dissenters are still imprisoned. China ranks highly in terms of human rights infringements, and the Fragile States Index shows that China has become less stable. Governance institutions are probably not yet sufficiently developed to cope with democratisation: although reforms of the central bank and fiscal system may have been the first step towards a more democratic regime, the Economic Freedom Index decreased, showing that there is still a long way to go. The country's GDP per capita is also now above the suggested US$10,000 threshold for a secure democracy.

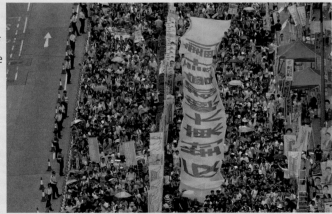

Figure 4.27: Pro-democracy demonstrations in Hong Kong, 2014

The mixed record of development aid
Aid, development, health and human rights

Since the links between aid, development, health and human rights are not clear, it is difficult to make correlations. There are inconsistencies in the delivery of aid, with different political choices being made, or in some cases aid removed too soon from a country, for example Peru. Certainly, well-managed humanitarian aid can make a significant difference to people affected by a disaster, whereas economic aid with conditions may be less successful if it creates dependency on the donor country or institution, or diverts recipient government funding away from health and education programmes. In a multipolar world, aid may become more regionalised, with groups of countries that share similar goals helping each other. Scale is another consideration: expensive, large top-down schemes may benefit the economy but disadvantage local people and the natural environment – such as the Santo Antônio Dam on the Madeira River in Brazil – while small bottom-up schemes are relatively cheap and aim to benefit local communities – such as micro-hydro in Chambamontera, Peru.

CASE STUDY: Haiti – development aid failure

Haiti has received aid for a very long time, sometimes linked to natural disasters and sometimes linked to poverty (see case study on page 225). Consequently there is a plethora of NGOs in the country with aid money totalling an estimated three times the national government budget (also see Thinking Synoptically on page 184). While there are some positives to the aid, such as helping to get things done that the government cannot afford to do, there are many negatives:

- local jobs have been lost to aid workers as little has been done to show local people how to help themselves

- dependency on aid has been created amongst people and the government, for example NGOs provide about 70 per cent of Haitian healthcare and 85 per cent of schooling

- local people have changed the way that they perceive things to the point that they cannot do things for themselves

- contrast between aid workers in new cars and clothes compared with the conditions that the poor Haitians live in

- locals not consulted about what they want and need so that aid organisations are perceived as meddling and deliberately keeping the people poor

- only small amounts of emergency aid money for the 2010 earthquake reached Haitians, for example, of US$10 billion pledged only about 2 per cent went to Haitian NGOs and government, and of the total from USAID for reconstruction only 1 per cent went to Haitian businesses while about 70 per cent went to US companies.

Yellow = Players, Orange = Attitudes and actions, Purple = Futures and uncertainties

CASE STUDY: The response to ebola in West Africa, 2014–16

Ebola is a virus disease transmitted from infected wild animals to people, and then from human to human via direct contact with bodily fluids. Mortality rates are usually around 50 per cent. It first appeared in 1976 in South Sudan and DR Congo, but in 2014 West Africa saw the largest outbreak, spreading quickly from Guinea to Sierra Leone and Liberia, and then on to Nigeria, Mali, Senegal and even reaching the USA and Spain. WHO estimated that over 28,600 people were infected and over 11,300 died. NGOs such as MSF were first in the area, offering immediate medical assistance. In the first year of response MSF set up 15 ebola management and transit centres (Figure 4.28), which dealt with 8,534 patients. MSF also shipped 1,400 tonnes of equipment, with 530,000 protective suits for health workers.

The EU provided nearly US$2.25 billion in aid, with the European Commission (EC) providing US$980 million. Of this EC money:

- 8 per cent was for urgently needed short-term humanitarian aid such as medical supplies and health workers

- 28 per cent was allocated to medical research projects and pharmaceutical industries to search for a vaccine

- 64 per cent was directed towards development aid, to help the countries affected recover after the outbreak.

The EU hosted conferences discussing the lessons to learn from the ebola outbreak and address criticisms – such as helping poor countries develop their health systems so that they are better prepared, reducing dependency on aid and reducing response times by IGOs so that the initial burden does not fall on NGOs. WHO aims to prevent further outbreaks through surveillance in at-risk African countries, help with laboratory testing, advice and training, and engaging communities.

Figure 4.28: An MSF ebola management and transit centre in Guinea

Economic inequalities and health and life expectancy

The purpose of development aid is to improve the economic and social wellbeing of people within a country. The traditional approach has been to improve the economy of a country to create wealth, which families or the government can then allocate to health and education. There is a strong positive correlation between increasing wealth and longer life expectancy and lower child mortality rates, for example. However, in some countries increased wealth may not be spread evenly among the population, creating inequalities such as in Brazil (see Table 4.5). While some countries in receipt of development aid have reduced inequalities, others have increased them.

Table 4.14 shows that the largest amount of aid or the largest increase in aid does not necessarily bring about the largest improvements. For example, Vietnam had the largest amount of official ODA in 2014, but the income shares of the lowest and highest deciles show an increased wealth gap; life expectancy improvement was below the UK and the child mortality rate was the fourth highest in 2015 out of the countries listed. Some of those receiving less aid, such as Bolivia, decreased inequalities while at the same time having the best increase in life expectancy and decrease in child mortality out of the countries listed. A correlation analysis of the sample data in Table 4.14 shows no significant correlation between an increase or decrease in development aid and any of the variables. There is a very small positive correlation between the change in ODA and higher income share of the lowest decile of population, and a very small negative correlation between ODA change and child mortality (health) (i.e. as ODA increases child mortality decreases). The correlation between a change in ODA and life expectancy is close to zero showing no correlation.

> **Maths tip**
>
> Data is often divided into groupings – such as quartiles that are used to calculate quartile deviation. Also commonly used are quintiles, dividing a database into fifths, and deciles, dividing a database into ten equal groups.

Table 4.14: Data on changes to aid, income inequalities, life expectancy and health for selected countries (source: World Bank)

Country	Net official ODA (US$m), 2000	Net official ODA (US$m), 2014	Change in net official ODA (US$m), 2000–14	Change in income share (%) held by highest decile, 2000–13	Change in income share (%) held by lowest decile, 2000–13	Change in life expectancy at birth (years), 2000–14	Change in child mortality rate (/000), 2000–15
Vietnam	2130.2	4304.3	+ 2174.1	29.1 to 30.1	3.3 to 2.6	+ 2.5	- 12.1
Turkey	487.6	3409.5	+ 2921.9	31.9 to 30.5	2.3 to 2.2	+ 5.2	- 26.1
India	1878.9	2985.6	+ 1106.7	27.0 to 30.0	2.0 to 3.5	+ 5.4	- 43.5
Colombia	282.0	1212.3	+ 930.3	47.0 to 41.9	0.1 to 1.1	+ 3.0	- 8.7
Haiti	308.0	1080.3	+ 772.3	47.8 to 48.2	0.6 to 0.6	+ 5.0	- 35.8
Brazil	294.9	909.7	+ 614.8	47.0 to 41.8	0.8 to 1.0	+ 4.4	- 15.6
Bolivia	744.2	669.7	- 74.5	49.0 to 35.6	0.1 to 0.9	+ 7.6	- 41.8
Thailand	707.5	359.5	- 348.0	33.7 to 30.4	2.6 to 2.8	+ 3.8	- 9.9
Chile	76.4	237.1	+ 160.7	45.3 to 41.5	1.3 to 1.7	+ 4.7	- 2.8
UK	**0**	**0**	**0**	**26.2 to 24.7**	**2.0 to 2.9**	**+ 3.4**	**- 2.4**

Synoptic link

The historian Niall Ferguson asserted that superpowers should use hard power, in the form of military and economic forces, to bring about the changes needed to ensure freedom around the world. However, this view is contested, as it emphasises one model of development, when other alternatives might also work. (See page 138.)

Superpowers and development aid

Economic and military superpowers often use development aid as an element of their foreign policies, to strengthen links within the globalisation process; this is supported by the involvement of TNCs. Trade conventions may be used to promote home companies and secure deals for resources or trade, a process sometimes described as neocolonialism. Superpowers may also judge their success by the amount of support they receive from the UN Security Council or General Assembly or similar regional alliances, or from military alliances such as NATO; or they may form new military alliances to gain influence in different world regions.

Yellow = Players, Orange = Attitudes and actions, Purple = Futures and uncertainties

CASE STUDY: China's development aid

Aid Data (2016) recorded 2,762 projects using Chinese development aid or other financial assistance. China initially focused on South America but recently has turned its attention to Africa where it is involved in 51 countries. Chinese overseas development aid (ODA) also features significantly in SE Asia.

China is a significant new donor of ODA, using bi-lateral aid and the Asian Development Bank. However, China has opted out of the global reporting mechanisms which were established to coordinate aid and guarantee transparency, so the aims of its aid are not always clear. Aid Data's analysis suggests that:

- China is less fussy than 'western' donors about where its ODA goes, but there is not enough evidence to show that it supports corrupt regimes to spread authoritarian rule
- when ODA is added to other forms of financial assistance from China, its influence in Africa is equal to that of the USA, with the addition of establishing commercial links
- China's aid has been concentrated on infrastructure projects, perhaps to facilitate access to resources, but there are also a wide range of social projects
- China provides more ODA to the African countries that support it in the United Nations
- Chinese aid is often concentrated in the home region of African leaders, consolidating their home support and opening up the possibility of future alliances
- Chinese aid has been a positive influence where 'western' aid has been withdrawn
- China's 'soft power' does not appear to have increased as a result of increased ODA in Africa, countries still look 'westward' for advice.

Example projects:

Social: Rwanda 2014–2016: 12 member medical team with medical supplies and equipment to Kibungo Hospital in Eastern Province.

Infrastructure: Zambia 2010–2016: US$930 million aid for 600 mW power plant at Kafue Gorge, southern Zambia.

Resource: Bolivia 2014–2016: US$344 million aid via a Chinese loan to cover 85 per cent of costs for iron ore processing plant to be run by Bolivian company El Mutun Steel (taking over from the Indian company Jindal Steel and Power).

In the 21st century are superpower interventions able to cope with the protracted issues of conflict (such as Afghanistan from 2001), huge refugee movements (over 1 million arrivals in Europe in 2015), climate change (average global temperature 1°C compared to pre-industrial level) and larger natural disasters (Indian Ocean tsunami in 2004)? The present humanitarian regime was devised half way through the 20th century and may no longer be 'fit for purpose'. For example, IDPs are not officially part of the UNHCR remit, yet outnumber refugees considerably. Perhaps response options need to be stronger and the burden of intervention needs to be shared more widely in the future.

The mixed record of military interventions
The costs of military interventions

Military interventions have been controversial because of the costs they may bring: the deaths of innocent civilians, the destruction of housing and infrastructure, the need to support and shelter those displaced, the disruption of livelihoods, the infringement of human rights during the conflict, and loss of sovereignty by the countries receiving the military intervention. The examples of Iraq, Afghanistan and Libya show that it is difficult to reunite and restructure countries after regime collapses. The short-term gains of preventing infringements of human rights, such as the persecution of minority groups and protecting resource pathways (for example oil supplies) need to be balanced against the long-term costs of lengthening conflicts, encouraging international terrorism and prolonging instability.

In 2012, UN General Secretary Ban Ki-Moon outlined three pillars of responsibility to protect (R2P), which he said were not necessarily in a sequence:

- **Pillar 1:** A state has legal responsibility to maintain stability and enforce human rights, such as reducing rhetoric that may target minority groups and inflame situations. International interventions could include preventative diplomacy and disruption of arms shipments that could be used for crimes against humanity.

- **Pillar 2:** A state is responsible for protecting its citizens and resolving tense situations before conflict breaks out. International intervention could include a UN commission to investigate situations, establish facts and identify the perpetrators of crimes.

- **Pillar 3:** Even though it is 'not the role of the UN to replace the state', there may be occasions when the state needs help to meet its legal responsibilities to protect its citizens. International intervention may include the use of peacekeeping missions with the consent of the legal **democratic government** of a state, in order to ensure a return to peace, social renewal and better institutions.

Non-military interventions

The response to the ebola outbreak in West Africa showed that non-military intervention can bring short-term benefits in saving lives, and also when linked with long-term development plans can improve a country (see page 239). It is important that development aid does not finish once there has been an improvement in a situation or a solution to a specific issue. Developing countries need continued help, especially with population growth and climate change increasing the difficulties faced by the poorest people. There is greater transparency now on where aid goes and what it is spent on, and some of the best schemes are those based on a donor's expert knowledge. Each possible intervention is unique, with different perceptions and nuanced understanding, which makes it difficult to have one approach; however, there will always be preventative and humanitarian response stages with the UN working with a variety of IGO and NGO partners.

Yellow = Players, Orange = Attitudes and actions, Purple = Futures and uncertainties

CASE STUDY: Non-military and military intervention in Côte d'Ivoire

Due to its colonial legacy, Côte d'Ivoire has a north-south ethnic division, with a Muslim immigrant north and a Christian south. For a long time those in the north were excluded from government, which led to their armed rebellion in 2002, one outcome of which was to allow them the right to vote. In late 2010 presidential elections were held between Laurent Gbagbo (incumbent president from the south) and Alassane Ouattara (from the north). The elections, overseen by the UN and the Electoral Commission, confirmed Ouattara as the winner but Gbagbo and the Constitutional Court quickly disputed this. Supporters of Ouattara were then attacked and the streets militarised by supporters of Gbagbo, making them unsafe for civilians. Random mass shootings and mortar attacks took place in broad daylight and it was clear that abuses of human rights were being committed and that democracy had broken down. This was enough to convince the UN that intervention was required, although initially this was delayed by concerns that the unrest would spill over into neighbouring West African countries.

During the conflict that followed in 2011, a million people were displaced from their homes and 147,000 refugees fled to Liberia. Armed forces loyal to Ouattara moved to the outskirts of the main city, Abidjan, where they met strong resistance from the better-equipped opposition. The battle lasted 12 days, killing 3,000 people, and ended when the UN Security Council authorised French forces to intervene. French air power attacked the main defences of forces loyal to Gbagbo, and a French armoured column secured the city and escorted Outtara's forces to Gbagbo's base. Gbagbo was arrested and within two weeks his forces surrendered or pledged their support to the new president. Within two months, the country was peaceful and normal life resumed.

UN agencies (Figure 4.29) brought in humanitarian aid for refugees and IDPs; the UN Food and Agriculture Organisation (FAO) provided seeds, tools and fertiliser for farmers. The new government pledged to investigate human rights abuses and produced disarmament plans. Despite criticism from Russia and China, who said that a power-sharing transitional government should have been established, rather than the use of force, most regard the speed of the intervention and resolution of the conflict as a success. In 2016 the UN mission was 'winding down', and a development objective of the country was to become an emerging economy by 2020. With an average annual economic growth rate of 9 per cent since 2011, this is a possibility, especially as structural reforms have created an environment for the international and private sector economy, and the growth of agricultural processing industries.

Figure 4.29: UN peacekeeping forces in Côte d'Ivoire

The consequences of lack of action

A lack of action can have consequences that may be as serious, or more serious, than those arising from action. For example, what might have happened if Saddam Hussein had been left in power in Iraq, or if Afghanistan had been left alone, or if IS had been allowed to expand its insurgency unchecked? Many believe that the peace and security around the world are better because of interventions, but not all agree. There have been times when intervention or action did not take place, when perhaps it should have.

CASE STUDY: The Srebrenica massacre, 1995

The Bosnian War (1992–95), part of the break-up of Yugoslavia, was an ethnic conflict between Muslims and Serbs. In 1993 the UN declared the besieged town of Srebenica a 'safe area', to be free from armed attack or any hostile act. After an evacuation of as many people as possible, French troops entered the town to observe the situation, and negotiations took place to allow aid convoys through Serb-controlled territory if the people in the town gave up their weapons to the UN. A Canadian UN peacekeeping force moved in to protect the civilians, but there was initial uncertainty about the use of force until 1994, when it was made clear that this was authorised. However, it was estimated that thousands of UN troops would have been needed to effectively carry out military intervention.

Aid convoys were allowed through but, increasingly, Serb forces restricted their contents. By now a Dutch UN peacekeeping force was in place, but reinforcements had not been allowed through and so a 600-strong force had halved by early 1995. After a minority of Muslims attacked Serbs for supplies, the situation deteriorated and it became clear that Srebrenica would be attacked even though the end of the war was near. The people of the town asked for their weapons back but the UN refused. The small UN force was powerless when in July 1995 Bosnian Serb forces moved into the town and took over 8,000 Muslim men and boys into Serb territory and systematically executed them, burying the bodies in an attempt to hide the genocide or 'ethnic cleansing'. This was the worst massacre in Europe since the Second World War and perhaps could have been avoided with stronger action by the UN.

The perpetrators of the crimes were convicted by international tribunals, and UN forces criticised for covering up what happened. The failure of the international community to fulfil its moral and legal duty led to the creation of the UN Peacebuilding Commission and raised awareness in democratic countries of the need for active intervention.

In July 2015 the UN tried to pass a resolution designating the massacre as genocide, but Russia used its veto to block this, saying that as the resolution only identified Serbs it would increase ethnic tensions.

Lack of action can also affect the natural environment. Climate change is having important global consequences (see Chapter 2), but action has been slow. In 1979 the first World Climate Conference (WCC) was held, in 1988 the Intergovernmental Panel on Climate Change (IPCC) was established, and in 1990 the WCC and IPCC called for a global treaty. In 1992 the UN Framework Convention on Climate Change (UNFCCC) international treaty was drawn up to achieve international cooperation to combat climate change and cope with its impacts. Negotiations started three years later, and two years after that the Kyoto Protocol was agreed, which required developed countries to reduce their greenhouse gas emissions in a two-stage commitment (2008–12 and 2013–20) using the 'polluter pays' principle. However, some countries never ratified the treaty – most notably the USA, which at the time was the largest polluter, and developing countries were excluded.

A level exam-style question

Explain why there may be long-term costs as a result of direct and indirect military intervention. (8 marks)

Guidance

Remember that the term 'costs' means problems and issues, and not just financial costs. In your answer, avoid just describing the costs; you need to link them to the action of military intervention. Refer to some examples, with a few facts.

Extension

Rwanda has been described as the worst genocide since the holocaust of the Second World War. Investigate this example of lack of international intervention to gather facts and opinions.

- To what extent did a lack of intervention lead to negative impacts on the natural environment and socio-political development in the country?

- What were the global consequences of this lack of intervention?

- Compare Rwanda with an example of successful intervention, such as Sierra Leone (1999–2006) or Burundi (2004).

Yellow = Players, Orange = Attitudes and actions, Purple = Futures and uncertainties

After two significant IPCC reports (2007 and 2014), the seriousness of global climate change impacts on the natural environment and socio-political development was clear. The Paris Agreement (2015) (see page 127) was an **international agreement** that increased the level of action to be taken around the world, with funding made available to help developing countries face the challenges. However, even though 177 countries signed the Paris Agreement, by June 2016 only 17 had ratified it. For the agreement to come into force, there need to be 55 ratifications accounting for at least 55 per cent of emissions. Unsurprisingly, some of the first to ratify the Paris Agreement were Fiji, the Maldives, Tuvalu, Palau and Nauru – small island states seriously affected by sea level rise. Fiji has already moved villages inland from its coast. This is just one negative consequence of the lack of intervention on climate change (Table 4.15).

Table 4.15: Summary of the environmental, social and political impacts of climate change

Environmental impacts	Social impacts	Political impacts
Degradation and loss of coral reefs	Spread of diseases in warmer and wetter climate	Loss of territory and communities to sea level rise or desertification
Coastal ecosystems threatened and processes changed	Fishing grounds move, with loss of income and food	Increase in numbers of environmental refugees
Biomes shifted by altitude and latitude	Indigenous groups lose their harmonious balance with nature	Traditional political balances threatened by change in resource availability, e.g. water
Glaciers and ice caps melting and water cycle balance changed	Small farmers struggle due to weather changes, diseases and pests	Increased poverty and hardship causing unrest and rebellion
Change to the thermohaline ocean conveyor	Extreme poverty for those reliant on their environment	Disputes arising over new frontiers, e.g. the Arctic
Increase in extreme weather hazard events	Reduced public spending on welfare and health, as funds are diverted to climate change adaptation and mitigation	Fragile states with poor governance unable to use climate change aid
	Increased health issues suffered by vulnerable groups	

ACTIVITY

Explain how a lack of action can have global consequences for the environment and socio-political development.

Summary: Knowledge check

Through reading this chapter and by completing the tasks and activities, as well as your wider reading, you should have learned the following and be able to demonstrate your knowledge and understanding of health, human rights and intervention (Topic 8 Option 8A).

a. Explain how human development can be measured and improved.

b. Describe and explain the variations in health and life expectancy in developing and developed countries.

c. Explain why the relationship between economic and social development is complex and dependent on the priorities of governments.

d. Explain the roles of IGOs in setting targets and goals for improving environmental quality, health, education and human rights.

e. Describe the purpose of the Universal Declaration of Human Rights, the European Convention on Human Rights and the Geneva Convention.

f. Assess the success of the UDHR, the ECHR and the GC.

g. Explain why countries have different priorities, ranging from prioritising human rights to prioritising economic development.

h. Explain the links between levels of democracy and levels of corruption, and the impact this has on human rights, freedom of speech and the rule of law.

i. Describe and explain the differences in rights of minority groups (gender/ethnicity) and how these are reflected in health and education levels.

j. Describe and explain the range of geopolitical interventions that can be used to address human rights issues and development.

k. Describe the forms of development aid, and explain how these may have positive and negative impacts on the natural environment and people.

l. Describe and explain the advantages and disadvantages of using military aid or intervention to solve human rights issues.

m. Describe the different ways of measuring the success of geopolitical intervention.

n. Explain why development aid has a mixed record of success, benefiting some countries but not others.

o. Explain why military or non-military interventions have a mixed record of success, benefiting some countries but not others.

As well as checking your knowledge and understanding of these subjects, you also need to be able to make links and explain the following ideas:

- The concept of human development is complicated, with different judgements about the best way forward; some people and groups emphasise economic development, some social development and some sustainability, with a balance between the natural environment and economic and social progress. Some say that the natural environment should have priority.

- A wide range of indicators can be used to measure human wellbeing; many of these overlap, and finding the best combination is not easy. Access to the basic human requirements of food, water, sanitation and shelter is an important measure, because these affect lifestyle as well as life expectancy, infant and child mortality and maternal mortality.

- The correlation between economic development and social development is not always clear; one important variable is the role of national governments, which make decisions that shape the country, including budgets for health care and education. The role of international organisations is also important: they may influence a government through trade and aid, and define certain conditions attached to aid or set goals for the future such as the MDGs and SDGs.

- Since the Second World War, international laws and agreements have been developed to prevent infringements of human rights. The UDHR provides the basis of many regional and national human rights conventions and laws, and even armed conflicts are moderated by the Geneva Convention. This has not prevented all crimes against humanity but it has allowed perpetrators to be brought to justice.

- Despite international conventions, countries still differ significantly in their actions on human rights. Authoritarian and totalitarian countries often have the worst records, partly because of the inherent corruption in their systems, but some developed countries are not without infringements. Human rights violations are frequently linked to conflicts but may also appear to be ethnically or gender-based. At times, human rights and freedom of speech may be restricted for reasons of national protection and sovereignty.

Yellow = Players, Orange = Attitudes and actions, Purple = Futures and uncertainties

- Countries, NGOs and IGOs can use a wide range of possible interventions, from trade sanctions to humanitarian aid to direct military action. Many forms of intervention are contested because it is believed that they bring negatives to the recipient country, such as creating dependency on aid money or goods, encouraging corruption through the misuse of aid money, the killing of innocent civilians, or damaging the natural environment.

- All forms of military involvement may be justified in terms of ensuring that human rights are upheld. However, such interventions may indirectly support human rights infringements. These types of action are therefore open to debate; for example, does the outcome justify the means? Success could be gauged by measures of development (e.g. gender equality) and democratic governance.

- Development aid is sometimes successful and sometimes not, and the correlation between the amount of aid and improvements in the measures of development is unclear. Some countries appear to have benefited while others have not, suggesting that development aid is only one aspect of bringing about improvements.

- Military intervention and other non-military IGO interventions are sometimes successful and sometimes not, and a main determining factor appears to be action or lack of action. There are benefits and problems associated with each situation, some short- and some long-term, and different situations dictate the order of the 'pillars of action'.

Preparing for your A level exams

Sample answer with comments

Explain why indigenous peoples experience lower health levels than non-indigenous populations. **(8 marks)**

Indigenous peoples, such as the Aborigines in Australia or the Indians in North America, appear to have poorer health than the non-indigenous white population. For example, the Aborigines suffer more from cardiovascular disease and cancer, and infant mortality rates are twice as high as in the non-indigenous population.

Australia has a per capita income of around US$35,000 a year and an average life expectancy of over 80 years, but not everyone is equal. Australia was ranked second on the 2015 Human Development Index and featured in the top ten for the Happiness Index (2015) and Freedom Index (2016). The country also gives lots of Overseas Development Aid to countries nearby, being one of the top ten donors in 2014, giving US$4.4 billion according to the OECD.

Aboriginal women and children are at a particular disadvantage, with type 2 diabetes three times more frequent in Aboriginal women than in white women, and children are very obese. This is because of poor diet, stress and poor education. There is a similar situation in North America and with the Inuit.

The first sentence names relevant indigenous groups but then just repeats what the question says, and this should be avoided – the candidate should be making an examiner credit every sentence. The second sentence is better and compares the two populations nicely in an accurate statement.

A factual paragraph, with a hint of comparison at the end of the first sentence. While facts are required in an answer, they must be relevant to the question, and here they are not. Perhaps Aborigines are poorer than the average Australian and this limits their access to health care, or Australia is giving too much aid abroad when more money is required 'at home' to improve health care services for Aborigines.

At last, the candidate gives some reasons, as required by the question, but these should have been the focus of the answer throughout. Other factors that could have been covered are smoking levels and their link to death rates, living and working conditions, the causes of obesity and stress, and how lower levels of education lead to poorer health levels. The introduction and conclusion both mention the North American indigenous population as well, but one example should be enough in an 8-mark question unless another example adds another major point, such as spending on health care by the Indian Health Service in North America.

Verdict

This is a below-average answer, mainly because the question is not answered throughout. There is a range of geographical ideas but they are not fully developed, and the way the knowledge is presented makes it mostly irrelevant.

Preparing for your A level exams

Sample answer with comments

Evaluate this statement: 'Development aid plays a significant role in improving the health and human rights of people in emerging and developing countries.' (20 marks)

There is a range of development aid that governments and their IGOs and NGOs may use to assist other countries. This includes humanitarian aid given in times of emergency which relieves suffering and reduces the number of secondary deaths. Aid may be given to help build infrastructure and other economic improvements, sometimes in the form of loans from the World Bank or financial aid from the IMF, and this form of aid may create debts or have conditions attached to it which means that it doesn't benefit poor people. Development aid is therefore contested.

> A sound introduction, which demonstrates that there may be two sides to development aid, and sets the tone for the rest of the answer. It could have been improved with the name of an NGO and a recognition of the breadth of human rights linked to health, freedom of speech and gender equality.

Charitable aid, such as from Christian Aid (CA), usually concentrates on the basics of life, such as water, food and shelter. After the 2010 earthquake in Haiti, CA, who were already operating in the country, helped poor people in rural areas by providing seeds and fertilisers so that people could grow their own food and be able to survive. CA also helped to reconstruct homes to make them more earthquake-proof. In total they helped about 180,000 people. CA said that there needed to be many years of continuing aid to Haiti, but the amount of aid has been falling and even CA themselves only allocated less than 1% of their funds to the country in 2015. A complication to the situation after the 2010 earthquake was that the humanitarian aid organisations did not really work with local people to find out what they needed, but imposed their own systems, and even worse there were additional deaths caused by a cholera outbreak, which was blamed on UN Nepalese peacekeepers who brought it into the country and spread it through contaminated river water in the area where they were based.

> This paragraph illustrates one example of development aid – helping recovery after a disaster. There is a range of accurate factual information and a balance of positives and negatives arising from development aid. Some consideration of 'significance' would raise the level of evaluation.

IGO loans from the World Bank to developing countries have been common to help poor countries up the development steps; unfortunately this resulted in large debts if countries did not make the progress expected and then could not pay back loans or make the interest payments. This forced countries to cut back on spending on health care and social programmes.

> The main problem with this paragraph is balance; recognises a weakness, but fails to expand or consider benefits.

Bolivia received US$670 million in ODA in 2014 and since 2000 life expectancy increased by 7.6 years and the child mortality rate declined by 41.8/000, showing that the aid was working. 40% of this ODA was allocated to reducing the gender equality gap: in 2010 there was a national conference for indigenous women, a female leader of the Senate, and in 2013 a domestic violence law made any form of abuse of women illegal. By 2015 53% of the parliament was made up of women, and the Gender Inequality Index improved from 0.576 in 2000 to 0.472 in 2013, and the 2015 Global Gender Gap Report (WEF) confirmed these improvements, with Bolivia moving up 36 places in the rank order in just one year (2014 to 2015), although there were still problems with equality within education.

> This paragraph adds to the breadth of development aid and provides a balanced evaluation with accurate factual content.

Development aid does play a significant role in improving health and human rights in poor countries. But since the global recession from 2008 the amount of money being donated has decreased, with many countries below the 0.7% of GDP target set. Aid does support some authoritarian regimes where rights may be abused, such as Australia supporting Indonesia, but that country has corruption and abuses the rights of minority groups and political opponents.

> Returns to the question, but makes points from notes rather than linking evidence together. An evaluation question needs a strong conclusion that brings together strengths and weaknesses, but also explores alternatives – the ways future aid could be improved, and what would happen if there were no ODA.

Verdict

This is a competent average answer, with some very good material. However, making a plan at the start would probably have ensured coverage of a wider range of development aid. In evaluation questions it is important to allow enough time for developing a conclusion. The candidate shows accurate knowledge and understanding and demonstrates some relevant connections and partial interpretations.

Human systems and geopolitics

Global development and connections: Migration, identity and sovereignty

Introduction

People have always migrated to improve their quality of life and to escape hardship and hazard. Modern technologies and globalised connections have increased the opportunities to do so. In 2015 the UN calculated that there were more than 243 million international migrants, of which about 19.5 million were refugees. The scale and speed of modern migration flows – particularly from rural to urban areas in developing or emerging economies – have supported dynamic economic growth in many countries. Migration of skilled workers promotes globalisation and creates unparalleled prosperity for some places, but also brings costs to both source and host regions, including increased inequality and challenges to ideas of national identity and sovereignty.

A sense of identity is important to human communities. Shared history and values have created countries and nation states, with nationalism calling for people's loyalty to a particular place. Understanding the role of nationalism and national culture helps geographers make sense of past and present geopolitics. However, globalisation is reshaping traditional ideas of nationalism and sovereignty, as some groups benefit more than others, creating divisions and diluting loyalty.

Since the Second World War intergovernmental organisations (IGOs) have managed global issues in an increasingly interconnected world. IGOs aim to mitigate the worst excesses of nationalism, but their interventions in geopolitics, economic management and environmental concerns have had mixed success. International cooperation and migration have raised questions about identity and loyalty, making national identity an elusive and contested concept. Geographers need to study these questions in order to investigate and understand the patterns of the 21st century.

In this topic

After studying this chapter, you will be able to discuss and explain the ideas and concepts contained within the following enquiry questions, and provide information on relevant located examples:

- What are the impacts of globalisation on international migration?
- How are nation states defined and how have they evolved in a globalising world?
- What are the impacts of global organisations on managing global issues and conflicts?
- What are the threats to national sovereignty in a more globalised world?

Figure 5.1: Migrants making the dangerous sea journey across the Mediterranean to Europe. To what extent is migration a symptom of globalisation or of local factors? What implications does migration have for sovereignty?

Synoptic links

Migration can have profound impacts on demography, politics and the economy of places. Movements of people are an important concern of national governments and intergovernmental organisations as they seek to facilitate some movements and limit others, such as into or within the European Union. The nature of the nation state is changed by this process, which has consequences for national identity and associated loyalties, such as in Ukraine. Patterns of economic and environmental regulation change as a result, especially in fast-growing cities such as Shanghai and Mumbai. The growth of intergovernmental decision-making bodies, such as the UN and NATO, affects many aspects of geography, as shown by their IGOs and their databases. The challenges to governance arising from changes in national identity and loyalty affect geopolitics, patterns of cooperation and conflict, regional and global economic networks and local and global environments such as the geopolitical situation in the Middle East.

Useful knowledge and understanding

During your previous studies of Geography (GCSE and AS Level), you may have learned about some of the ideas and concepts covered in this chapter, such as:

- the process of migration
- migration patterns
- the processes of globalisation
- political geography and the differences between countries
- patterns of inequality
- intergovernmental organisations
- political, economic and environmental problems
- transnational ownership and multiculturalism.

This chapter will reinforce this learning and also modify and extend your knowledge and understanding of migration, identity and sovereignty.

Skills covered within this topic

- Use of flow lines to show the direction and size of migrant flows.
- Interpreting oral accounts from migrants to investigate the causes of migration.
- Interpreting opinions on the contribution of migrants to the culture and social life of contrasting nations.
- Use of divided bar graphs to compare the ethnic diversity of countries.
- Comparison of maps of languages and colonial histories to analyse the relationship between them.
- Use of the Gini coefficient to describe inequalities within and between nation states.
- Evaluation of source material to determine the impact of IGOs on managing global environmental issues.
- Use of proportional circles to show the scale of foreign ownership in certain economic sectors.
- Critical analysis of a variety of source material to assess the costs and benefits of foreign ownership.
- Critical analysis of source material to identify possible misuse of data in the assessment of role of the state and the success in promoting national identity.

What are the impacts of globalisation on international migration?

Learning objectives

8B.1 To understand how globalisation has led to an increase in migration both within countries and between them.

8B.2 To understand the varied, complex causes of migration that are subject to change.

8B.3 To understand the varied and disputed consequences of international migration.

Globalisation and increasing migration

The IMF defines **globalisation** as 'the growing economic interdependence of countries worldwide through increasing volume and variety of cross-border transactions in goods and services, freer international **capital** flows, and more rapid and widespread diffusion of technology'. However, the idea of growing interdependence can also be applied to people, because globalisation has created opportunities for them to migrate, as well as incentives (pull factors) and pressures (push factors) that help explain these migrations. An emigrant is a **migrant** from the point of view of the country they are leaving, and an immigrant from the point of view of the country to which they are moving.

A migrant is defined as someone who moves their 'permanent' residence from one country to another for at least one year:

- An **economic migrant** is someone emigrating for better employment opportunities or an improved financial position.

- A **refugee** is a person who has left their home country because they have suffered (or fear) persecution on account of their race, religion, nationality or political opinions. They may be a member of a persecuted social or ethnic group, or fleeing a war. Such a person may become an '**asylum seeker**' if the host country grants them this status.

- An **irregular migrant** is a person who enters a country illegally or remains in a country without a valid visa or permit from that country, or who has overstayed the duration of a visa or whose visa has been cancelled.

Globalisation has caused significant changes to the global economic system, creating the push and pull factors required to drive migration. The two main trends are rural-to-urban migration within emerging economies and developing countries, and international migration between interconnected developed economies.

Figure 5.2 shows bilateral flows of people from 2005 to 2010 and the estimated flow between the 50 most significant host and source countries during this period. Adjusted for population growth, the global migration rate has been constant since about 1995. It is not the developing countries that

ACTIVITY

TECHNOLOGY/ICT SKILLS

Investigate the recent migration patterns between Mexico and the USA. What is the scale of the movement? Has the pattern changed in recent years? What role do the Mexican and US governments play in this international migration?

Yellow = Players, Orange = Attitudes and actions, Purple = Futures and uncertainties

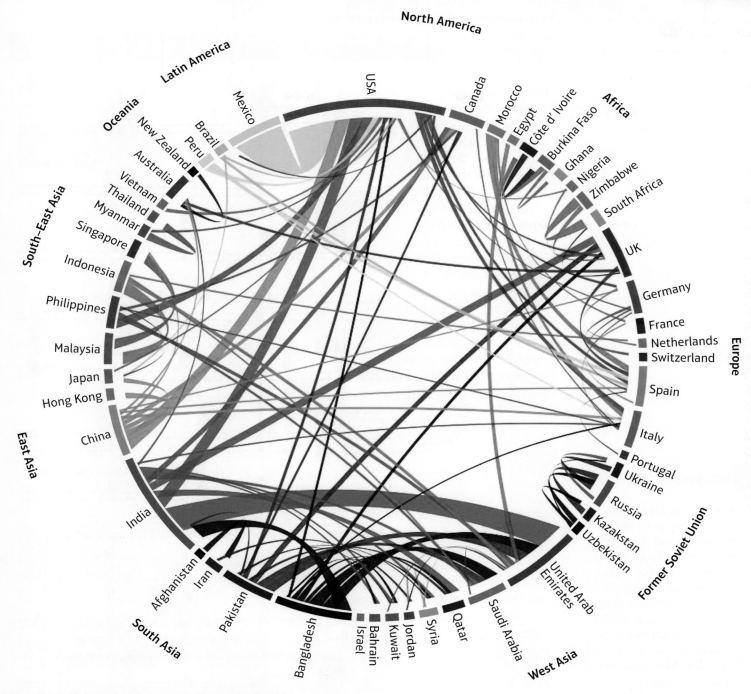

Figure 5.2: Global migration patterns from 2005 to 2010
Flows between the 50 countries that send/receive at least 0.5% of migrants, 2005–10. (Colour of movement indicates source country.)

send the most people to the developed countries; it is the emerging economies whose people are still relatively poor but have education and mobility. The largest regional migration is from South-East Asia to the Middle East, driven by oil wealth and construction booms, as well as a demand for house servants, although this may be slowing due to the recent economic recession and lower oil prices. The biggest flow between individual countries is the steady stream from Mexico to the USA. There is also a significant circulation of migrants among sub-Saharan African countries. This migration within the continent is much larger than the number leaving Africa, although mentions in the Western media create an impression that it is the other way around because immigration is a contentious issue in the EU.

Migration within countries

CASE STUDY: Rural–urban migration in China

According to the International Labour Organisation, since 1979 China has experienced the largest internal migration ever recorded. China has a total of 229.8 million rural migrant workers. Around 70 per cent of migrant workers are employed in China's eastern areas, with two-thirds of them working in large or medium-sized cities. Approximately two-thirds of these migrants work in manufacturing and construction industries, as indicated in Figure 5.3. It is clear that the global shift in manufacturing to South and East Asia has created demand for this **labour**.

Higher wages in secondary industries and the possibility of a better quality of life act as the pull factors for this migration. Also, the transfer of improved agricultural techniques and technologies made possible by China's increasing openness to the wider world has reduced the need for rural labour, depressing incomes and quality of life in rural areas, acting as a push factor for the migration. This sector of population is known as the 'floating population'. Chinese citizens are registered in their place of birth by the Hukou system: rural citizens do not have an urban Hukou, and cannot easily regularise their relocation to a city. They are often denied access to housing, health care and education as a result. Despite this official barrier, the pressures to migrate engineered by globalisation are so strong that it has been estimated that by 2025, a further 250 million Chinese will have migrated from their rural homes, taking the Chinese urban population beyond 1 billion.

Figure 5.3: Migrant workers at a building construction site in Chengdu, China

Literacy tip

A kleptocracy is a country that suffers systematic corruption and thievery by the state. Austerity refers to spending cuts and tax rises designed to reduce government debt. Peri-urban growth creates hybrid landscapes on the margins of developing cities that have rural and urban characteristics. Dysfunctional states and economies behave in abnormal and undesirable ways.

Yellow = Players, Orange = Attitudes and actions, Purple = Futures and uncertainties

CASE STUDY: Rural–urban migration in the Democratic Republic of the Congo

Rural to urban migration has also grown significantly in developing economies. The Democratic Republic of the Congo is the second largest nation in Africa, with a population of about 70 million, growing by nearly 3 per cent annually. It is a desperately poor country, decimated by three decades of civil war and poor governance. The United Nations Development Programme ranks the country 176th out of 188 countries in human development. The country has seen significant rural to urban migration since the 1990s, not as a result of of pull factors, but rather because the quality of life in rural areas has become so bad. The capital city, Kinshasa, has doubled in size every five years since 1950 and now has an estimated population of between 11 and 14 million. The absence of a reliable population census illustrates the low level of development in the country.

According to the African Development Bank, 72 per cent of rural households in DR Congo are poor. Nearly 40 per cent of children under five suffer from chronic malnutrition, and most of the population lives under conditions of moderate to serious food insecurity. The country received huge loans from the West during its early years of independence from Belgium (the early 1960s). However, the kleptocracy of President Mobuto Sese Seko squandered much of this on grand projects, and much of the revenue from the country's considerable mining wealth disappeared into bank accounts in tax havens abroad. The country could not repay the debts, and was forced into drastic austerity by structural adjustment programmes imposed by the IMF. The domestic economy collapsed and civil war ensued. Rural to urban migration has therefore been driven almost exclusively by push factors. Migrants are fleeing poverty and conflict in their home provinces, moving to peri-urban areas or cities in search of work. There is little or no formal employment in the cities, due to the dysfunction of the state and economy. The majority of migrants survive through informal employment such as street hawking. Wages are so low that researchers have coined the phrase 'the wage puzzle', because it is unclear how these populations are able to maintain themselves.

Migration between countries

Between 3 and 4 per cent of the global population lives outside their country of birth, but this proportion varies greatly between countries according to their different migration policies and levels of engagement with the global economy. For example, Singapore, an island state located between Malaysia and Indonesia, is the fifth most globalised country in the world according to the 2015 KOF Index of Globalisation. It has attracted large international migration flows, which it seeks to manage through regulations. By contrast, Japan is the 54th most globalised country, and has placed stricter limits on migration: restrictive immigration laws stop Japan's farms and factories from employing foreign labour, and stringent qualification requirements with complex rules and procedures shut out skilled foreign professionals. In 2008 Japan's Liberal Democratic Party called for Japan to accept at least 10 million immigrants, but opinion polls showed that most Japanese were opposed to this, and the Party went on to lose the election. A survey by the newspaper *Asahi Shimbun* showed that 65 per cent of respondents opposed a more open immigration policy.

ACTIVITY

Compare and contrast the causes of rural-to-urban migration in emerging economies and developing countries.

CASE STUDY: Singapore, international migration and the global shift

Singapore was established as a trading **colony** of the British Empire in 1819 and grew quickly, attracting labour from mainland China, India and the Malay Archipelago. However, Singapore's road to self-governance in the 1950s and 1960s saw the passing of new laws to limit immigration, including strict citizenship laws imposed after independence from Malaysia (1965); as a result the city-state's migrant population fell to just 3 per cent of the total. Since the 1980s Singapore has industrialised as part of the global economic shift, becoming one of the four East Asian Tiger economies, drawing fresh waves of migration. As a result, the population of Singapore can be divided into two categories of people:

- citizens (including naturalised citizens) and permanent residents who are regarded as 'residents'
- temporary immigrants (such as students and certain workers) who are considered 'non-residents'.

The non-resident population increased at an unprecedented pace in the first decade of the 21st century, according to the 2010 Singapore census: It accounted for 25.7 per cent of the total, up from 18.7 per cent in the previous decade. The majority of immigrants were born in Malaysia (386,000), China, Hong Kong and Macau (175,200), South Asia (123,500), Indonesia (54,400) and other Asian countries (90,100).

Two-thirds of Singapore's non-resident workforce are low-skilled migrants working in construction, domestic labour, services, manufacturing and marine industries. Since 2008 some have also been admitted to work specifically in bars, discotheques, lounges, nightclubs, hotels and restaurants. The remaining third are skilled, along with a small number of entrepreneurs. The size of this group has increased rapidly due to intensive recruitment and liberalised immigration criteria, while low-skilled migrants are maintained as a transient workforce by the visa regime.

Figure 5.4: The Singapore skyline

The termination of employment of a foreign-born worker results in the immediate termination of their work permit, in which case the immigrant must leave Singapore within seven days. Work permit-holders are also subject to a regular medical examination that includes testing for HIV/AIDS. They may not marry Singaporeans or other permanent residents without the approval of the controller of work permits, and failure to get approval may result in repatriation. Female work-permit holders (typically domestic workers) who, through the compulsory medical screening process, are found to be pregnant are also subject to repatriation without exception.

In contrast, highly skilled workers are encouraged to migrate to Singapore, which has rebranded itself as a culturally vibrant 'Renaissance city'. The skyline shown in Figure 5.4 reflects this image. In 2015 skilled workers and professionals accounted for 22 per cent of Singapore's total non-resident workforce, due to 1990s policies of recruiting from the 'non-traditional' source countries of the United States, UK, France, Australia, Japan and South Korea. The majority of skilled workers are from China and India. These workers have less restrictive work permits and can apply for permanent residency after two years.

As Singapore has globalised, so its own citizens have sought opportunities abroad. An estimated 192,300 Singaporeans live overseas, mainly in Australia (50,000), the UK (40,000), the USA (20,000) and China (20,000). Many Singaporean migrants are highly skilled and are employed in specialist sectors such as banking, information technology, medicine, engineering, and science and technology. Additionally, a significant proportion are students, some of whom have been sponsored by government scholarships and are obliged to return home upon finishing their studies.

Yellow = Players, Orange = Attitudes and actions, Purple = Futures and uncertainties

CASE STUDY: Japan, non-immigration policy

Japan has an ageing and shrinking workforce, which is likely to lead to a lower standard of living and reduced economies of scale. It has been suggested that the country needs 200,000 immigrants a year as well as a fertility rate above replacement level. However, the Japanese culture is based on a homogeneous population and racial unification and government policies reflect this. Many politicians and citizens believe the restrictions on immigration have brought harmony and cooperation to their society and that an introduction of large numbers of foreigners would disrupt society and increase crime.

The top answers to a 2014 opinion poll asking people how to solve the future labour shortage were to increase the number of working women and encourage the elderly to work. It has also been suggested that robots and robotics could help to fill the labour shortage. Encouraging immigration is well down the list, and so the long lasting government policy of restricting foreigners and access to Japanese nationality is still widely supported. In 2015 there were 790,000 foreign workers in Japan with about 40 per cent from China, and the foreign trainee programme has been expanded – especially to help with construction for the 2020 Tokyo Olympics. The latter has been criticised for exploiting a cheap workforce rather than fulfilling the original aim of giving skills to those from developing countries.

Patterns of international migration

The Lee model of migration (Figure 5.5) is one of several migration models. This model explains migration in terms of push and pull factors and intervening obstacles. Patterns of migration may change in response to environmental, economic and political change in both the home and host areas. Environmental factors might include natural hazards or the pollution of water sources; economic factors might relate to reduced employment in rural areas as agriculture is mechanised, or to an increased demand for labour as urban areas industrialise; political change can reduce barriers to migration, creating a pull factor, or may impose a barrier – such as the tiered levels of entry to the UK.

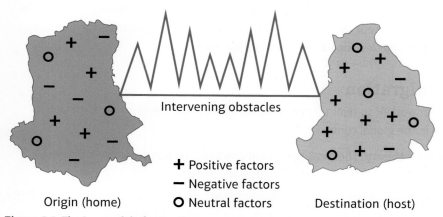

Figure 5.5: The Lee model of migration

The Lee model suggests that the decision to migrate is based on a balance of positive (push) and negative (pull) factors. When the balance of push and pull factors is strong enough to overcome any intervening obstacles, individuals will migrate. Both voluntary and forced migrations follow this pattern.

ACTIVITY

CARTOGRAPHIC SKILLS

Flow-line maps show how something moves from one place to another, and the scale of that movement. The flow lines are arrows showing the direction of movement, with the width of the arrow drawn to a scale to represent the volume of movement.

1. Investigate the United Nations Refugee Agency (UNHCR) website and review the latest interactive flow-line map showing migration within the Mediterranean region. Use search words – UNHCR Mediterranean refugee map 2016 (or current year).

2. On an outline map of the Middle East and Europe, draw a flow-line map to show Syrian migration using the data in Table 5.1.

    a. Draw a proportional arrow between the source and host countries in the appropriate direction. Choose a scale, such as 1 mm = 100,000 people. Be careful to place the arrows so that they do not overlap.

    b. Describe and explain the pattern shown by your completed map, referring to the reasons outlined in this section.

Table 5.1: Selected migrant flows from Syria to destination countries in 2015

Host country (migrant flow in '000s)		
Turkey (2,200)	Jordan (600)	Egypt (80)
Lebanon (1,000)	Iraq (250)	Sweden (80)

International migration patterns have changed over time. Some significant historical migrations included Europeans colonising the 'new world' of the Americas and Oceania, the forced migration of Africans to the Americas and the economic migration from developing countries to European countries. Economic motives for voluntary migration are always likely to be the strongest factor, such as Mexicans into the USA, but environmental and political factors are increasing in importance.

- Environmental factors include the impacts of climate change; which is making some areas drier so that water and food supplies are affected, while low lying coastal areas and small island states will flood as sea levels rise. Environmental refugee movements are already taking place, such as from Kiribati to New Zealand, and will increase greatly by the end of the century.

- The world is currently experiencing an unprecedented level of regional conflict, with fighting in several failed states. This unrest has caused a large increase in the number of internally displaced people; by the end of 2014 the top five countries were Syria (7.6 million), Colombia (6 million), Iraq (3.4 million), Sudan (3.1 million) and DR Congo (2.8 million). In addition the number of refugee movements to other countries had increased, such as from Syria to Europe (Figure 5.7). Some refugees are escaping persecution due to their ethnicity or religious beliefs, and therefore seek asylum in a new stable host country. Many make risky journeys, for example approximately 10,000 deaths occurred in the Mediterranean Sea between 2011 and 2015.

Most movements are from developing countries to developed countries, and this is increasing. In 1960 developed countries had 44.4 per cent of the migrant stock but by 2013 this had increased to 58.6 per cent. As more countries emerge economically this movement may slow as the push and pull factors will not be as strong, even though modern transport and communications has decreased the number of intervening obstacles. During economic recessions developed countries have been more reluctant to accept immigrants, however, with ageing populations further immigration will be needed by developed countries in the future.

The causes of migration

Most migrants move for work or to rejoin family members. Five main theories help explain the reasons for migration in the global context:

- **Neoclassical economic theory**: The most significant push/pull factors are wage differences, which cause migration flows from low-wage to high-wage areas.
- **Dual labour market theory**: Pull factors in developed countries bring migrant workers to fill the lowest-skilled jobs because the home population does not wish to do this work.
- **The new economics of labour migration**: Migration flows and patterns cannot be explained solely at the level of individual workers and the push and pull factors that cause migration; it is more complex, for example a household or social group in the source country may improve their living conditions by using remittances sent by family members working abroad.
- **Relative deprivation theory**: Awareness of income differences between neighbours in a source community is an important factor in migration. Successful migrants can afford better schooling for their children and better homes for their families, and so may serve as examples for others and encourage them to move.

Yellow = Players, Orange = Attitudes and actions, Purple = Futures and uncertainties

- **World systems theory**: Trade between countries may cause economic decline in some, encouraging people to move to a more successful country. Even after decolonisation, former colonies may remain economically dependent on the former colonial superpower, which may encourage migration along trading routes. However, **free trade** may encourage people to stay in their home country if they can earn higher wages from new production processes.

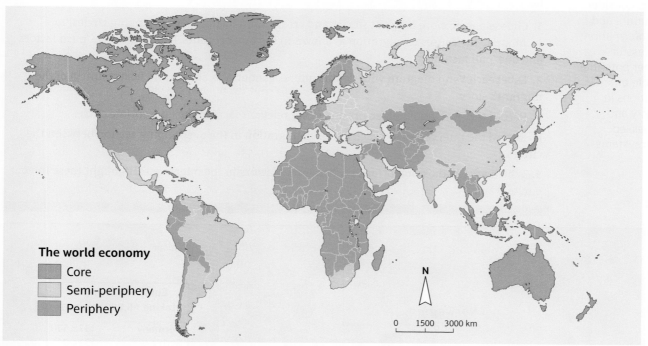

Figure 5.6: World systems theory: core, semi-periphery and periphery (2000)

Another significant cause of migrations is the enforced **displacement** of people as a result of conflict and poverty in their home regions. According to the UNHCR, in 2016 there were 16 million people of concern (refugees, internally displaced persons and asylum-seekers) in West, Central and East Africa, a further 15 million people of concern in the Middle East and 8 million in South Asia.

CASE STUDY: Displaced person migrations from Africa and the Middle East

The Syrian civil war, which began in 2011, has pushed millions of people to leave their homes (Figure 5.7). The countries bordering Syria remain at the forefront of the crisis, with more than 4 million Syrian refugees hosted by Egypt, Iraq, Jordan, Lebanon and Turkey. Inside Syria, a total of 12.2 million people remain in need of humanitarian assistance, including 7.6 million internally displaced people. Nearly 1.1 million Syrians have applied for asylum in Europe.

Continued pressure on host countries, and the difficulties of finding safety, have resulted in a rise in the number of Syrians seeking refuge beyond the Middle East and North Africa (MENA) region. By early 2016 more than a million refugees had undertaken the dangerous sea routes from Egypt, Libya and Turkey to Europe, and not all made it to their destination with many perishing at sea. However, this migration is part of a larger complex migration movement with growing numbers of asylum-seekers and refugees from the Middle East, sub-Saharan Africa and the Horn of Africa, with people journeying to the Gulf region, North Africa and the Mediterranean region. Failed states like Libya in North Africa, where fighting continues in many areas, make this situation worse.

In 2015 there were about 435,000 internally displaced people, and UNHCR provided humanitarian assistance to nearly 60,000 refugees. Escalating conflict in Yemen caused 2.3 million people to become internally displaced during 2015; 25 million are in need of humanitarian assistance and more than 100,000 have fled the country. Migrants are joining these flows from other countries, such as Nigeria, Pakistan and Eritrea. In 2014 Eritreans were reportedly among the most numerous of those attempting the risky crossing from North Africa to Europe by boat, and the UNHCR reported that nearly 40,000 Eritreans applied for asylum in 38 European countries in 2014, compared with about 13,000 in the same period in 2013. It puts the Eritrean refugee population at more than 321,000, which is almost 7 per cent of the country's total population.

Extension

Consider the extent to which migrant stories are based on perception or factual evidence. Consider especially the nature and importance of information flows from successful migrants to other potential migrants, the role of the living conditions in the home country and the role of globalised communication systems.

ACTIVITY

'I am a migrant' is a campaign run jointly by the International Labour Organisation and the Joint Council for the Welfare of Migrants to allow migrants to tell their migration stories in their own words. You can read these stories on their website, which can be researched by the home and host countries.

1. Choose one or more migrant stories and create Venn diagrams (overlapping circles) to summarise the causes of migration as shown in the oral accounts. Label one circle 'push factors and one 'pull factors'.

2. Label each Venn diagram with details (name, gender, age, source and host country) of the migrant's story.

3. Record the push and pull factors and other relevant facts in the appropriate circle.

4. Record the intervening obstacles of each migration in the overlapping section between the circles.

5. Which are the most important push and pull factors in the migration? Highlight these in your circles.

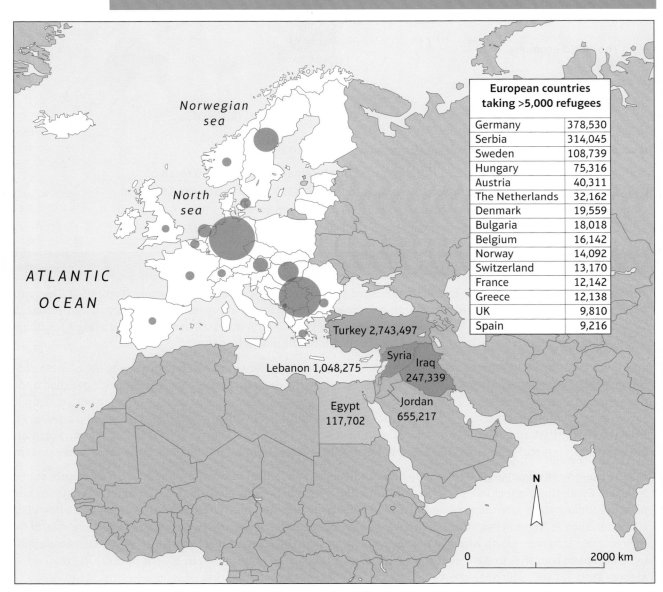

Figure 5.7: Syrian refugees and the countries taking them, cumulative figures to June 2016

Yellow = Players, Orange = Attitudes and actions, Purple = Futures and uncertainties

Challenges to national identity and sovereignty

The global population has almost tripled since 1950 but the growth rate is slowing as countries develop and average age increases, which means that in the future there could be labour shortages. The World Bank and the IMF reported in 2015 that half the world's population is in developed countries experiencing ageing populations and slower population growth; Germany has a declining population. Even countries like China will face ageing population issues by the middle of the century. World fertility rates are declining, but many developing countries face a legacy of late 20th-century population growth. In 1950 32 per cent of the global population lived in developed countries and just 7 per cent in sub-Saharan Africa: in 2015 developed countries accounted for only 17 per cent, while sub-Saharan Africa's share of the global population had doubled to 14 per cent. By 2050, the share of children in the global population is expected to have fallen to 21 per cent, in contrast to a peak of 38 per cent in the late 1960s, while the share of over-65s will triple from just 5 per cent in 1960 to 16 per cent in 2050, so migrations will be smaller.

Economic theory suggests that maximum efficiency comes from free trade, **deregulated financial markets**, and open borders for movement of workers: people should be able to move freely to meet skilled labour needs, minimise wage inflation, and enable a flexible and dynamic economy. Countries with an ageing population therefore require the immigration of younger skilled workers, who can also add **cultural heterogeneity**. In 2015 the World Bank President Jim Yong Kim said: 'If countries with ageing populations can create a path for refugees and migrants to participate in the economy, everyone benefits.'

Costs for developed countries from immigration may include increased population density, traffic congestion and higher taxes for spending on public services, housing, education and health. The increased supply of labour may depress or stagnate wages, leading to a fall in the standard of living. Some believe that families of immigrants just bring their own culture, rather than adopting the host culture. Community groups may exist separately, without shared loyalties, posing challenges for national identity and **sovereignty**.

The movement of labour within countries

The movement of labour is unrestricted within many countries, to ensure efficient allocation of this resource. Migration flows within countries are usually far greater than immigration from other countries. For example, according to ONS census data, of the total UK population in 2011 (63.2 million), 7.5 million (12 per cent) had a different address compared to a year previously, but for only 1 per cent (687,200) had this been outside the UK. Two-thirds of internal migrants did not move out of their home local authority area. In 2014 the Centre for Cities reported that 35 per cent of jobs in London required candidates to be educated to degree level or higher, and that more than a third of graduates from all UK universities migrate to London in their 20s. This internal migration within the UK enables London employers to recruit from a deep pool of skills and maintain the capital city's economic dynamism.

Supranational groupings such as the European Union have encouraged migration between member countries for the benefit of the European community.

ACTIVITY

Explain how immigration and the free movement of labour may pose challenges for national identity.

Synoptic link

Migration has an important role in determining the diversity of places, not only in cities such as London but also in rural towns such as Boston, in Lincolnshire. Globalisation has given migrants a new perspective on the world, with internet access, mobile communications and the media providing information to allow migrants to make decisions. London is a world city, especially globalised in the financial sector and as a home for wealthy groups, both of which create other job opportunities. (See Book 1, pages 193–194 and this book page 277.)

CASE STUDY: Migration within the EU and the Schengen Area

With modern transport and communications, globalisation has increased migration within and between wealthier nations. The EU has a 'single market' based on the free movement of goods, capital, services and people – the 'four freedoms' – between its 28 member states. The increasing interconnection of EU countries has created many incentives for people to migrate for economic and family reasons. There are also significant temporary migration and tourism flows within the EU. According to a 2010 study by Oxford University:

- EU migrants accounted for 35 per cent of the total migrant stock in EU countries as a whole
- in Luxembourg 80.7 per cent of all migrants were from within the EU
- the EU countries with the largest numbers of EU migrants were Germany (3.7 million), Spain (2.5 million), France (2.4 million), the UK (2.2 million) and Italy (1.2 million)
- the EU countries with the largest number of people living in other EU countries were Romania (2.3 million), Poland (1.9 million), Italy (1.7 million), Germany (1.5 million) and the UK (1.4 million)
- the EU countries hosting the largest number of British people in 2010 were Spain (411,074), Ireland (397,465), France (172,836) and Germany (154,826)
- the EU states with the largest numbers of migrants in the UK were Poland (521,446), Ireland (422,569), Germany (299,753) and France (128,010).

Migration is even easier in the 26 countries of the Schengen Area, which have abolished passport and border controls so that they function as a single country for travel purposes. However, since February 2016 Austria, Denmark, France, Germany, Norway and Sweden have imposed controls on some or all of their borders with other Schengen states, as a result of increased migration from Syria, Iraq and Afghanistan and in the aftermath of the Paris (2015) and Brussels (2016) terrorist attacks.

The consequences of international migration

Migration changes the cultural and ethnic composition of countries. According to the ONS, the UK population increased by 4.1 million (nearly 7 per cent) between 2001 and 2011, when 50 per cent of the current foreign-born population of the UK arrived, many from the eastern European countries. In 2011 only a quarter of the foreign-born UK population had been resident in the UK for more than 30 years.

The success of assimilation of migrants varies from nation to nation, and often depends on ethnicity. One measure of assimilation is proficiency in the native tongue of the host nation. The ONS (2011 census) identified that 89 per cent of the non-UK-born population were proficient in the English language. However, proficiency was lower among some ethnic groups, for example the Bangladeshi-born population (70 per cent), and varied little by length of residence. The lowest level of proficiency was among Chinese-born who had been resident in the UK for more than 30 years (62 per cent).

The amount of time that migrants are resident in the host nation also has an impact on assimilation. For example, the ONS also identified that 77 per cent of those resident in the UK for more than 30 years reported having a UK identity, compared to only 10 per cent for recent arrivals (2007–11). Family reunification is permitted in many migration schemes, although this can act as a barrier to assimilation, by giving families incentives to use marriage to work around the European immigration system. In Norway, for example, the proportion of cousin-marriages within the Pakistani immigrant community is greater than in Pakistan itself.

Yellow = Players, Orange = Attitudes and actions, Purple = Futures and uncertainties

CASE STUDY: Labour flows across the Mexico–US border

In 2015 the US Census Bureau estimated the US population to be 323 million, of which 45 million were born abroad, about 14 per cent of the total. Hispanic migrants from Mexico are the largest proportion of these, at 47 per cent. It is estimated that 11 per cent of Mexicans reside abroad, 98 per cent of them in the USA. Migration from Mexico to the US has increased significantly since the 1960s, mainly because of the better economic opportunities across the border. Traditionally, most migrants were young men who found manual work in the USA, often in agriculture. They sent remittances (in US dollars) to their families back in Mexico.

Despite border controls to stem the flow of illegal migration, between 2000 and 2010 the US's Hispanic population increased by 43 per cent, while the non-Hispanic population rose by just 4.9 per cent. This is partly due to migration, but also to higher natural growth within this population group.

Figure 5.8: Migrants attend a mass by the Pope by the border fence separating US and Mexico on 17 February 2016

A level exam-style question

Explain how supranational groups manage migration both within and between nations. (6 marks)

Guidance

'Explain how' means that you need to identify the methods that groupings of countries such as the EU are able to use to encourage or restrict movements of people. This is a short-answer question, so avoid extended case study examples, but do include a few facts.

ACTIVITY

TECHNOLOGY/ICT SKILLS

Compare and contrast opinions on the contribution of migrants to the culture and social life of two contrasting countries.

Information sources could be migration stories in different newspapers, chosen for contrasting political perspectives: in the UK, compare coverage of a migration story from the *Daily Mail* and *The Guardian*; for Australia, use *The Australian* and *Green Left Weekly*; or for the USA, use *The Wall Street Journal* and *National Review*. Crowd-sourced data sites could be used as well as or instead of newspapers.

Contrasting responses to migration

Migration causes **political tensions** because of different perceptions of its impacts (social, economic, cultural and demographic). For example, Ipsos MORI conducts a monthly poll asking respondents to name the most important issue facing the UK and any other 'important' issues. Respondents are not prompted and reply purely from their lived perception. Immigration ranks consistently among the top five issues: in June 2015 it was the issue given most often by respondents (45 per cent). The 2013 British Social Attitudes Survey found that 77 per cent of respondents believed that immigration should be reduced either 'a lot' or 'a little'. Though as Figure 5.9 shows, worldwide people perceive the number of migrants to be much higher than it is.

Globalisation could be considered to have eroded the traditional power of the state and players such as unions to protect citizens from rapid social change. This may lead to insecurity at work, inequality within communities, and a loss of confidence in governments to manage the impacts of migration. However, a 2010 survey by Transatlantic Trends found that 72 per cent of the UK public

1. Suggest what Figure 5.9 tells us about peoples views of immigration.

2. Explain any differences between different world regions.

supported admitting more doctors and nurses from other countries to cope with increasing health care demands, and 51 per cent supported admitting more care workers to help manage the ageing population. These views form part of multiculturalism. In *The Guardian* in January 2005, Professor Bhikhu Parekh wrote:

'No culture represents the last word in human wisdom. It articulates a particular vision of human life, develops a particular range of human capacities and emotions, and marginalises others… Multiculturalism is … not about shutting oneself up in a communal or cultural ghetto and leading a segregated and self-contained life. Rather it is about opening up oneself to others, learning from their insights and criticisms, and growing as a result into a richer and tolerant culture.'

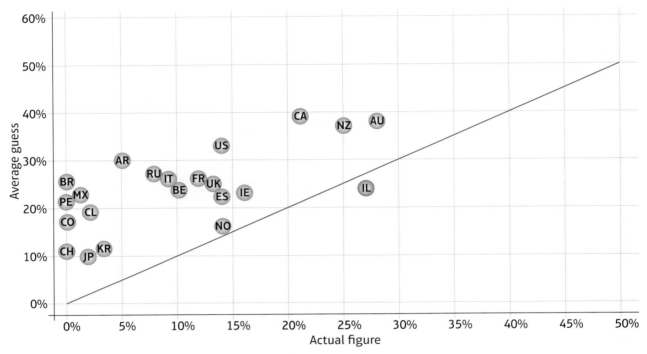

Key: BR = Brazil (0,25), PE = Peru (0,21), CO = Columbia (0,17), CH = China (0,11), MX = Mexico (1,22), CL = Chile (2,19), KR = South Korea (3,11), JP = Japan (2,10), PL = Poland (2,9), AR = Argentina (5,30), SA = South Africa (5,29), RU = Russia (8,27), IT = Italy (9,26), BE = Belgium (10,24), NE = Netherlands (12,25), DR = Germany (12,26), UK = United Kingdom (13,25), US = United States (14,33), ES = Spain (14,22), NO = Norway (14,16), SW = Sweden (16,25), IE = Ireland (16,23), CA = Canada (21,39), NZ = New Zealand (25,37), AU = Australia (28,38), IL = Israel (27,24)

Figure 5.9: Perceptions of migration, 2015, 'What percentage of our population are immigrants?'

Unequal controls on international migration

The ability of people to migrate across national borders varies according to their levels of skill and income. For example, the UK regulates migration from countries outside the European Economic Area with a points-based system. There are five tiers:

- Tier 1 relates to entrepreneurs, investors and highly skilled workers. This category promotes migration to the UK of wealthy entrepreneurs (investors require at least £2 million, for example) and highly skilled persons.

- Tier 2 relates to migrants with a job offer with a UK-based employer. Migration under this tier is currently limited to 21,700 people per annum. Level of salary is seen as a proxy for the level of skill required to do the job, and extra points are awarded to migrants expected to earn the highest salaries.

Yellow = Players, Orange = Attitudes and actions, Purple = Futures and uncertainties

- Tier 3 was designed for low-skilled workers filling specific temporary labour shortages, but it is currently suspended because of the numbers of eastern Europeans able to do these jobs.

- Tier 4 relates to foreign students studying at a UK-based higher education institution. Migration under this tier is limited to the duration of the course of study.

- Tier 5 relates to temporary workers and youth mobility schemes such as gap-year students and is normally limited to six months, or at most one year.

The USA has a similar system of limiting international migration that advantages applications based on the wealth and skills of potential migrants, and Australia also has a points-based migration system that prioritises elite migration and discourages low-skilled migration.

However, some states do not have full control over their borders, which allows irregular migration flows, often facilitated by people traffickers. One reason may be internal conflict: since the defeat of the Gaddafi regime in 2011, Libya has been divided by rival armed militias affiliated to regions, cities and tribes, while the central government has been unable to exercise authority over the country. As a result, Frontex estimated that in 2015 half a million migrants and asylum-seekers attempted to travel from Libya to the EU. Another reason may be the physical characteristics of a border that make it difficult to secure, as in Greece, which has the longest coastline in the Mediterranean, at 13,676 km. This coastline includes thousands of islands, of which only 227 are inhabited; Greece has found it impossible to control this border in the face of the migration by those escaping the instability and conflict in the Middle East. Many countries have erected border fences or walls to try to control migration (Table 5.2).

A level exam-style question

Explain why migration in the 21st century is increasing both within and between countries. (8 marks)

Guidance:
Identify the reasons why migration is increasing, perhaps under categories such as economic, socio-cultural, globalisation forces and government policies. Use named examples to support the points you wish to make, but remember that you will not have space to develop these into full case studies.

Table 5.2: Selected examples of worldwide border controls, 2015

Country	Against	Announced	Begun	Completed	Stated reasons
Brazil[1]	Bolivia		2007		immigration, trafficking
	Paraguay		2007		immigration, trafficking
	Argentina	2013			immigration, trafficking
	Colombia	2013			immigration, trafficking
	French Guiana	2013			immigration, trafficking
	Guyana	2013			immigration, trafficking
	Peru	2013			immigration, trafficking
	Suriname	2013			immigration, trafficking
	Uruguay	2013			immigration, trafficking
	Venezuela	2013			immigration, trafficking
China	North Korea		2006	2012	territorial
Greece	Turkey		2012	2012	immigration
India	Bangladesh		1989		immigration
	Pakistan		1990	2004	security
	Myanmar	2003			security, smuggling
	Bhutan	2015			security

Country	Against	Announced	Begun	Completed	Stated reasons
Israel	Palestinian Territories (Gaza)		1993		security
	Lebanon		2001		immigration
	Egypt		2010	2013	immigration
	Palestinian Territories (West Bank)		2012		security
	Syria		2013		immigration, security
	Jordan	2015			immigration, security
Saudi Arabia	Yemen		2003	2004	immigration, security
	United Arab Emirates		2005		security, smuggling, territorial
	Oman		2005		immigration
	Iraq		2006	2014	security, trafficking
	Qatar	2014			immigration
UK	France (Calais & Coquelle)		2015	2015	immigration
USA	Cuba		1959	1964	territorial
	Mexico		2006		immigration, smuggling

[1] The majority of Brazil's proposed barrier comprises a non-physical 'virtual' wall, monitored by drones and satellites.

Yellow = Players, Orange = Attitudes and actions, Purple = Futures and uncertainties

How are nation states defined and how have they evolved in a globalising world?

Learning objectives

8B.4 To understand that nation states are varied and have different histories.

8B.5 To understand how the modern world has been shaped by nationalism.

8B.6 To understand how globalisation has led to the deregulation of capital markets and the emergence of new state forms.

What is a nation state?

A **nation state** is a political entity that has **sovereignty** over its territory (clear boundaries), authority to govern without outside interference, and recognition by other countries. Historically, states have taken many forms, such as self-governing cities, kingdoms, empires and confederations of urban and rural areas. However, the idea of a nation state is a cultural one, a collective identity

---

CASE STUDY: Iceland and Singapore compared

Iceland and Singapore are both island states. Singapore has a population of 5.5 million crammed on to a territory of only 720 km², whereas Iceland has a much larger territory at 103,000 km² but a population of only 330,000.

Singapore is considered a city-state (a city with dependent territories) rather than a nation state, with a global commercial, financial and transport hub (entrepôt trade). Its founding principles are meritocracy, multiculturalism and secularism. Singapore is culturally and ethnically diverse: the 2010 census recorded that 48 per cent of the population were foreign-born. Linguistic differences further emphasise this diversity: 50 per cent of the population speak Mandarin as their native tongue, 32 per cent English, 12 per cent Malay and 3 per cent Tamil.

Iceland, on the other hand, is considered a nation state because its national culture and language are homogenous and unique. The inhabitants are ethnically related to other Scandinavian groups, but there are a few cross-border minorities. Mainland Europe is too far away to have a strong influence; in fact there is greater influence from the USA. An example of cultural homogeneity is the custom of using patronymic or matronymic surnames – patronyms being more common. Patronymic last names are based on the first name of the father, while matronymic names are based on the first name of the mother, and these follow the person's given name, for example Elísabet Jónsdóttir ('Elísabet, Jón's daughter' (Jón being the father)) or Ólafur Katrínarson ('Ólafur, Katrín's son' (Katrín being the mother)). Consequently, Icelanders refer to one another by their given name, and the Icelandic telephone directory is listed alphabetically by first name rather than by surname. All new names are decided by the Icelandic Naming Committee.

derived from a shared history or ethnicity. Nation states vary greatly in their ethnic composition and cultural and linguistic heritage, which all are the outcome of their unique history, demography, location and physical geography.

The United Kingdom was formed by the union of England, Scotland, Wales and Northern Ireland, but separate treaties and agreements, such as devolution, mean that the state is not uniform. The Treaty of Union (1707) between England and Scotland ensured the continuation of separate legal systems and separate national churches, and in 2003 the British Government described the United Kingdom as 'countries within a country'. Japan is the world's largest nation state, with 127 million people. Its four large islands (Honshu, Hokkaido, Kyushu and Shikoku) contain about 97 per cent of its land area, and there is little cultural or ethnic diversity.

Figure 5.10: Traditional Icelandic turf houses, an example of a unique culture

> **Maths tip**
>
> Divided bar graphs can be constructed using percentages or original raw data. The Activity on this page asks for percentages to be used, but you could use the population numbers, however, with such a large range you could use a logarithmic scale with single log graph paper. This would give a logarithmic scale for the vertical axis, where each level multiplies by a factor of 10 (e.g. 10; 100; 1,000; 10,000; 100,000; 1,000,000; 10,000,000; 100,000,000). This is perhaps more common with line graphs where there is a large change over time, such as the world's population growth, and here you can use double log graph paper with log scales on both axes.

ACTIVITY

GRAPHICAL SKILLS

In a divided bar graph, the length of each section is proportionate to the value of an item of data. The value of the data in each section is found by measuring the length of the section and comparing it to the relevant axis. When divided bar graphs are placed side by side, it is possible to compare the data sets and see several differences or changes.

1. Study Table 5.3. Calculate the percentage of the total population represented by each ethnic group for the UK and Japan.

2. Using graph paper, create a bar for each country that has a scale from 0 to 100, with a vertical scale alongside in percentages.

3. Create sections within the bars that represent the proportion of the population from each ethnic group.

4. Label each section (if the section is too small, put the label alongside). Title the graph and bars.

5. Using your completed graph, compare the ethnic diversity of Japan and the UK.

Yellow = Players, Orange = Attitudes and actions, Purple = Futures and uncertainties

Table 5.3: Population of Japan and the UK in 2016 by major ethnic group

Japan (total population in '000s: 127,000)		UK (total population in '000s: 64,716)	
Ethnic group	**Population**	**Ethnic group**	**Population**
Japanese	125,095	White	56,368
Korean	635	Asian	4,530
Chinese	508	Black	1,941
Filipino	254	Mixed	1,294
Brazilian	254	Other	583
Vietnamese	127		
Other	127		

> **Literacy tip**
>
> Meritocracy is a political philosophy that believes that those with the best skills should hold power. Multiculturalism is the presence, acceptance and promotion of multiple cultures within a country. Secularism is the independence of government from religious influences or power.

The origins of national borders

National borders are very often linked to physical geography, such as lakes, rivers, coastlines and the watershed (divide) in mountains, because these are more easily definable than open land. For example, the border between the UK and France is the middle of the English Channel, and long lengths of the Franco-German border are in the middle of the River Rhine. However, borders may also be the legacy of historical events. One example is the border between the Republic of Ireland and Northern Ireland, which does not follow natural boundaries in the physical landscape but is a 499-km border from Lough Foyle to Carlingford Lough that was negotiated on the partition of Ireland in the early 1920s.

Many borders are the result of colonial history, drawn by foreign colonial powers that took no account of the location of local ethnic or religious groups. Many boundaries ignored physical and social geography and were straight lines drawn on a map, for example parts of the USA–Canadian border, Guatemala in Central America and Mali in West Africa. These create difficulties for nation states and often lead to ethnic tensions and conflict.

CASE STUDY: The colonial heritage of Iraq's borders

The region between the Tigris and Euphrates rivers, historically known as Mesopotamia, has been called the cradle of civilisation. It was here that mankind first began to read, write, create laws, and live in cities under organised government – notably Uruk, from which Iraq was derived. The area has been home to successive civilisations since the sixth millennium BC. At different periods in its history, Iraq was the centre for the indigenous Akkadian, Sumerian, Assyrian and Babylonian empires. It was also part of the Median, Achaemenid, Hellenistic, Parthian, Sassanid, Roman, Rashidun, Umayyad, Abbasid, Mongol, Safavid, Afsharid and Ottoman empires. Iraq's modern borders were mostly demarcated in 1920 by the League of Nations when the Ottoman Empire was divided by the Treaty of Sèvres. Iraq was placed under the authority of the UK as the British Mandate of Mesopotamia. A monarchy was established in 1921 and the Kingdom of Iraq gained independence from Britain in 1932, inheriting its present borders from this historical process.

The decisions that led to these borders have their origins in the Sykes-Picot Agreement of 1916, a secret agreement between the UK and France that had the assent of the Russian Empire (Figure 3.25). These countries were at war with the Ottoman Empire at this time and wished to carve up the Middle East into their respective spheres of influence. Since independence in 1932 Iraq has suffered coups, civil conflict, authoritarian rule and regional wars. Some see the incorporation of three distinct ethnic groups within these colonial borders as a root cause of this conflict (Figure 5.11). Ongoing intermittent conflict between Sunni, Shiite and Kurdish factions has led to increasing debate about the splitting of Iraq into three autonomous regions, including Kurdistan in the north-east, a Sunnistan in the west and a Shiastan in the south-east. There have also been two modern wars involving the superpowers, in 1990 and 2003, and the different ethnic groups continue to fight both within the country and beyond.

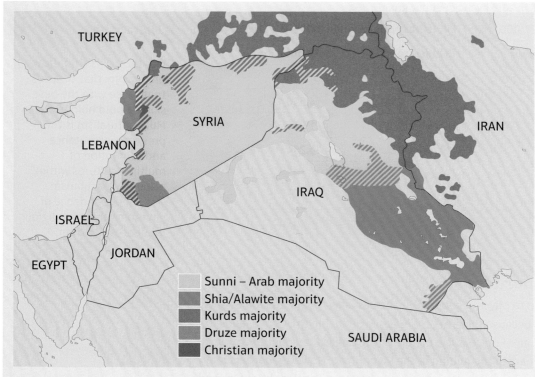

Figure 5.11: Syria and Iraq's main ethnic groups

Map legend:
- Sunni – Arab majority
- Shia/Alawite majority
- Kurds majority
- Druze majority
- Christian majority

CASE STUDY: Rwanda

The divisions within Rwanda pre-date colonisation but were exacerbated by it. Early in the 19th century the Tutsi king established the borders, controlling the majority Hutu through cattle ownership and fighting skills, before the first Europeans from Germany arrived in 1894. Colonisation of this part of Africa was disputed by Germany, Belgium and the UK but in 1910 it was agreed that Germany would control Rwanda and they allowed the existing structure to continue. During the First World War Belgium gained control of the country with official supervision status from 1923. After the formation of the UN, Belgium was given trusteeship and they restricted Tutsi power but a 1952 development plan for Rwanda left the Tutsi minority in control. Unrest between the Hutu majority and Tutsi minority increased, necessitating a state of emergency to be declared and Belgium used military forces to restore order. It was thought by Belgium and the UN that Rwanda could be merged with Burundi, but it soon became apparent that this would lead to a high level of conflict. In 1962 the UN terminated Belgium's trusteeship effectively making Rwanda independent with the Hutus gaining control. Many Tutsis fled the country because of fear of persecution.

In 1973 there was a coup led by Habyarimana, who became President, and the constitution was suspended and only one political party allowed. A civil war began in 1990 and in 1992 talks began on peace and a political restructure between Habyarimana (Hutu) and the Rwandan Patriotic Front (RPF) (Tutsi). When the President's plane was apparently shot down in 1994 there was a genocide, with between 0.5 and 1 million people killed (mostly Tutsis). The superior forces of the RPF ended the genocide by taking over the country. Unrest continues today, with Human Rights Watch reporting many human rights abuses with many refugees and asylum seekers.

Yellow = Players, Orange = Attitudes and actions, Purple = Futures and uncertainties

Contested borders and conflict

There are many contested borders in the modern world, including those of Western Sahara, Kashmir and the seabed under the Arctic Ocean. This may be due to the desire of one state to absorb the territory of another, the desire to unite a culturally and ethnically similar population from neighbouring countries, or to gain access to valuable resources. Any threats to territorial sovereignty are likely to lead to instability and possibly conflict, especially if superpowers take a strong interest.

Not all nation states are universally recognised. For example, Taiwan is officially called the Republic of China. The original Republic of China was established in 1912 in mainland China; however, by 1949 it had lost all of its mainland territory in the Chinese civil war, when the Communist Party took full control of mainland China and founded the People's Republic of China (PRC). The ROC continued to represent China at the United Nations until 1971, when the PRC assumed China's seat and Taiwan (ROC) lost UN membership. International recognition of the ROC gradually eroded as most countries switched their 'China' recognition to the PRC. Twenty-one UN member states and the Holy See currently maintain official diplomatic relations with the ROC, while many other countries maintain unofficial ties through offices at institutions that function as de facto embassies and consulates, and through trade. However, the PRC continues to claim that it represents the sole legal government of 'China' and that Taiwan is China's 23rd province. This stance denies Taiwan recognition as a sovereign state, and the PRC has threatened to use military force against any formal declaration by Taiwan of national independence.

ACTIVITY

Assess the extent to which contested borders are a major cause of conflict in the modern world.

CASE STUDY: Conflict in Ukraine

Ukraine gained independence from the Soviet Union in 1991. Ukraine has a Slavic culture, its own language and Orthodox Church. However, the territory itself has been contested for hundreds of years. Many empires have ruled and divided the current land, including Lithuania, Poland, the Ottoman Empire, Austria-Hungary and Russia. Within modern Ukraine's borders there are peoples with strong cultural, ethnic and linguistic connections to neighbouring countries, especially Russia. This partly explains why Ukraine is currently in territorial dispute with Russia over the Crimean peninsula and eastern Ukraine, which Russia annexed in 2014, but which Ukraine and most other countries recognise as part of the Ukraine.

Following independence, Ukraine declared itself a neutral state, but nonetheless formed a limited military partnership with the Russian Federation and a partnership with NATO. In the 2000s, the government began leaning towards NATO, signing a NATO-Ukraine Action Plan in 2002. In 2013 protests broke out in Kiev after President Yanukovych's government suspended the Ukraine-European Union Association Agreement and sought closer economic ties with Russia. The street protests escalated into the 2014 Ukrainian revolution, resulting in the overthrow of the government and the establishment of an EU-friendly government. These events precipitated the annexation of Crimea by the Russian Federation in February 2014, and the war in Donbass in March 2014 (Figure 5.12) – a conflict that was still not resolved midway through 2016.

Figure 5.12: Ukrainian border guards patrol, July 2015, along the barbed wire fence on the Senkivka border post, around 200 km north of the Ukrainian capital Kiev. The fence marks the start of a project to seal the 2,000-km frontier with Russia.

Nationalism and the modern world

Nationalism is the shared feeling for a special, significant geographical area; this may be expressed by political identification with and a sense of belonging to a nation. Nationalism may

CASE STUDY: The British Empire and nationalism

The British Empire had its roots in piracy, conflict and trade, rather than in nationalism. In the 17th and 18th centuries wars with the Netherlands and France left England (and then, after the union of England and Scotland in 1707, Britain) the dominant colonial power in North America and India. When the Thirteen Colonies in North America gained their independence in 1783, Britain lost some of its oldest and most populous colonies and resources. British attention then turned towards Asia, Africa and the Pacific. After its defeat of France in the Revolutionary and Napoleonic Wars (1792–1815), Britain emerged as the principal naval and imperial power of the 19th century.

At the same time in Britain, an integrated, country-wide economy was developing. People began to identify with the country at large, rather than with the smaller units of their family, town or region. National symbols, anthems, myths, flags and narratives were adopted and the construction of the Empire came to be seen as a national mission. British dominance was later described as Pax Britannica (Figure 5.13). The abolition of the slave trade in the Empire in 1807, followed by the abolition of slavery in the Empire in 1833, were examples of the civilising aspects of the British mission to the world.

Figure 5.13: An elaborate map of the British Empire in 1886

As the Industrial Revolution began to transform Britain, the Empire expanded to include India, large parts of Africa and many other territories. A missionary movement to spread Christianity was an element of the emigration from Britain to her colonies. Rudyard Kipling captured the sense of national mission in his 1899 poem 'The White Man's Burden' – which reflects how **colonialism** was seen as improving and civilising the wider world.

Yellow = Players, Orange = Attitudes and actions, Purple = Futures and uncertainties

be linked to a particular ethnic, cultural or religious group, or may embrace multiple minorities that express and exercise a collective identity. In the 19th century, a wave of romanticised nationalism swept through Europe, linked to the wealth that was the result of industrial revolutions transforming some countries and overturning established structures of power. An exponent of the modern nation state was the German philosopher Friedrich Hegel (1770–1831), who said that a sense of nationality was the cement that held modern societies together in an age when dynastic and religious allegiance was in decline. New countries such as Germany, Italy and Romania formed by uniting various regional states with a common national identity, providing security and stability. The identities of Greece, Poland and Bulgaria were formed after they won independence from multinational empires, although initially their nationalism was a source of conflict.

In the 19th century some European countries built global empires, and competing nationalism was an important factor in this process. This expansionist nationalism was demonstrated in an extreme form by Nazi Germany, when it expanded into parts of Eastern Europe, believing that the German people needed more 'living space' at the expense of Slavic populations. After 1918 a new variant of nationalism emerged, liberal nationalism, as shown by US President Woodrow Wilson's 14-point plan for a post-war order in Europe. The main principle of liberal nationalism was the right of national self-determination, which then led to anti-colonial nationalism in the late 1940s and the process of decolonisation, which sometimes involved armed insurrection and sometimes peaceful protest.

Independence movements since 1945

By the 20th century the ideals of European nationalism had been exported worldwide, perpetuated through the education of the colonial elites, but after 1945 the idea of national self-determination in the colonies began to threaten the imperial powers. At first they resisted decolonisation but the human and financial cost of fighting two world wars removed their capacity to maintain their empires against local and international opposition.

The post-war world was dominated by the United States of America (USA) and the Union of Soviet Socialist Republics (USSR), neither of which was sympathetic to old Europe's empires. The USA, which had fought its own war of independence against the UK, regarded itself as the leader of the free world, bringing liberty to oppressed peoples through decolonisation. The communist USSR regarded empires as a form of oppression of the proletariat, so they also were in favour of colonies achieving independence, especially as it would allow them to exert their new superpower status more easily around the world.

Independence movements came in two broad categories:

- those won by non-violent protest, as in India
- the result of violent revolutions or wars of independence, as in Angola (1961–74) against Portugal.

CASE STUDY: India's road to independence

India had had a movement for home rule or independence since at least the early part of the 20th century. Gandhi, the main leader of this movement, taking advantage of the post-war chaos in Europe encouraged many Indian citizens to view the British as the cause of India's problems. He inspired a newfound nationalism among its population, gathering the support needed to push the British out and create an independent India in 1947.

Timeline of independence and the creation of India, Pakistan and Bangladesh

1858: India comes under direct rule of the British crown after failed Indian mutiny

1885: Indian National Congress founded as forum for emerging nationalist feeling

1920–22: Nationalist figurehead Mahatma Gandhi launches anti-British civil disobedience campaign

1942–43: Congress launches 'Quit India' movement

1947: End of British rule and partition of subcontinent into mainly Hindu India and Muslim-majority state of Pakistan

1947–48: Hundreds of thousands die in widespread violence after partition

1951–52: Congress Party wins first general elections under leadership of Jawaharlal Nehru

1962: India loses brief border war with China

1965: Second war with Pakistan over Kashmir

1966: Nehru's daughter Indira Gandhi becomes prime minister

1971: Third war with Pakistan over creation of Bangladesh, formerly East Pakistan.

A level exam-style question

Suggest how nationalism has shaped the identity of modern nations. (6 marks)

Guidance

Identify briefly what nationalism is but concentrate on its features and make links to examples of how the identity of selected countries has been shaped. Try to choose examples from more than one continent.

The 'wind of change'

Most British colonies in Africa became independent nations in the 1960s. The historical process of creating the new nation states has become known as the 'wind of change', from an address made in February 1960 by British Prime Minister Harold Macmillan to the Parliament of South Africa in Cape Town. He had spent a month in Africa visiting a number of British colonies, and his speech signalled that the British Government intended to grant independence to many of these territories. Seventeen African countries achieved interdependence in 1960. In his speech, Macmillan said:

'The wind of change is blowing through this continent. Whether we like it or not, this growth of national consciousness is a political fact.'

The speed and extent of decolonisation can be seen in the growth of the United Nations: from 51 member states in 1945 to more than 120 by the end of the 1960s, mostly newly independent former colonies. The newly independent states derived moral purpose and political direction from the events of 1960 with far-reaching effects on Africa.

Yellow = Players, Orange = Attitudes and actions, Purple = Futures and uncertainties

Post-colonial conflict

The process of decolonisation led to conflicts that were costly – environmentally, economically and in human life. For example, West Africa alone has experienced seven post-colonial conflicts in the last 50 years: the Nigerian civil war (1967–70), two civil wars in Liberia in 1989 and 1999, a decade of fighting in Sierra Leone (1991–2002), the Guinea-Bissau civil war (1998–99) and a recent conflict in Côte d'Ivoire (2002–07 and 2010–11).

Extension

Investigate the current situation in South Sudan, and the roles of IGOs such as the UN and the African Union. and the internal groups.

CASE STUDY: South Sudan

South Sudan is the world's newest country, having been formed from the ten southernmost states of Sudan. The country was established in 2011 as an outcome of a 2005 agreement that ended the civil war in Sudan. Sadly, independence has not brought peace: conflicts between the country's 60 major ethnic groups has led to a civil war that erupted in 2013, displacing 2.2 million people. A long road to negotiated peace may be necessary, reflecting the independence struggles in other African nations.

CASE STUDY: Post-colonial conflict in Vietnam

The Indochina Peninsula was colonised by France in the mid-19th century and then briefly occupied by Japan in the 1940s. The Vietnamese people fought French rule in the First Indochina War, eventually expelling the French in 1954, but this divided the country into two rival states, North and South Vietnam. The North allied itself with the communist states of USSR and China, seeing its post-colonial future as a communist country. The South saw its post-colonial future as a capitalist economy and received support from the USA to prevent communist influence from the North. Conflict between the two sides intensified, with heavy intervention from the United States, in what became the Vietnam War. The war ended with a North Vietnamese victory in 1975. Vietnam was then unified under a communist government but remained impoverished and politically isolated. In 1986 the government initiated a series of economic and political reforms, which began Vietnam's path towards integration into the world economy, greatly assisted by the global economic shift to East Asia (Figure 5.14).

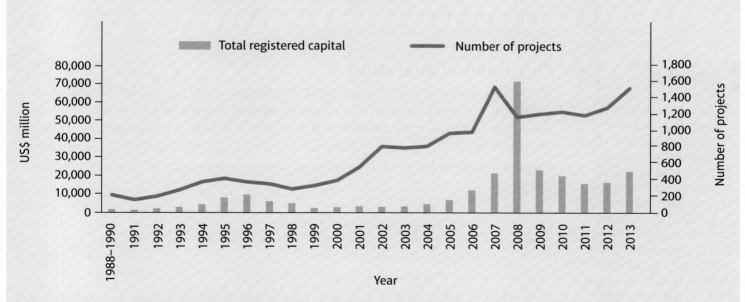

Figure 5.14: FDI projects in Vietnam, 1988–2013, Vietnam became a member of WTO in 2005

Synoptic link

Energy security is a key goal for every country, most of which rely on fossil fuels. The main players include OPEC and the International Atomic Energy Authority, which promotes the development of civilian nuclear energy under the auspices of the UN. National players include the Renewable Energy Association, which represents and promotes providers of renewable energy in the UK. (See page 96.)

A new international economic order

The Cold War made identity and sovereignty in the new nations more complicated. In 1960, the UN General Assembly passed the Declaration on the Granting of Independence to Colonial Countries and Peoples; in 1961 the Non-Aligned Movement was created in Belgrade to give new nations a common forum for resisting US and Soviet influences. In 1964 the UN Conference on Trade and Development (UNCTAD) promoted a New International Economic Order (NIEO). The NIEO was opposed to the 1944 Bretton Woods system because emerging economies needed fairer international policies to allow for their development. The main tenets of the NIEO were that developing countries must be:

- entitled to regulate and control the activities of multinational corporations operating within their territory
- free to nationalise or expropriate foreign property on conditions favourable to them
- free to set up associations of primary commodities producers similar to the Organisation of the Petroleum Exporting Countries (OPEC). All other states must recognise this right and refrain from taking economic, military or political measures to restrict freedoms
- able to benefit from stable and equitable prices for raw materials, generalised non-reciprocal and non-discriminatory tariffs, and transfer of technology. Economic and technical assistance should be provided without conditions.

UNCTAD failed to implement the NIEO, which was opposed by several superpowers, and socio-economic inequalities between developed and developing countries, especially newly independent states, continue to exist. An example is the Latin American debt crisis of 1982, when countries such as Brazil found that they could not pay their debt interest, which later spread to African states. Developed countries were concerned that the world economy could be destabilised and the Washington Consensus response was structural adjustment, which further widened the development gap between colonial powers and their former colonies. Some debts have been 'written off' gradually, or 'debt-for-nature' swaps arranged, to keep world economic stability but perhaps affecting economic sovereignty.

Synoptic link

In some former colonies, significant groups have fewer rights than the dominant group; this encourages migration to the former colonial power (e.g. the Karen people in Burma). The UN High Commission for Refugees acts to protect the rights of internally displaced persons. NGOs such as the International Committee of the Red Cross also act to protect the interests of disadvantaged groups, especially those facing violence. (See page 235.)

Post-colonial migration

The conflicts, poor governance and economic struggles associated with decolonisation created strong push factors for migration from newly independent states to the former imperial powers. The economic boom in many developed countries in the 1960s also introduced strong pull factors. The same migration flows are still evident today, and have been important in changing the ethnic composition and cultural homogeneity of some former colonial powers, and consequently their identities.

Yellow = Players, Orange = Attitudes and actions, Purple = Futures and uncertainties

CASE STUDY: Immigration and the changing face of the UK

After 1945 the UK, especially London, received over 500,000 migrants from the Caribbean, because workers were needed to help rebuild industry and services. There were no intervening obstacles because visa restrictions did not apply at this time to members of the British Commonwealth, and some had their passage paid for by the government. Push factors included poverty in the Caribbean associated with the economic transition after independence.

During the 1960s, Britain's textile industries in the Midlands and north-west were booming, and needed labour; about 750,000 Pakistanis and a million Indians were attracted by this work, and few visa restrictions and cheaper air transport helped this migration.

Conflicts in other former British colonies also led to refugee immigration: for example, about 30,000 Ugandan Asians (1972) and 20,000 Vietnamese boat people from Hong Kong (1975) settled in cities around the UK.

The emergence of new state forms

Globalisation has encouraged the movement of capital between countries, including more foreign direct investment as governments have reduced restrictions on foreign ownership. Consequently businesses and entrepreneurs have been able to relocate to countries with low-tax regimes, and these act as havens for their profits. Most **tax havens** also offer services to wealthy individuals and companies that may be more significant than the tax reductions: they permit business transactions to take place in secret, encouraging tax avoidance and evasion of financial regulations and inheritance dues in other countries.

For example, the British Virgin Islands specialise in incorporating offshore companies, so a business in another country can reinvest profits in the subsidiary offshore company and defer taxes indefinitely. Ireland offers companies 12.5 per cent corporation tax, compared to 20 per cent in the UK. Switzerland and Luxembourg offer secure banking, corporate tax avoidance and a wide range of offshore services. While the UK does not offer secret banking, it does offer a wide range of offshore services. Tax havens such as Anguilla, Bermuda, the British Virgin Islands, the Cayman Islands, Guernsey, Montserrat and the Turks and Caicos Islands are British territories with constitutions that were established by the UK parliament, and while some believe that these overseas territories are independent of the UK and govern themselves, others point out their strong links with the City of London financial hub.

Tax havens and tax avoidance

Most governments and **IGO**s have accepted the emergence of tax havens and are sympathetic towards foreign investors because of the productivity gains and technology transfer that arise from foreign direct investment, and so agree that their profits should not be taxed multiple times. However, there are concerns that some large businesses are deliberately seeking ways of avoiding taxes, which has consequences for the tax income of countries and knock-on effects on their spending power for services, for example.

The release of documents by a 'whistle-blower' from the Panamanian law firm Mossack Fonseca in 2016 highlighted this issue, with documents revealing the actions of some wealthy and famous individuals who had secret assets in Panama. For example, the Prime Minister of Iceland was forced to resign when the leak revealed that he had been hiding assets in an undeclared offshore company. Collectively these offshore havens have a major impact on the global economy: they account for about US$13 trillion of personal wealth – an amount equivalent to the annual US

Extension

The Nations Online Project makes maps of indigenous languages available online. Research the patterns of non-European language groups in Africa, and then analyse how the languages have changed since the countries gained their independence.

gross national product – and are the legal home to 2 million corporate entities and half of all international lending banks. Tax havens may also distort the distribution of global costs and benefits, to the detriment of developing economies.

CASE STUDY: The impact of tax havens on the poorest countries in Africa

A 2014 report by Christian Aid (an **NGO**) identified that Africa has the highest proportion of assets held abroad of any world region. The secrecy surrounding global financial flows makes the amounts involved difficult to quantify, but a report by the African Development Bank (ADB) and Global Financial Integrity (GFI) in 2013 found that illicit financial flows were draining US$1.2 to US$1.3 trillion a year from Africa. Estimates by GFI suggested average annual outflows between 2000 and 2008 to be about US$50 billion. In comparison, foreign direct investment flows into Africa in 2008 were US$38 billion, US$62 billion in 2009 and US$52.3 billion in 2011.

Illicit financial outflows undermine development and poverty reduction in two ways:

- They reduce the amount of money available to invest in the country. Studies show that if all the capital 'flight' over the period 2000–08 had been invested in Africa, the average rate of poverty reduction would have been 4 to 6 per cent higher each year.

- Because money is often kept offshore in secret, it avoids being taxed, dramatically reducing the money available for government spending on public services and other investments. Raising tax revenues is largely impossible, contributing directly to increasing inequality. As most income and assets offshore are out of reach, governments are forced to ignore them.

Maths tip

Remember that the Gini index is a number between zero and one, which measures the degree of inequality in the distribution of income or wealth. The index would be 0 for a society in which each member received exactly the same amount. An index of 1 would mean that one member got everything and the rest got nothing. Sometimes the Gini coefficient is scored out of 100 instead of 1.

Growing inequality: a threat to the world economy

Global inequalities are recognised as a major threat to the sustainability of the global economic system. The World Economic Forum (WEF) is focusing on the problem of increasing income inequality within developed and emerging economies, whereby small elites now enjoy unprecedented economic rewards while many are squeezed out of the middle-class category into poverty. For example, Oxfam calculated in 2016 that the world's 62 richest people share a combined wealth of US$1.76 trillion, as much as the poorest 3.6 billion of the world's population combined.

The severe consequences of this situation include unequal access to quality education, which reinforces inequality and reduces social mobility. Chronic diseases, such as heart disease and diabetes, are linked to poor diet and 'food deserts', where it is hard to buy affordable fresh fruit and vegetables, so people are eating high-calorie food low in nutrients. In developing countries, where much of the population is under 30, the lack of access to jobs is likely to increase the risks of political and social strife. However, research by the Joseph Rowntree Foundation in 2016 revealed that people aged 25 to 29 in eight rich countries – the US, UK, Australia, Canada, Spain, Italy, France and Germany – have also become poorer over the last 20 years compared with the average population, and unattached young adults are finding it hard to afford their own home.

Inequality appears to be an inevitable outcome of capitalist activity. **Capitalism** has three key principles:

1. Land, resources and capital are privately owned.
2. The majority of people work for a wage.
3. Markets mediate between producers and consumers.

Alternative economic models place businesses in the hands of workers and ordinary people, with democratic principles applied to all levels, including finance, jobs, participatory governance, budgets and planning, as used in Porto Alegre, Brazil. **Economic democracy** (Professor David Schweickart) is where workers control most enterprises democratically through labour trusts, coupon-based markets and sharing 'levies' on corporate profits. Examples of this model include

Yellow = Players, Orange = Attitudes and actions, Purple = Futures and uncertainties

Spain's Mondragon Co-operative and the UK's Co-operative Group and John Lewis Partnership. Workers control (but do not own) the enterprises they work in and, after paying a 'capital assets tax' on revenues generated, share any surplus equally between them.

CASE STUDY: Ecuador and Bolivia, an alternative economic model?

Ecuador and Bolivia established new constitutions based on an idea called *Buen Vivir*, which means to live well in a community with a healthy environment. Capitalism promotes individual rights: the right to own, sell, keep, and have. *Buen Vivir* subjugates the rights of the individual to those of the larger community and to nature. For example, *Buen Vivir* argues that humans are not owners of the Earth and its resources, only stewards. Ecosystem services, where a monetary value is allocated to environmental goods such as water provision from rivers or carbon sequestration by forests, are opposed, and instead *Buen Vivir* promotes collaborative consumption and a sharing economy.

A level exam-style question

Explain how globalisation processes have influenced the emergence of new forms of nation state. (8 marks)

Guidance

In your explanation, identify which aspects of globalisation are involved, as well as the actual forms of nation states. However, do not cover these as separate paragraphs: justify the links between them with a few examples.

ACTIVITY

GRAPHICAL SKILLS

1. Go to the World Bank website and find the database with Gini coefficients for most countries.

a. Select ten or more countries with a range of economic development, as shown by their gross national income (GNI).

b. Plot these countries, using the horizontal (x) axis of a scatter graph for GNI and the vertical (y) axis for the Gini coefficient.

c. Describe and explain the pattern of inequality shown between nations.

What are the impacts of global organisations on managing global issues and conflicts?

Learning objectives

8B.7 To understand that global organisations are not new, but have been important in the post-1945 world.

8B.8 To understand how IGOs established after the Second World War have controlled the rules of world trade and financial flows.

8B.9 To understand that IGOs have managed environmental problems facing the world with varying success.

Important global organisations (post-1945)

IGOs have come to play a significant role in **global governance**. The United Nations was the first post-war IGO, and was established by the victors of the Second World War in October 1945. At first the UN had 51 member states; this increased to 193, following widespread decolonisation and independence in the 1960s. The UN's founding charter stated that its purposes were:

- to save succeeding generations from the scourge of war, which twice in our lifetime has brought untold sorrow to mankind
- to reaffirm faith in fundamental human rights, in the dignity and worth of the human person, in the equal rights of men and women and of nations large and small
- to establish conditions under which justice and respect for the obligations arising from treaties and other sources of international law can be maintained
- to promote social progress and better standards of life in larger freedom.

The UN established several bodies to put these principles into practice. Five are still in operation:

- the General Assembly (the main meeting for debates)
- the Security Council (for deciding specific resolutions for peace and security)
- the Economic and Social Council (ECOSOC) (for promoting international economic and social cooperation and development)
- the Secretariat (for providing studies, information and facilities needed by UN decision-makers)
- the International Court of Justice (the main judicial organisation).

The UN has established various specialist agencies with particular responsibilities, including the World Bank Group, the World Health Organisation, the World Food Programme, UNESCO and UNICEF. The UN's most prominent officer is the Secretary-General, an office held by South Korean Ban Ki-moon in 2016. Non-governmental organisations also contribute to the UN's work.

Yellow = Players, Orange = Attitudes and actions, Purple = Futures and uncertainties

Figure 5.15: The United Nations General Assembly Hall in New York City

---

CASE STUDY: The United Nations Security Council

Of all the UN's bodies, the Security Council is the most significant. It is charged with maintaining peace between countries and has the power to make decisions binding on all member states. The Security Council has five permanent members – China, France, Russia, the UK and the USA – and ten non-permanent members with places for two-year terms (member states serve on a world region basis). The five permanent members hold veto power over UN resolutions, allowing one of them to block adoption of a resolution, although considerable debate will still take place.

As a result of this veto power, the UN's role in global governance has been affected by the differing geopolitical visions of Security Council members. For example, in the early days of the UN, the Soviet Union Minister for Foreign Affairs, Vyacheslav Molotov, vetoed resolutions so many times that he became known as 'Mr Veto'. In fact, the Soviet Union is responsible for nearly half of all vetoes ever cast, including 79 in the UN's first ten years. Molotov regularly rejected bids for new membership because of the US's refusal to admit all the Soviet republics. Since the dissolution of the Soviet Union, Russia used its veto power sparingly until the 2014 conflicts in Ukraine and Syria. These are some recent examples of permanent members exercising their veto powers:

- July 2015: Russia vetoed a draft resolution seeking to set up an international criminal tribunal for the MH17 air disaster in Ukraine
- May 2014: China and Russia vetoed a resolution condemning the state of Syria
- February 2011: the US vetoed a draft resolution condemning Israeli settlements in the West Bank.

The UN's role in global governance

The UN aims 'to achieve international cooperation in solving international problems of an economic, social, cultural, or humanitarian character'. It was under the auspices of the UN that member states agreed to achieve eight Millennium Development Goals (MDGs) by 2015. These have now been superseded by the 2030 Agenda for Sustainable Development, in which 17 Sustainable Development Goals (SDGs) and 169 targets are designed to build on and expand the Millennium Development Goals. Numerous UN bodies are working towards the SDGs:

- **The UN Development Programme (UNDP)** publishes the Human Development Index (HDI), a comparative measure of levels of poverty, literacy, education, life expectancy and other factors.
- **The Food and Agriculture Organisation (FAO)** promotes agricultural development and food security.

Extension

Investigate one example of the work of each UN organisation and assess how it helped to improve a socio-economic, socio-cultural, socio-political or environmental situation. You may wish to set this out in a table with the following column headings: UN organisation; Example; Contribution.

- **The United Nations Children's Fund (UNICEF)**, created in 1946 to aid European children after the Second World War, has since expanded its mission to provide aid around the world and to uphold the UN Convention on the Rights of the Child.
- **The World Health Organisation (WHO)** focuses on international health issues and has largely eradicated polio, river blindness and leprosy.
- **The UN Population Fund**, which also dedicates part of its resources to combating HIV, is the world's largest source of funding for reproductive health and family planning services.
- **The World Food Programme (WFP)**, along with the International **Red Cross** and **Red Crescent** movement, provides food aid in response to famine, natural disasters and armed conflict, and currently feeds an average of 90 million people in 80 nations each year.
- **The Office of the United Nations High Commissioner for Refugees (UNHCR)** works to protect the rights of refugees, asylum-seekers and stateless people.
- **The UN Environment Programme (UNEP)** sets a global environment agenda by assessing environmental trends at all scales, developing strategies and encouraging wise management of the environment (Table 5.4).

The work of the WHO, FAO and WFP is often regarded as among the best contributions that the UN has made to 'social progress and better standards of life in larger freedom'. However, there have been criticisms of UN agencies: in 1984 US President Ronald Reagan withdrew funding from UNESCO (the United Nations Educational, Scientific and Cultural Organisation) over allegations of mismanagement. Boutros Boutros-Ghali, Secretary-General from 1992 to 1996, and his successor Kofi Annan (1997–2006) initiated many reforms of the institutions in the face of further threats from the US to withhold its UN dues.

Table 5.4: The top 15 contributors to UNEP's Environment Fund in 2013

Country	Amount (US$ million)	Country	Amount (US$ million)
Netherlands	10.25	Denmark	4.60
Germany	9.89	Finland	4.36
United States	6.25	Norway	3.00
Belgium	5.93	Canada	2.97
France	5.85	Japan	2.78
United Kingdom	5.57	Russian Federation	1.50
Sweden	4.79	Australia	1.12
Switzerland	4.66		

ACTIVITY

Evaluate the ways in which the UN can intervene in global environmental, socio-economic and political situations in the 21st century.

UN geopolitical interventions

Interventions by the UN through the use of **economic sanctions** and direct military intervention have been made in defence of **human rights,** but with mixed degrees of success. In the decade after the Cold War ended the number of Security Council resolutions doubled and the peacekeeping budget increased more than tenfold. The UN negotiated an end to the Salvadorean civil war, had a successful peacekeeping mission in Namibia, oversaw democratic elections in post-apartheid South Africa and post-Khmer Rouge Cambodia, and in 1991 a UN-authorised, US-led coalition repulsed the Iraqi invasion of Kuwait. However, there were some less successful interventions, especially in the 1990s: the UN mission to keep peace in Somalia was viewed as a failure after heavy US casualties in the Battle of Mogadishu; the peace-keeping mission to Bosnia

Yellow = Players, Orange = Attitudes and actions, Purple = Futures and uncertainties

was unable to prevent ethnic cleansing; and the indecisiveness of the Security Council led to failure to intervene altogether in the Rwandan genocide.

Interventions authorised by the UN have often been supported or delivered by third parties. For example, the UN mission in the Sierra Leone civil war (1991–2002) was supplemented by British Royal Marines, and NATO oversaw the invasion of Afghanistan in 2001. The UN Charter stipulates that to help maintain peace and security around the world, all UN member states should make available armed forces and infrastructure (Figure 5.16). Since 1948 nearly 130 nations have contributed police personnel to peace operations. However, some peacekeeping forces have been drawn from developing countries, which may not have the capacity to deliver the objectives of a mission. Some world regions operate their own international peacekeeping forces, such as the African Union.

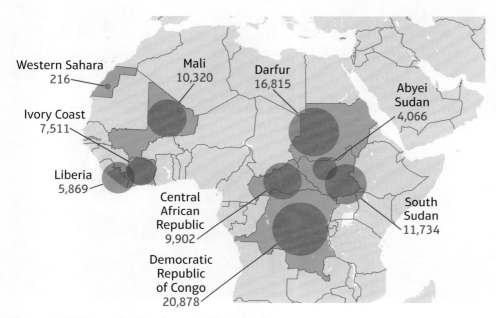

Figure 5.16: UN peacekeeping forces in Africa (2015)

CASE STUDY: UN sanctions against Iran

The UN's use of sanctions to limit Iran's nuclear programme is regarded as a successful intervention to resolve potential conflict. In 2006 the Security Council passed a resolution demanding that Iran suspend all uranium enrichment-related and reprocessing activities because it was believed that Iran was enriching uranium in order to develop a nuclear weapon – although Iran insisted it was simply working to develop civilian nuclear power. The Security Council threatened sanctions against Iran if it did not cease enrichment activity, and by the end of 2006 imposed limited sanctions banning the supply of nuclear-related materials and technology and freezing the assets of individuals and companies related to the programme.

Iran did not respond to the UN resolutions to the satisfaction of the Security Council and so sanctions were extended until 2010. Eventually, they included an arms embargo, an extended asset freeze and travel bans on individuals involved with the programme. Countries were also permitted to inspect Iranian cargo, prohibit the servicing of Iranian vessels involved in banned activities, prevent the provision of financial services for sensitive nuclear activities, closely watch Iranian individuals and organisations, prohibit the opening of Iranian banks outside Iran and doing business with them, and prevent financial institutions making investments in Iran.

This pressure brought Iran to the negotiating table and its government agreed to suspend enrichment activity. A 2015 UN resolution set out a schedule for suspending and eventually lifting UN sanctions, with provisions to reimpose sanctions in the event of non-performance by Iran.

The 'war on terror', geopolitical relations and global stability

Some countries, such as the US, UK and Russia, occasionally decide to operate independently of the UN. Their actions have included intervening in **'failed states'** or conducting the **'war on terror'**, and may strain **geopolitical relations** and affect global stability. US President George W. Bush first used the term 'war on terror' on 20 September 2001; it refers to the global military, political, legal and conceptual struggle against identified terrorist groups and the states that support them.

ACTIVITY

Explain, with the use of examples, why global stability in the early 21st century is insecure.

Extension

Investigate the impacts that the UK has had on geopolitical relations and global stability in the 21st century.

A level exam-style question

Suggest how the 'war on terror' may bring challenges for intergovernmental organisations. (6 marks)

Guidance

Consider the obstacles that may be placed in the way of IGOs, both in terms of policies and practical action. Consider the relationships between countries and IGOs, and how the war on terror may change cooperation and trust and encourage countries to work outside organisations such as the UN.

Figure 5.17: Images from the 'war on terror'

- In 2003 the US invaded Iraq, despite failing to pass a UN Security Council resolution specifically authorising intervention. The invasion began with an air campaign followed by US-led ground troops. The Bush administration justified the invasion by saying that Iraq possessed 'weapons of mass destruction', but the evidence is widely debated.

- China has engaged in its own war on terror, predominantly in response to violent actions by Uighur separatist movements in Xinjiang. This campaign has been criticised for unfairly targeting China's predominantly Muslim Uighur population.

- Russia has also been engaged in counter-terrorism, including the Second Chechen War, the insurgency in the North Caucasus, and military intervention in the Syrian civil war. Like China, Russia has also focused on separatist and Islamist movements that use violence to achieve their political goals.

A new conflict in Africa arose with the advance of Boko Haram in northern Nigeria in 2009 and Al-Qaeda in the Islamic Maghreb in the Sahara and parts of West Africa in 2012. France and the US are increasing their support to governments in the region, but these groups continue to operate in the many remote lawless areas.

CASE STUDY: Russia's impact on geopolitical relations and global stability

Russia has had recent significant impacts in eastern Europe, eastern Asia and the Middle East. Assisted by China's will to create stability to its west, Russia has been able to reach agreements with China such as the 2015 Silk Road Economic Belt which involves investment and resources from China and security and geopolitical stability from Russia. Through such geopolitical relations Russia effectively secures the Eurasian region from USA influence.

Russia wishes to regain some of its influence in eastern Europe and in the early 21st century forged greater ties with Germany, which allowed it to ensure that all eastern European countries did not join NATO and the EU. Using its energy supply system as a means of intimidation Russia tried to prevent eastern European countries looking to the 'west', this created instability, especially with Russian forces entering Ukraine. The latter led to increasingly hostile rhetoric between Russia and the 'west' and considerably strained geopolitical relations. Sanctions against Russia were imposed which severely restricted its income from energy sales to Europe.

Russia's intervention in Syria was aimed to reduce the terrorist threat to Russian territory and to strengthen its foothold in the Middle East region. But this may add to global instability when any political ramifications spread beyond the region, such as the sanctions imposed by Russia on Turkey. The geopolitical nature of the Middle East is likely to provide future issues for Russia as well as other superpowers.

IGO control of world trade and financial flows

After 1945 the Western allies viewed the resumption and growth of international trade as crucial for rebuilding shattered economies and also as a way of preventing future conflicts. The Bretton-Woods Agreement (1944) created a system of rules for managing the international monetary system for the USA, Canada, Western Europe, Australia, New Zealand and Japan. All the countries linked their currencies to gold and the US dollar. The agreement also established the IMF and the International Bank for Reconstruction and Development (IBRD). These institutions developed a free market, or a neoliberal approach, known as the 'Washington Consensus', and its legacy is the Western dominance of global economic management and free trade policies.

The International Monetary Fund (IMF)

The purpose of the IMF is to foster global financial stability and help governments balance their payments in times of economic difficulty, by making loans to member countries that cannot repay their debts. The idea is that if a country receives support during hard times, it is less likely to resort to the protectionist policies of the 1930s. IMF members are assigned financial ratings, reflecting their relative economic power, and pay a financial 'subscription' matching their rating. This creates a fund that is used to stabilise the international monetary system through a system of exchange rates and international payments that enables countries (and their citizens) to continue to make transactions.

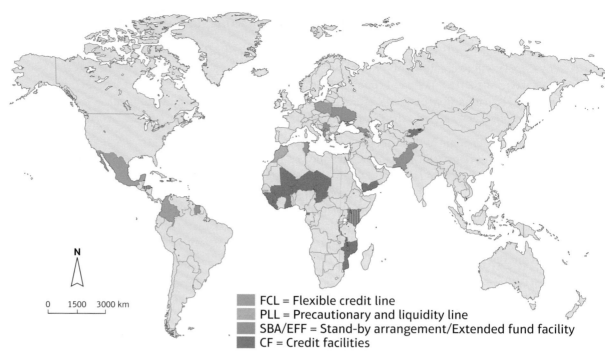

FCL = Flexible credit line
PLL = Precautionary and liquidity line
SBA/EFF = Stand-by arrangement/Extended fund facility
CF = Credit facilities

Note: Loans similar to a line of credit are approved by the IMF Executive Board to support a country's adjustment programme. The arrangement requires the member to observe specific terms in order to be eligible to receive a disbursement. The IMF lends under various arrangements as shown in the key and, at reduced rates, under Poverty Reduction and Growth Trust and Exogenous Shocks Facility arrangements.

Figure 5.18: IMF loans (April 2016)

The International Bank for Reconstruction and Development (IBRD)

The aim of the IBRD was to provide loans for rebuilding economies after 1945 and to alleviate poverty. (The communist Soviet Union did not participate until the 1990s.) Now part of the World Bank Group, the IBRD has changed its focus to tackling extreme poverty. It aims to decrease the percentage of people living on less than US$1.25 a day to no more than 3 per cent of the world's population by 2030, by encouraging income growth for the poorest 40 per cent in a country. It provides low-interest loans, grants and technical assistance to developing countries, and often works in partnership with governments and multilateral institutions and commercial banks to deliver projects.

GATT and the World Trade Organisation (WTO)

In 1947, 23 countries signed the General Agreement on Tariffs and Trade (GATT) to restart global trade, which continues through negotiations, or 'trade rounds'. Over time, more countries have joined and agreed to further reductions in tariffs and more standardisation of products. In 1995 this process led to the establishment of the WTO, which is where countries seek a 'substantial reduction of tariffs and other trade barriers and the elimination of preferences, on a reciprocal and mutually advantageous basis'.

Yellow = Players, Orange = Attitudes and actions, Purple = Futures and uncertainties

Structural adjustment and poverty reduction

Structural adjustment programmes (SAPs) consist of loans provided by the IMF and World Bank to countries that have experienced economic crises. To obtain such loans, countries must implement specific policies or reduce interest rates on existing ones. The Washington Consensus set out ten policy recommendations for SAPs:

1. Fiscal policy discipline, with avoidance of large fiscal deficits relative to GDP.
2. Redirection of public spending from subsidies towards broad-based provision of key pro-growth, pro-poor services like primary education, primary health care and infrastructure investment.
3. Tax reform, a broader tax base and moderate marginal tax rates.
4. Market-determined interest rates with positive effects.
5. Competitive exchange rates.
6. Trade liberalisation of imports, with any trade protection using low and relatively uniform tariffs.
7. Liberalisation of inward foreign direct investment.
8. Privatisation of state enterprises.
9. Deregulation – the abolition of rules that impede market entry or restrict competition (except for those justified on safety, environmental and consumer protection grounds) and prudential oversight of financial institutions.
10. Legal security for property rights.

The World Bank report 'Learning from a Decade of Reform' looked back on the impact of the Washington Consensus in the 1990s. Most SAPs had an initial 'transition' phase during which removal of subsidies and closure of state-owned enterprises led to an initial worsening of economic problems. Many sub-Saharan African economies failed to take off (Rostow model) despite policy reforms, changes to political and external relations and the continued influx of foreign aid. Uganda, Tanzania and Mozambique were among the countries that showed some success, but they remain fragile. Some argue that the imposition of reforms has destabilised developing countries and decreased the quality of life; for example, the reforms have not led to a great economic boom in Latin America, but instead to economic crises, the accumulation of external debts and dependence on core world regions. Others point to the success of China and India since the 1990s, despite their economic policies being the opposite of the Washington Consensus's recommendations – with high levels of protectionism, little privatisation, extensive industrial planning and lax fiscal and financial rules.

In response to criticism that SAPs were implementing generic free-market policies without involvement from the borrowing countries, those countries are now encouraged to develop poverty reduction strategy papers (PRSPs) to replace SAPs. The idea is that increasing participation in creating policy leads to greater ownership of loan programmes and better internal fiscal management.

ACTIVITY

Assess the impact of structural adjustment and poverty reduction strategies on emerging economies in the late 20th and early 21st centuries.

Another alternative to SAPs is the Highly Indebted Poor Countries Initiative (HIPC), introduced by the IMF and World Bank in 1996. It provides debt relief and low-interest loans to cancel or reduce external debt repayments to sustainable levels. To be considered for the initiative, countries must face a debt burden that they cannot manage through traditional means, meet a range of economic management and performance targets, and undertake economic and social reforms. The HIPC initiative identified 39 eligible countries (33 in sub-Saharan Africa), but by 2016 only 36 of these had received full or partial debt relief.

CASE STUDY: Jamaica's structural adjustment programme

Jamaica became independent from Britain in 1962, but it was only in the 1970s that the government of Michael Manley initiated policies to support health and education, nationalise industries, increase taxation on foreign investment and encourage agricultural self-sufficiency. But these projects struggled because the oil crisis of the 1970s increased costs. As the price of imports rocketed and exports fell, Jamaica accumulated debts. When interest rates rose at the start of the 1980s, debt payments increased exponentially, from 16 per cent of exports in 1977 to 35 per cent in 1986. This gave the IMF and World Bank the opportunity to impose large-scale SAPs in return for loans.

The austerity measures had a big impact: during the 1980s the number of registered nurses fell by 60 per cent, food subsidies were removed, currency devaluation made the cost of food increase, and wages were kept low. Spending on health, education and housing provision was also cut. Oxfam described the consequences as 'a grim daily struggle to pay for food, clothing and transportation – even on the part of people who ten years ago would have been considered middle class'. Ten years later the government tried neoliberal policies to solve the crisis, but these did not work either and there has been little progress in reducing hunger or increasing basic water and sanitation provision. In 1990, 96 per cent of children completed primary school; in 2014 only 88 per cent did. In 1990, 79 mothers died in childbirth for every 100,000 live births; in 2015 it was 89. Figure 5.21 shows the poor quality of life for many citizens of Montego Bay, Jamaica.

Jamaica has now repaid more money (US$19.8 billion) than it has borrowed (US$18.5 billion), and yet the government still owes US$7.8 billion because of accumulated interest payments. Annual government foreign debt payments (US$1.2 billion) are double the amount spent annually on education and health combined (US$600 million). Since Jamaica is classified as an upper middle income country, it is not eligible for debt relief. It has tried negotiating deals with domestic private lenders to reduce interest rates, but with little impact on government debt. In 2013 the IMF announced another US$1 billion loan so that Jamaica could meet its huge debt payments, but this was linked to four more years of austerity, including a pay freeze, amounting to a 20 per cent cut in real wages.

Yellow = Players, Orange = Attitudes and actions, Purple = Futures and uncertainties

Figure 5.19: A shanty town in Montego Bay, Jamaica

Economic IGOs and trading blocs

Almost all countries are involved in global trade and financial IGOs. However, members of IGOs do not have equal influence within them. For example, the USA contributes about 18 per cent of the capital of the IMF and has about 17 per cent of the votes on the Board of Governors, while smaller economies provide less capital and therefore have less voting power. Austria provides 0.84 per cent of the capital and receives 0.82 per cent of the votes. To consolidate power, regional groups of countries have formed **trading blocs**, with mutual trade agreements and cooperation. These allow countries with smaller economies to participate more equally in the global economy. Within a trade bloc, barriers to trade within a world region are reduced or eliminated; these can be discrete agreements between several countries, such as the Association of SouthEast Asian Nations (ASEAN) or a collection of agreements such as the European Union (EU). Some military alliances have also brought countries closer together, such as South East Asia Treaty Organisation (SEATO) (1954–77), which was a political union to block communist expansion but also led to joint cultural and educational programmes.

If governments can achieve better trade terms for their secondary and tertiary businesses, their trade should increase and they can employ more people, which will help their economy further. Countries benefit from a larger tax base, higher skill levels and lower unemployment, and more international links should increase foreign direct investment (FDI). The majority of FDI will come from members of the trade bloc, but countries outside the trade bloc will also invest inside it to avoid tariffs and gain access to a larger market.

CASE STUDY: The North American Free Trade Association (NAFTA)

NAFTA is a trade bloc formed in 1994, involving Canada, the USA and Mexico. These three economies account for about 21 per cent of global GDP. The main aim of NAFTA has been to reduce tariffs on goods traded between the countries, but it also protects the intellectual property rights of companies trading within the bloc, coordinates environmental regulation, and promotes the construction of railways, roads and pipelines between the countries.

Many studies show the economic benefits to all three countries, with increased trade and FDI. However, specific groups have experienced problems: for example manufacturing workers in north-east America, whose jobs were relocated to Mexico, and Mexican farmers who are unable to compete with US agribusinesses. Since 1994 the North American economy has doubled in size: trade in goods has more than tripled to over US$1 trillion, with an exchange of about US$2.6 billion in goods a day, and employment levels have increased by 23 per cent, representing a net gain of 39.7 million jobs.

IGO management of environmental problems

Biomes, the atmosphere and the hydrosphere are transboundary, so the wellbeing of the environment requires cooperation between all countries. IGOs have been at the forefront of action to protect and conserve the environment, but they still require the support of all countries, especially the wealthiest and most powerful states. It has often taken many decades to achieve coordinated action on environmental issues, but there have been some successes.

CASE STUDY: The Montreal Protocol on the depletion of stratospheric ozone

In 1973 scientists discovered that the Earth's stratospheric ozone (O_3) layer at the poles was thinning, and that the emission of chlorofluorocarbons (CFCs) and other substances was responsible for this depletion. The ozone layer prevents harmful ultraviolet (UV) light from passing through the Earth's atmosphere; increased exposure to UV light is known to cause cancer and other diseases.

Public opinion in the developed world was alarmed by this potential health risk, and governments acted quickly; within 14 years of the discovery the Montreal **Protocol** on Substances that Deplete the Ozone Layer was signed. This international **treaty** phased out the production of chemicals responsible for ozone depletion. It entered into force on 1 January 1989 and has since been revised several times. As a result of the protocol, scientists now predict that the ozone layer will return to 1980 levels by 2070. The protocol has been ratified by 197 countries, making it the first universally ratified treaty in United Nations history. This protocol is therefore regarded as an example of timely and effective international action to reverse damage to the environment.

Yellow = Players, Orange = Attitudes and actions, Purple = Futures and uncertainties

CASE STUDY: The Convention on International Trade in Endangered Species of Wild Flora and Fauna (CITES)

CITES is an intergovernmental agreement to manage international trade in specimens of wild animals and plants so that their survival is not threatened., The international wildlife trade is estimated to be worth billions of dollars annually, and includes live animals and plants and a vast array of wildlife products derived from them. This trade, together with other factors such as habitat loss, is capable of bringing some species close to extinction. Tigers, elephants (Figure 5.20) and rhinos are three high-profile animals affected by illegal trade.

The convention came into force in 1975 after 12 years of negotiations, and by 2016 182 countries had signed it. They adhere voluntarily to the rules, passing their own laws to implement CITES and protect the natural environment. All imports, exports, re-exports or introduction of species covered by the convention have to be authorised through a licensing system – especially those threatened by extinction, for example monitoring illegal trade in elephant products.

More than 35,000 species of animals and plants are now given varying degrees of protection. CITES introduced meaningful protection for several rare reptile and amphibian species that were seeing an increase in trade over the internet, and governments have agreed strong protection for tigers and rhinos from the medicine and fashion trade in Asia. However, CITES is criticised by some for allowing its decisions to be influenced by commercial interests: for example Atlantic bluefin tuna are not protected, despite concerns of conservationists over their sustainability.

Figure 5.20: An African elephant in the Ngorongoro Conservation Area, Tanzania

Management of the oceans, rivers and biodiversity

IGOs have helped develop laws for managing the oceans and international rivers as well as monitoring the state of the global environment and levels of biodiversity.

CASE STUDY: The UN Convention on the Law of the Sea (UNCLOS)

UNCLOS defines the rights and responsibilities of nations over their use of the world's oceans, establishing guidelines for businesses, the environment and the management of marine natural resources. The Convention came into force in 1994; by 2016 166 countries and the European Union had ratified it. The UN manages ratification and accession and supports its meetings, but has no direct operational role.

Among numerous ocean management strategies are the International Maritime Organisation, the International Whaling Commission and the International Seabed Authority and Marine Protection Areas (MPAs). The main provisions of UNCLOS define the boundaries of coastal zones where countries have exclusive use of marine and mineral resources, and provide general obligations for safeguarding the marine environment and protecting freedom of scientific research in international waters. There is a legal regime for controlling mineral resource exploitation in deep-seabed areas beyond national jurisdiction, through the International Seabed Authority and the 'common heritage of mankind' principle. Landlocked states are also given a right of access to and from the sea, without taxation of traffic through transit states.

A key achievement of the Convention is that over 90 per cent of international trade takes place by sea, as well as 95 per cent of global internet traffic through submarine cables. It also imposes a responsibility on all countries to ensure the long-term sustainability of fish and other marine resources.

ACTIVITY

Compare and contrast the views of different players on the management of global environmental issues by IGOs and international agreements.

Your research could include reviews of magazines that present contrasting perspectives, such as *The Ecologist*, which supports regulation to protect the environment and a greater role for IGOs, *The Spectator*, which publishes original and often controversial opinions, *National Geographic*, which provides opinions and stories, and *Nature*, which covers ecological and biological perspectives.

CASE STUDY: The Helsinki Rules

The Helsinki Rules on the Uses of the Waters of International Rivers (1966) are guidelines on the use of rivers and connected groundwaters that cross national boundaries. There is no mechanism in place to enforce the rules, but they led to the creation of the UN Convention on the Law of Non-Navigational Uses of International Watercourses. The main principle established by the Helsinki Rules was that international treaties managing whole drainage basins must be based on 'equitable use' or 'equitable share'.

Often, international water treaties do not achieve 'equitable use' or 'equitable share' because one or more parties in the negotiations has disproportionate political, economic or military power and can strongly influence other countries, such as Egypt in the Nile basin. Some governments – for example Brazil, Belgium and China – have objected to the drainage basin management approach, stating that it infringes national sovereignty. In 2004 the Helsinki Rules were superseded by the Berlin Rules on Water Resources, which are wider in scope.

CASE STUDY: The Millennium Ecosystem Assessment (MEA)

The MEA, launched in 2001, aims to assess how changes in ecosystems have affected human wellbeing, and how to conserve and use them sustainably. It reported on conditions and trends in the world's ecosystems and the benefits they provide, such as clean water, food, forest products, flood control and natural resources, and the options for their sustainable use. The Great Barrier Reef in Australia (Figure 5.21) is an example of a vulnerable ecosystem studied in detail for the MEA.

Figure 5.21: A Blue Starfish resting on hard Acropora coral, Great Barrier Reef

The UN Environment Programme (UNEP) reviewed the MEA and drew four main conclusions:

- From 1950 to 2000 humans changed ecosystems more rapidly and extensively than in any other 50-year period, largely to meet increased demands for food, water, timber, fibres and fuel. This has resulted in a substantial and often irreversible loss of biodiversity.

- While ecosystems have helped to enhance human wellbeing and economic development, there has been a cost in terms of degraded ecosystems, which could substantially diminish their ability to benefit future generations.

- The degradation of ecosystems could grow significantly worse during the first half of the 21st century and be a barrier to achieving the SDGs.

- Reversing ecosystem degradation is possible but will need significant changes in policies, institutions and practices.

ACTIVITY

Evaluate the success of IGOs in managing different global concerns – cultural, political and environmental.

Extension

Investigate the 2015 to 2030 Sustainable Development Goals. Make notes on those associated with managing oceans, river basins and biodiversity. Read the latest SDG annual report and add information on the levels of success being achieved.

Yellow = Players, Orange = Attitudes and actions, Purple = Futures and uncertainties

The management of Antarctica

The cooperation between countries in managing Antarctica as a continent of peace and science is perhaps the best example of successful IGO management, with countries claiming territory being given 'slices' of the continent. However, there is some disputed territory, especially on the Antarctic Peninsula; however, because it is remote and has the coldest climate on Earth, there is reduced potential for exploitation and so lower interest in it.

CASE STUDY: The Antarctic Treaty system

The International Geophysical Year was an international scientific project that lasted from 1957 to 1958, with over 50 Antarctic research stations established by Argentina, Australia, Belgium, Chile, France, Japan, New Zealand, Norway, South Africa, the Soviet Union, the UK and the US. These 12 countries signed the Antarctic Treaty in 1959, which was a diplomatic expression of the operational and scientific cooperation they had achieved. By 2016 there were 53 signatory countries. For the purposes of the treaty system, Antarctica is defined as all the land and ice shelves south of 60° latitude (Figure 5.22).

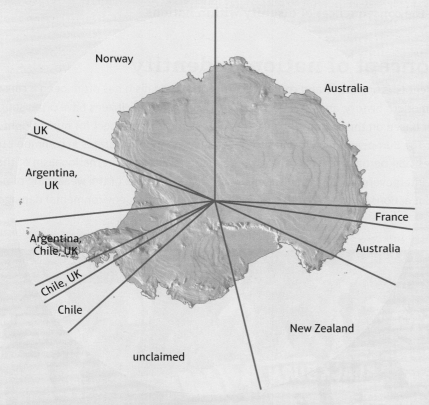

Figure 5.22: Satellite image of Antarctica with territorial claims

The Antarctic Treaty broadly makes Antarctica a scientific preserve, with freedom for scientific investigation and bans on military activity. Officially there is no citizenship or government of Antarctica, and it is governed through the Antarctic Treaty system including the Treaty itself and an Environmental Protocol (1998), a Convention for the Protection of Antarctic Seals (1972), and a Convention on the Conservation of Antarctic Marine Living Resources (1982) – and these are incorporated into national laws that apply to the citizens of the 53 nations wherever they are in Antarctica. All parties to the Treaty meet annually to discuss issues and manage the continent by consensus. Their decisions create protocols for protection of the Antarctic environment, conservation of plants and animals, preservation of historic sites, designation and management of protected areas, management of tourism, collection of meteorological data and hydrographic charting.

Extension

The British Antarctic Survey website is a good source of information. Review their science briefings to find out about the latest research in Antarctica and consider its relevance for managing global environmental concerns.

A level exam-style question

Explain the role of IGOs in the management of global environmental problems. (8 marks)

Guidance

Use your knowledge of examples of IGO management of global environmental problems to help explain how they manage them, perhaps recognising successful and less successful strategies.

CHAPTER
5

What are the threats to national sovereignty in a more globalised world?

Learning objectives

8B.10 To understand why national identity is an elusive and contested concept.

8B.11 To understand the challenges to national identity.

8B.12 To understand the consequences of disunity within nations.

The concept of national identity

National identity is an elusive and contested concept. Nationalism has often been a cause of conflict in modern history, and after each conflict politicians and citizens have often sought a new world order based on the common values of liberty, democracy, respect for human rights and the rule of law. The European Union is one such legacy, linking member states in common pursuit of these values, and perhaps reducing the power of nationalism. The internationalism of the UN and associated intergovernmental organisations is another expression of the weakening of nationalism through joint actions to solve common problems such as conflict, environmental damage and migration.

Figure 5.23: England football supporters with St George's flags

Yellow = Players, Orange = Attitudes and actions, Purple = Futures and uncertainties

Despite these changes, nationalism remains a powerful force; for example the Balkans war of the 1990s showed that nationalistic ideas could still cause destructive conflict. The concept of nationalism is reinforced through education, sport and politics, where loyalty to institutions, groups and common ideals is emphasised.

Displays of nationalism are often seen in public places or at public events (Figure 5.23), but may be interpreted in different ways. They may be associated with extremism, such as the National Front, British National Party and English Defence League, or linked to the desire to show pride in national achievements and culture, as in the opening ceremony of the Olympics in 2012.

Identity, loyalty and national 'character'

Nationalism inspires feelings of loyalty and devotion to the institutions and ideals of the nation state, often through symbols and slogans. Symbols of nationalism – flags, works of art, national anthems, architecture, currency, postage stamps, passports and many other forms of media – reinforce a national consciousness and create a sense of pride in national culture and interests. This has been seen in the UK in the Scottish independence referendum (2014) and the referendum on the UK's membership of the EU (2016). In his book *Nationalism* (1967), K. R. Minogue states that

> 'flags and anthems can be used to create members of a nation by developing new habits and emotions; the Star-spangled banner [US flag] with its stars increasing as a new state joined the Union was an important symbol of America for the millions of immigrants to the United States.'

Identity and loyalty are often tied to legal systems and methods of governance, to the idea of a national 'character' and to physical landscapes. For example, the essence of English common law is judges applying legal precedent to the facts before them, while other European countries such as France and Germany have a civil law system, where the law is interpreted rather than developed or made by judges, and only legislative enactments are considered legally binding. This contrast is more than a technical detail; it creates different interpretations of what is meant by justice. Such differences create a sense of identity: a loyalty to one approach is reinforced through opposition and criticism of another approach.

National identity and multinationalism

As a result of globalisation, most developed countries are now highly interconnected, and emerging countries are also strengthening their connections to the global economy. Inevitably, these processes have resulted in the migration of workers and entrepreneurs, ranging from elite wealthy groups to refugees. The legacy of these movements is that most developed countries are multinational, with many contrasting ethnic groups. As a result, questions of national identity and loyalty have become complex. For example, in the 2011 UK Census, only 13 per cent of the population over 75 years of age described themselves as being British; people tended to identify first with their home nation before the UK. Furthermore, across all age groups, only 14 per cent of those whose ethnicity is 'white British' describe their identity as British. In contrast, about half of people with black or Asian ethnicity selected British. The religious group least likely to describe themselves as British are Christians (15 per cent) and the most likely are Sikhs (62 per cent). In England, populations on the coasts, particularly on the eastern side of the country, self-identify most strongly with an English-only identity, while in Wales, younger people are more likely to describe themselves as Welsh than their parents. There is a complicated cultural pattern.

Global terrorism has created tensions between Muslim communities in secular (religiously neutral) countries. The Islamic Society of Britain recognises that first-generation migrants may have felt culturally displaced from their home countries, but second- and third-generation Muslims regard

Synoptic link

There is an increasing role for IGOs in shaping the economic and social characteristics of places. For example, UNESCO creates World Heritage Sites and the EU schemes PDO (protected designation of origin) and PGI (protected geographical indication) protect specialist quality farm products. Some NGOs such as the National Trust in the UK are active in protecting distinctive places, often using a heritage catalyst. Urban regeneration can be delivered by partnerships of local authorities and local bodies, such as Bournemouth 2026 Trust. (See Chapter 5 pages 253–265.)

A level exam-style question

Explain the historical and contemporary influences on national identity. (8 marks)

Guidance

Consider the range of factors that influence a culture such as images of the past, migration, and globalisation connections. Try to keep a balance between historical and recent influences.

CASE STUDY: Englishness and the English countryside

'Of all the things that are sacred to us in England, the countryside is one of the most precious of all.' So said former Poet Laureate Sir Andrew Motion, when speaking to the Council for the Protection of Rural England in 2014. He argued that ever since the Industrial Revolution, the English countryside has held a central place in the idea of what it means to be English. The 18th-century poet William Wordsworth tried to capture the romantic ideal of the English countryside in the face of the reality of industrialisation. He was part of the Romantic Movement across Europe that emphasised emotion and individualism as well as glorification of the past and nature.

The Haywain by John Constable (1821), shown in Figure 5.24, is an iconic image of the lowland British countryside, depicting a scene on the River Stour between the English counties of Suffolk and Essex. Constable's representation of rural England contains the emotion of the countryside experience, a sentiment that endures today: it was voted the second most popular painting in any British gallery (second only to Turner's *Fighting Temeraire*) in a 2005 BBC Radio 4 poll.

Figure 5.24: John Constable's The Haywain (National Gallery, London)

Nationalistic images are often conjured up at times of change or difficulty, and the English countryside as a meme for a timeless and unchanging England was sometimes used in First World War recruitment posters.

In 1945 the government established National Parks and Green Belt legislation to restrict development in the countryside, in order to preserve English heritage and landscape. However, today many contest that England is a predominantly urban nation with diverse ethnic communities, and that any generalisation about Englishness based on white British and rural areas is incorrect. Some would go as far as to say that planning laws to protect the countryside are out of date and have become a way of protecting the affluent from urbanisation.

ACTIVITY

With reference to at least two countries, explain why national identity may vary within a country.

the UK as their home. The Society has said that when tension exists between the UK and a Muslim nation, British Muslims should support the UK, since they are citizens of the country and that is the right thing to do.

Yellow = Players, Orange = Attitudes and actions, Purple = Futures and uncertainties

Challenges to national identity

'Made in Britain' is an increasingly complex idea, because many UK-based companies are foreign-owned. In 2011 a UK-based manufacturer of ovens, Stoves, commissioned independent market research of 1,000 British adults to discover how knowledgeable they were about which brands were actually British-made. The results showed that 48 per cent of those surveyed admitted confusion: 40 per cent thought HP Sauce (Netherlands), 43 per cent thought Royal Doulton (China and Thailand), 32 per cent thought Dyson (Malaysia), and 23 per cent thought Raleigh bikes (Vietnam, South Korea, Bangladesh) were British-made. This research led to the formation of an independent, non-profit organisation called the Made In Great Britain Campaign in 2013. It aims to educate the public about brands that promote themselves as British but have little or no production in the UK.

CASE STUDY: Foreign ownership of UK utilities

The UK's water and electricity supply may appear to be national service industries, but in 2011 the UK Office of Fair Trading (OFT) found that foreign investors owned 40 per cent of infrastructure assets in the energy, water, transport and communication sectors. Foreign ownership of UK utility assets is now widespread, most frequently in the form of foreign-listed public limited companies, but also through various private ownership methods such as infrastructure funds.

According to the OFT, the increasing globalisation of corporate investment activity has diminished the relevance of nationality in respect of utility ownership. They hold this benign view because of the legal protections for consumer interests, the legal separation of network and retail functions, and the requirement for licence-holders to protect core assets and maintain an investment credit rating. The legal ability to put a failing utility operator into special administration, to protect continuity of service while sorting the finances of a company, is also a major safeguard. This has enabled London's electricity network to be owned by Cheung Kong Infrastructure Holdings, a Hong Kong Investment company for a Chinese billionaire.

Three-quarters of the UK's energy generation and 92 per cent of the energy supply is provided by six large companies. Two of them, Centrica and SSE, have their shares listed on the London Stock Exchange, indicating British ownership, but foreigners may own shares listed in London. The other four large energy companies are foreign-owned: EDF Energy (owned by French firm Électricité de France and listed on the Paris Stock Exchange), npower and E.ON UK (both German-owned) and Scottish Power (Spanish-owned).

Table 5.5: The output of the UK car industry by manufacturer

Manufacturer	Nationality of owner	Output in thousands (2015)
Nissan	Japanese	500
Jaguar Land Rover	Indian	450
Honda	Japanese	250
Toyota	Japanese	200
Vauxhall (GM)	US	200
BMW Mini	German	200
Bentley	German	10

'Westernisation'

Fortune magazine publishes a list of the world's largest companies called the *Fortune* Global 500. In 2014 the world's 500 largest companies generated US$31.2 trillion in revenues and US$1.7 trillion in profits, and employed 65 million people worldwide. Although these companies come from 36 different countries, most are based in USA/Canada, Europe or South East Asia and all project their influence globally through ownership of foreign companies. In 2013 the *Harvard Business Review*

ACTIVITY

GRAPHICAL SKILLS

Use proportional circles to represent the level of foreign ownership of UK car manufacturing by output. The data is shown in Table 5.5. To calculate the radius or diameters of each circle, use the square root of each piece of data and develop a manageable scale (in mm or cm). You could draw the circles on plain paper or graph paper, alongside each other, or add the circles to a world map, placing the proportional circles over the country of ownership.

found that the US/Canada-based companies had 8,131 majority-owned affiliates, European-based ones had 15,143, and Asian-based ones had only 2,920.

TNCs from North America and Europe, the 'Western world', have much larger networks of FDI than non-Western countries. Their collective influence has been described as **Westernisation**, which can also be seen in other aspects of globalisation and in the adoption of Western cultural values, of which one element is the promotion of the benefits of the capitalist model. This is not necessarily a benign influence; in 2000 Naomi Klein criticised brand-oriented consumer culture and the operations of large corporations, and raised concerns about the unethical exploitation of workers in the world's poorest countries.

Capitalist theory and national identity

Capitalism is a system of mostly private ownership in which innovation and new ideas are encouraged, creating new firms, new owners and new capital. Perhaps its greatest strength is its dynamism, the ability to change regardless of economic conditions and replace older outdated businesses. However, this dynamism is also capitalism's greatest weakness, as the dividing line between success and failure is often thin. Growth and economic decline can occur at the same time, and instability and job insecurity in one region may cause its identity to differ from another area of the same country.

The economies of East Asia grew fastest when they opened up to global markets while maintaining centralised control over key industries and investment decisions. However, recent slower growth is being linked to state intervention, with bureaucracy, large subsidies and guaranteed markets leading to overinvestment and then insolvency. These economic models are now being challenged as future economic growth is considered. However, US and European capitalist models, in which capital is free to flow in new directions without interference from the state, have performed less well than East Asia so far in the 21st century. Some are considering that more central planning and control may be necessary to support economic dynamism.

CASE STUDY: Indian TNCs

There are an estimated 38,000 transnational corporations with over 200,000 foreign subsidiaries, although these are not all large in size. In the past, TNCs from developed countries have invested and established themselves in developing countries, however, in the 21st century this is changing. Emerging economies such as India have expanded their FDI through the activities of their TNCs. The Indian approach has been to acquire existing companies rather than set up new operations; in 2013 there were 8 Indian TNCs in the Fortune top 500 and the top 15 Indian TNCs had a combined overseas assets total of US$63 billion. The Tata Group is the largest and most evolved internationalised Indian corporation, acquiring 25 foreign companies in 10 years with 58 per cent of its revenues earned abroad: for example, Corus in UK (Tata Steel), Jaguar Land Rover (Tata Motors UK), and Tetley (Tata Global Beverages).

The IMF has stated that India's economic data has problems. Data availability is an issue, for example not all corporations submit their tax returns on time so each year GDP is estimated, India uses the 'wholesale price index' in its GDP calculations which makes data look more positive, and there are many small 'shell' corporations which make it difficult to estimate the actual number of legitimate corporations. Ambit Capital (Mumbai) have used an alternative measure of economic change in India which showed decline between 2014 and 2016 rather than growth.

Yellow = Players, Orange = Attitudes and actions, Purple = Futures and uncertainties

International ownership patterns

As a result of the spread of foreign investment in a globalised world, ownership of property, land and businesses is also increasingly non-national (Figure 5.25). For example, between 2011 and 2013, estate agents Chesterton Humberts reported that 70 per cent of newly built homes in prime London locations were sold to overseas purchasers, increasing the multinational presence and creating diverse communities that affect national identity. This process has contributed to fast-rising rents and house prices, leading to questions about whether British people can afford to live in their own capital city. Similar debates surrounded the sale to foreign owners of iconic British brands such as Cadbury and the Mini. Can they still be considered British, and is national identity changed when iconic brands or locations come under foreign ownership and decision-making?

ACTIVITY

1. Explain how the various globalisation processes are challenging national identities.

2. Assess the ability of national identities to survive in the 21st century.

CASE STUDY: The Qatar Investment Authority (QIA)

The QIA is Qatar's sovereign wealth fund. It invests the surplus from Qatar's oil and gas wealth into economic sectors other than energy, in order to secure the long-term economic future of the emirate. It is estimated to have over US$180 billion of assets, which are invested in major Organisation for Economic Co-operation and Development (OECD) economies. For example, in the UK the QIA owns more than 12 per cent of Barclays Bank and is the largest shareholder in Sainsbury's. In Germany the QIA owns more than 17 per cent of Volkswagen, the car manufacturer, and in France 4 per cent of the Total oil company and 12 per cent of the Lagardère media company. In 2013 the QIA invested over US$200 million in Indian real estate, in 2014 it launched a plan to invest over US$10 billion in China, and in 2015 it announced a strategy to invest over US$30 billion in US assets over five years.

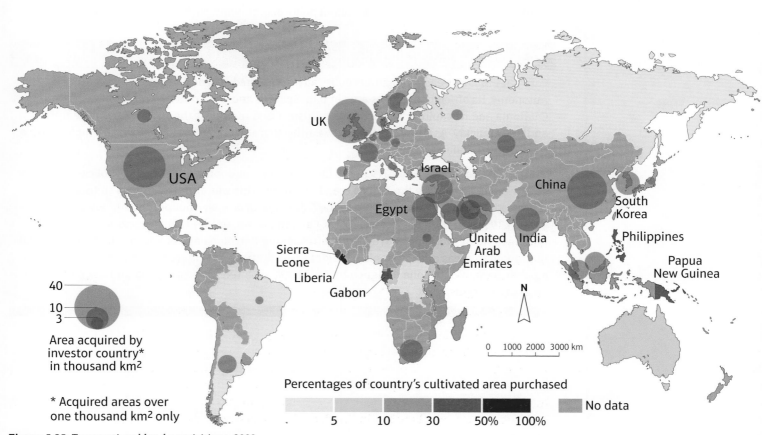

Figure 5.25: Transnational land acquisitions, 2009

Extension

Explore how the costs and benefits of foreign ownership are presented in different media.

- Choose two or three contrasting types of TNC and make notes from their official websites on the benefits that the corporation claims to bring to a foreign country.
- Then search news sites and crowd-sourced data sites, and make notes on any benefits or problems brought about by that corporation.

CASE STUDY: Facebook – challenging national identities

In 2015 Facebook, the online social networking service based in California, had a market capitalisation greater than US$300 billion, with 1.59 billion registered users. Facebook allows anyone with access to the internet to stay in touch with friends, relatives and other acquaintances, wherever they are in the world. The company's revenue comes from selling information about its users to advertisers and other companies.

It is a US company selling a global service, but since users generate their own content they do not always think of the company as American. However, the company itself has little connection to most of the countries in which it operates. All users outside the US and Canada have a contract with Facebook's Irish subsidiary, Facebook Ireland Ltd, an arrangement which allows Facebook to avoid US taxes for all users in Europe, Asia, Australia, Africa and South America. Facebook is making use of the 'Double Irish' arrangement, which allows it to pay only between 2 and 3 per cent corporation tax on all international revenue. Facebook generates as much revenue and profit as it can globally, while paying back as little back as possible into the societies in which it generates its revenue.

The consequences of national disunity

Many modern states contain strong nationalist movements that seek to create independent, smaller states separate from the larger country. Examples include Scotland in the UK, Catalonia in Spain and Quebec in Canada. However, the advocates of these new independent states still often see their future within larger supranational political groupings, such as the European Union.

CASE STUDY: Catalonia in the EU

Catalonia in north-eastern Spain is politically designated as a nation state by its Statute of Autonomy. Its official languages are Catalan, Spanish and the Aranese dialect of Occitan. Catalonia and its capital city Barcelona industrialised rapidly at the end of the 19th century and grew quickly. As Catalonia became wealthy, there was a cultural renaissance coupled with a burgeoning Catalan nationalism.

During the Spanish civil war (1936–39) Catalonia had autonomy under the Republican government, but during the Franco era after the war this was annulled and its culture severely repressed. For example, official use of the Catalan language was not permitted. However, from the late 1950s through to the early 1970s, Catalonia again enjoyed rapid economic growth, pulling in many workers from across Spain, making Barcelona one of Europe's largest industrial metropolitan areas and Catalonia a major tourist destination. Figure 5.26 shows Parc Güell, a public park in Barcelona designed by Antoni Gaudí, a renowned architect and a symbol of Catalan modernism.

Yellow = Players, Orange = Attitudes and actions, Purple = Futures and uncertainties

Figure 5.26: Parc Güell in Barcelona

Since Spain's transition to democracy (1975–82), Catalonia has regained some political and cultural autonomy and is now one of the most economically dynamic communities in Spain. This process has occurred alongside Spain's integration into the European Union. The twin processes of economic growth and security and cooperation within the EU have emboldened Catalan nationalism. As a result, on 9 November 2015, Catalan lawmakers approved a plan for secession from Spain by 2017, with a vote of 72 to 63. However, the Constitutional Court suspended the plan. It is not clear whether Catalonia's future lies within Spain or as an autonomous state within the EU, or whether Spain would allow Catalonia to join the EU if it legally seceded from Spain.

Political tensions in emerging economies

Jim O'Neill, a Goldman Sachs economist, coined the term BRIC in his 2001 report, 'The World Needs Better Economic BRICs'. BRIC stands for Brazil, Russia, India and China (many now extend the grouping to include South Africa, creating BRICS). He grouped these countries together because they were undergoing similar rates of rapid economic growth and their size – collectively 25 per cent of the world's land area and 40 per cent of its population – meant that this growth was of global significance. If the rate of growth had been maintained, their combined economies would have been larger than the combined economies of the current richest countries of the world by 2050. The countries have now formed a loose voluntary grouping to exert influence in developing countries. However, growth rates slowed in China, Russia, Brazil and South Africa as the 2008 recession reached them, especially with the decline in commodity values (raw materials). China's government plans to cut steel production by up to 150 million tonnes, which could see the loss of up to 400,000 jobs in this sector alone as they attempt to restructure China's slowing economy. Officials point to excessive industrial capacity, a slump in demand and falling prices.

Political tensions have been rising in all BRICS countries. Russia has also been involved in conflicts in Ukraine and Syria. Brazil and South Africa have seen allegations of corruption at the highest levels of government. As their economies have slowed, the uneven pattern of the costs and benefits of globalisation has become more apparent within their societies. These countries are vulnerable because they lack strong democratic institutions to mediate between the people directly affected by an economic downturn and those governing them. This is made worse by the lack of independent media scrutiny of official government, such as in China where a fearful ruling Communist Party has cracked down on public debate, labour activists, independent lawyers, civil society organisations, and access to the worldwide web. In Brazil, China and Russia, there is a fear of spreading social unrest. This risk is increasing in China, where the government is seeking to restructure unprofitable state-owned industries, such as coal and steel, with the potential loss of millions of jobs (Figure 5.27); economic growth targets of between 6.5 and 7 per cent for 2016 to 2020 are unlikely to be met.

Figure 5.27: A 'zombie' factory in China's steel belt (an abandoned Qingquan Steel plant in Tangshan)

Failed states

The role of the government varies from country to country. Some countries, like Russia and Saudi Arabia, have powerful, authoritarian governments. Western countries like the UK and the US are powerful states but there are checks and balances on authority through democratic processes. However, many emerging countries have governments that work with powerful local elites, rather than operating independently. Identity in these countries often has greater allegiance to family, tribe or religion than national identity. This is especially true in 'failed states' where the government is weak or non-existent, and stark differences are apparent between the politically and economically powerful elite, foreign investment groups and the wider population (Figure 5.28).

A failed state means that government has broken down, and is characterised by:

- loss of control over its territory, or lack of physical dominance (military)
- erosion of legitimate authority in the decision-making process
- inability to provide public services
- inability to interact with other states as a full member of the international community.

Yellow = Players, Orange = Attitudes and actions, Purple = Futures and uncertainties

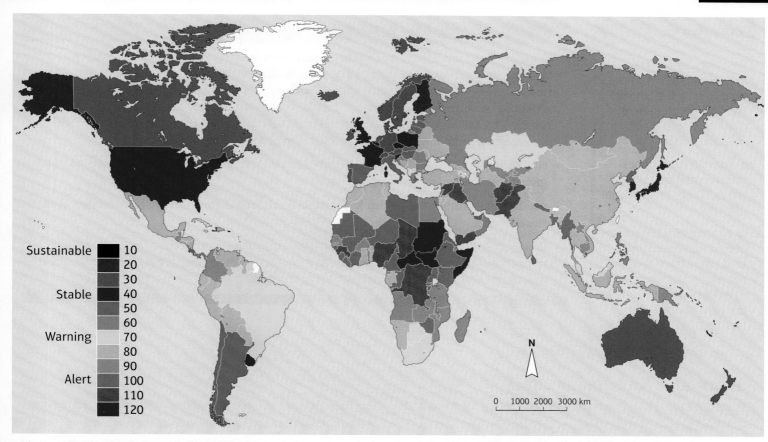

Figure 5.28: The Fragile States Index 2015

Legend:
Sustainable — 10, 20, 30
Stable — 40, 50, 60
Warning — 70, 80, 90
Alert — 100, 110, 120

0 1000 2000 3000 km

CASE STUDY: Somalia, a failed state

Somalia has been considered a failed state since 1991, when it collapsed into anarchy following the overthrow of the military regime of President Siad Barre. This breakdown of government occurred as rival warlords tore the country apart into clan-based fiefdoms. An internationally backed unity government was formed in 2000, but struggled to establish control. As a result, the two relatively peaceful northern regions of Somaliland and Puntland became self-governing. The seizure of the capital Mogadishu and much of the country's south by a coalition of Islamist Sharia courts in 2006 prompted an intervention by Ethiopian and African Union forces. These forces slowly began to reintroduce some form of control to the country. Since 2012, when a new internationally backed government was installed, Somalia has been getting closer to stability, but the new authorities still face a challenge from Al-Qaeda-aligned Al-Shabab insurgents. Somalia's disintegration is reflected in its intimidation of domestic media by government security agencies and in President Hassan Sheikh Mohamud's decision to halt plans for the next presidential election, citing a lack of security and infrastructure.

A level exam-style question

Explain the possible causes of disunity within developed and emerging nations. (8 marks)

Guidance

Don't spend long on definitions, but concentrate instead on demonstrating your understanding of the factors that may cause unrest in a country. Refer briefly to examples, and perhaps consider whether these factors are similar in different types of country.

Extension

Research the current impressions and messages produced by the Somali national government or clans. Use official and media websites.

a) To what extent does this information suggest the existence of a national identity?

b) Analyse how the sources of information use data to support the role of the national government and demonstrate that Somalia is a nation with a clear identity.

CASE STUDY: Somalia, a national identity?

The majority of Somalis have a similar ethnicity, heritage and culture, language, religion and laws. Nowever, there are also strong allegiances to clans and sub-clans, and so while Somali has many features of a national identity these are threatened by competing clans and the failure of the national government to protect and provide for its citizens. This has certainly been the case since the 1991 collapse of government and Figure 5.28 shows that the country is in the highest 'alert' category. Some success has been achieved in promoting national identity through:

- civil society organisations - which provide discussion forums in communities, especially giving a voice to marginalised groups such as women

- poetry and cultural song - which have national unifying themes (but these have been weaker since 1991)

- dividing the tasks of national and regional governments – with a central government co-ordinating national themes and resolving disputes between clans, and regional government providing services (but this model not yet operative).

Summary: Knowledge check

Through reading this chapter and by completing the tasks and activities, as well as through your wider reading, you should have learned the following and be able to demonstrate your knowledge and understanding of migration, identity and sovereignty (Topic 8 Option 8B).

a. Explain how globalisation has changed the global economic system.

b. Explain migration types and the processes that control movements of people.

c. Explain how environmental, economic and political events shape the pattern of international migration.

d. Assess how economic theory explains the movements of resources, goods, capital and labour.

e. Suggest how migration changes the cultural and ethnic composition of nation states, and how tensions arise as a result of these changes.

f. Explain why national sovereign states vary greatly in their ethnic, cultural and linguistic unity.

g. Suggest why borders are sometimes contested and how these reasons are linked to historical and contemporary situations.

h. Explain how the forces of nationalism have shaped nations today through the colonial period and independence.

i. Explain how TNCs and IGOs have helped develop new forms of states.

j. Suggest how global inequalities threaten the sustainability of the global economic system and encouraged the emergence of alternative economic models.

k. Explain the role of the United Nations and its affiliated IGOs in managing global environmental, socio-economic and political problems.

l. Explain why some countries operate independently on the world political stage.

m. Explain the different economic strategies, including trading blocs, for achieving economic development.

n. Explain the different management strategies for achieving environmental sustainability.

o. Explain the concept of nationalism and all of the ways in which a national identity may be changed.

Yellow = Players, Orange = Attitudes and actions, Purple = Futures and uncertainties

As well as checking your knowledge and understanding of these subjects, you also need to be able to make links and explain the following ideas:

- Globalisation forces, such as communication improvements, fast transport, the spread of TNCs and historical migrations have greatly increased the ability of people to move from one country to another and also within a country, as a result of FDI, for example.

- The causes of migration are rooted in push and pull factors, and the economic motive linked to jobs and money is still the strongest factor, but increasingly people are displaced by conflicts and this increases refugee movements and also the numbers of internally displaced people.

- The consequences of international migration are varied and disputed because, while some impacts are immediate, such as filling important jobs, the need for more housing and services or the creation of tension within communities, others are longer-term – such as conflict, integration to create a multicultural society, or changes in migration policy.

- Today's nation-states have arisen as a result of historical processes based on physical or human geography, but often not both, which has led to contested borders.

- Nationalism has played a role in the development of the modern world through the formation of many new countries as empires have disintegrated, with significant individuals shaping the future of these countries.

- In Western capitalist markets the profit motive is strong, and new financial flows have been created, such as to tax havens. While inequalities between countries are slowly diminishing, inequalities within countries have grown between a rich elite and large numbers of poorer people, prompting alternative approaches to be tried.

- Global organisations have proliferated since 1945 and now play an important part in every aspect of managing the world, including covering socio-economic, environmental and political issues, with some of these leading to military conflicts and much suffering. Often there have been major nations on opposite sides, which can have serious implications for global stability.

- While many people link themselves strongly to a nation, international migration and globalisation have perhaps modified the definition of national identity. Symbols remain but what people feel and think is changing.

- External influences on countries are stronger today than they have ever been, as shown by the spread of Westernisation, perhaps centred on American culture, and the migration and investment by a global elite and their businesses in foreign countries.

- A counter-reaction to globalisation is perhaps the small but significant movement to gain independence by small states, and popular uprisings against the corruption and power of elites in emerging countries

Preparing for your A level exams

Sample answer with comments

Explain the role of national governments in managing migration flows. (8 marks)

Developed countries may try to limit migration flows; for example, the UK regulates immigration from non-EU countries through a tiered and points system (2008). Tier 1 enables entrepreneurs, investors and exceptional talent to enter the UK in order to further its development and Tier 2 allows migrants with a UK job offer to enter. The EU has an agreement that allows free movement of goods, capital, services, and people between its 28 member states. The increasing interconnection of EU economies has created incentives for people to migrate for economic reasons. There are also temporary migration and tourism flows within the area.

In developing or emerging countries there is a migration from rural to urban areas. For example, China has strong control over these flows. It uses the Hukou system to try to control the movement of people and their families. Despite this official barrier, the pressures to migrate are strong and overwhelm this system.

Some states do not have full control over their borders. For example, the 2011 Libyan Civil War, with numerous rival armed militias affiliated to regions, cities and tribes, has caused refugee movements internally and externally, which the government cannot control.

> The UK is a useful example but is largely descriptive, apart from one explanation of the policy. The reference to the Schengen area does not clearly connect to the role of the national governments, and tourism is not migration. It is good practice with 8-mark questions to get straight to the point of the question.

> Developing countries need mentioning and China is a suitable example. Singapore could also have been mentioned, or developing countries with weaker policies. The role is stated but could have been explored further.

> The answer finishes with another example rather than a brief summary answer to the question.

Verdict

This is a competent average answer. There is general understanding and relevant case studies, but the amount of explanation is limited and undeveloped, and more accurate data could have been included.

Preparing for your A level exams

Sample answer with comments

Evaluate this statement: 'In a globalised world, nationalism remains a powerful force.' (20 marks)

Globalisation has involved a broad spectrum of international economic, cultural and even environmental processes, including trade, capital flows, and diffusion of technology. International players are therefore very important, and the UN agencies play a part in international political decisions and bringing countries together, as does the WTO. However, there are about 220 nations in the world and these have their own cultures, political systems and economies. Which is stronger in a globalised world – the national identity or the global identity?

> *This introduction avoids the trap of just repeating basic definitions; it does link ideas together within the context of the question and it sets the evaluation parameters for a comparison of nationalism and internationalism.*

Nationalism in its various forms has been seen as a driving force for many conflicts in modern history, which has prompted politicians and the general public within nations to build a new world based on common values of liberty, democracy, human rights and civil liberties, and the rule of law. This has created stronger international links such as the European Union and the United Nations. The aim was to protect the nationalism of individual countries, but it also started the political and economic globalisation process. The United Nations and associated intergovernmental organisations of the post-1945 world show international attempts to weaken the worst aspects of nationalism – territorial expansion – by working together in global conflict areas. The Montreal Protocol on depletion of stratospheric ozone was agreed by 197 countries, making it the first universally ratified treaty in United Nations history. The protocol shows the importance of international agreement to tackle a global problem.

> *This paragraph illustrates the start of greater international cooperation and political globalisation, but doesn't take the opportunity to explore how individual nations work within these organisations to represent their own interests, such as vetoes in the UN Security Council, or the UK 2016 referendum on EU membership.*

The IMF and the World Bank are IGOs that promote international financial movements and trade. However, their attempts to support development in the poorest countries suffering from balance of payments crises through Structural Adjustment Programmes (SAPs) have been criticised for worsening the problem. Their reforms have destabilised developing countries and worsened their quality of life. SAPs did not lead to an economic boom in Latin America, but large external debts. Economic cooperation between nations has not been beneficial for the poorest nations, but served to reinforce the advantages of the most powerful.

> *This paragraph lacks direction. Is it trying to say that international links are weaker because they can cause problems, encouraging countries to focus inwardly, or that international links are stronger than a single nation? Wording of answers must always directly answer the question.*

Nationalism inspires feelings of political and cultural loyalty, often through symbols or sport. Despite the actions of IGOs and supranational groups such as the EU, nationalism is strong because it is reinforced by education, experience and by political movements stressing loyalty. For example, the essence of English common law is that it is made by judges sitting in courts, applying legal precedent to the facts before them, while other European countries follow different systems. Migrations, for example from the Caribbean and South Asia to the UK, or from Eastern Europe to the UK, spread culture, which may weaken the national culture of the host country, creating complex national identities and loyalty.

Currently, nationalism is strong despite the world having globalised, as shown by the KOF index, although the dominant 'Westernisation' from the USA and TNCs plays an important part in weakening it.

> *This adds to the breadth of forces – i.e. culture. There is also some evaluation, although it is not fully developed.*

> *A very short conclusion that does offer an answer to the question but with points not previously made. It is important that the key points made in the answer are pulled together in the conclusion.*

Verdict

This is a sound average answer. There are a lot of relevant ideas and the candidate shows a good understanding of connections and links. However, making a plan at the start would have ensured greater coherence and especially identification of the key evaluation points – i.e. the points that support nationalism still being powerful and the points that suggest that it is not powerful. These points are implied rather than made clear. A plan would have also helped timing and the balance of themes within the answer.

Appendix: Maths and Statistical answers

Independent Investigation
Activity page 14

Chi squared test (two or more data sets)

Male primary school: $(49 - 46.97)^2 \div 46.97 = 0.0877$

Male lower secondary school: $(26 - 31.47)^2 \div 31.47 = 0.9508$

Female pre-primary school: $(48 - 51.44)^2 \div 51.44 = 0.2300$

Female primary school: $(51 - 53.03)^2 \div 53.03 = 0.0777$

Female lower secondary school: $(41 - 35.53)^2 \div 35.53 = 0.8421$

Chapter 2: Carbon and Energy
Activity page 81

Net ocean flux per year: $80 - 78.4 = 1.6$ PgC/yr into oceans

Net terrestrial flux per year = $123 - 118.7 = 4.3$ PgC/yr into vegetation

Activity page 89

Total anthropogenic emissions: $375 + 180 = 555$ PgC

Atmosphere: $240 \times 100 \div 555 = 43.24\%$

Oceans: $155 \times 100 \div 555 = 27.93\%$

Terrestrial ecosystems: $160 \times 100 \div 555 = 28.83\%$

Activity page 96
Table 2.5: Global oil consumption and prices 1955 to 2015

Year	Average oil price (US$ per barrel, 2014 values)	Index	Oil consumption (millions of barrels a day)	Index
1955	17.05	100	15.377	100
1960	15.17	88.97	21.471	139.63
1965	13.49	79.12	30.806	200.34
1970	10.97	64.34	45.348	294.91
1975	50.74	297.60	54.327	353.30
1980	105.81	620.59	61.233	398.21
1985	60.64	355.66	59.247	385.30
1990	42.97	252.02	66.737	434.01
1995	26.43	155.01	70.322	457.32
2000	39.17	229.74	76.868	499.89
2005	66.09	387.62	84.411	548.94
2010	86.31	506.22	87.867	571.42
2015	50.75	297.65	93.700	609.35

Spearman's rank correlation analysis shows a positive correlation of +0.643, suggesting that as oil consumption increases, so does the price. This is significant at the 95% (0.05) level where the critical value is 0.560.

Activity page 121

(a) The mean for CO_2 levels from 1960 to 2015 is 352.80 ppm
(4233.58 ÷ 12 = 352.7983)

(c) Rate of change 1960 to 2015 is +26.48% (400.83 x 100 ÷ 316.91
= 126.48)

Chapter 5: Migration, identity and sovereignty
Activity page 268

a) Calculate percentages
Table 5.3: Population of Japan and the UK in 2016 by major ethnic group

Japan (total population in '000s: 127,000)			United Kingdom (total population in '000s: 64,716)		
Ethnic group	Population ('000s)	Percentage	Ethnic group	Population ('000s)	Percentage
Japanese	125,095	98.5	White	56,368	87.1
Korean	635	0.5	Asian	4,530	7.0
Chinese	508	0.4	Black	1,941	3.0
Filipino	254	0.2	Mixed	1,294	2.0
Brazilian	254	0.2	Other	583	0.9
Vietnamese	127	0.1			
Other	127	0.1			

Glossary

Aquifer: a permeable or porous rock which stores water.

Asylum-seeker: a refugee who has made a request for protection and residence in a host country and is awaiting a decision by the state where they made the request.

Bi-polar: a world where two countries either share or compete to have the greatest global influence.

Biogeochemical: the transfer of elements and compounds, such as carbon, between living organisms and the physical environment through chemical processes that create new compounds and elements in the atmosphere, hydrosphere and lithosphere.

Capital: assets, goods or finance that can be used for economic activity.

Capitalism: an economic and political system in which trade and industry are controlled by private owners for profit, rather than by the state.

Capitalist society: a culture based on an economic system of enterprise and private ownership of all parts of the system with minimal interference from governments so that economic conditions are competitive.

Carbon fluxes: the movement or transfer of carbon, in different compounds, between stores in atmosphere, biosphere, hydrosphere and lithosphere.

Carbon-neutral: a process or activity that results in no net release of carbon into the atmosphere, perhaps through using renewable energy or planting trees.

Carbon pathway: the steps involved in moving carbon to a store where it is fixed.

Carbon store or sink: as part of the natural carbon cycle carbon accumulates in places within the cycle, often for a very long time period. These stores include vegetation, atmosphere, oceans and rock.

Channel flow: the water flowing in a rivulet, stream or river contained within banks.

Child mortality: the number of deaths of children under five years of age, compared with the total number of live births in one year in an area. Usually expressed out of 1,000, but sometimes given as a percentage.

Closed system: a sequence of linked processes with a transfer of energy but not matter between the parts of the system (the inputs and outputs happen within the system). An example is the global hydrological cycle.

Colonialism: the policy or practice of acquiring full or partial political control over another country or territory, occupying it with settlers and using its resources for trade and economic gain.

Colony: a territory occupied as part of an empire.

Contested borders: national borders that are not recognised by different players because of disputed ownership of territory.

Convectional precipitation: occurs when intense insolation (solar radiation) reaches the ground and lower atmosphere and causes convection in humid air. As the warmed moist air rises, it cools and water vapour condenses into clouds, forming precipitation – often in the form of thunderstorms.

Convention: an international agreement between countries to cover a specific matter, which may then become a source of international law.

Corruption: the abuse of power for private gains, usually by those who hold political power, and undermines trust and restricts freedom; the scale may vary depending on the amount of money involved.

Counterfeiting: creating an imitation or fake product or component that is significantly cheaper and inferior than the original, but breaks patent laws and seriously undermines trade in legitimate goods.

Critical threshold: the level at which there is a sudden or very rapid change.

Cultural heterogeneity: the presence of multiple groups of people from different backgrounds within a population of a place.

Democratic government: a system of governing where the people have power, usually through 'one person, one vote' in free elections of representatives as determined by the historical development of the country.

Democratic institution: a government organisation responsible for ensuring that democratic processes take place, from a parliament to the courts of law to a government agency or department.

Deprivation: where people lack things that they need. In a developing country this may be food, water, shelter and human rights, while in a developed country it may be lack of a job, below-average access to health care or education, or substandard housing.

Deregulated financial market: the removal or reduction of rules governing financial transactions.

Deregulation: the reduction or elimination of government control or support over a part of the economy (business and industry), especially removing barriers to competition.

Deregulation of capital market: the removal or reduction of rules governing investments in business and other assets.

Desalination plant: the process of converting saltwater to freshwater suitable for human consumption and industrial purposes.

Development aid: the assistance given by a wealthier country or group of countries to a poorer country to help it develop, which can take various forms such as humanitarian aid from NGOs who receive money from public donations, or official development assistance (ODA) from governments who allocate a proportion of tax income.

Development: the process by which a country evolves over time for the better in terms of economic, social, political and sustainable change.

Direct control: the imposition of power by a more powerful group, such as through military rule.

Direct military intervention: the use of military force by one state in the territory of another state.

Direct runoff: water flowing over the surface of the ground after precipitation, snowmelt or throughflow reaching the surface.

Drainage basin: the area of land drained by a river and its tributaries (river system) and separated from neighbouring drainage basins by a ridge of highland called a watershed or divide.

Drainage density: the total length of all the streams and rivers in a drainage basin divided by the total area of the drainage basin.

Drought: the definition varies internationally. According to the UN, drought is an extended period (a season, a year, or several years) of deficient rainfall relative to the statistical average for a region measured over a very long period of time.

Economic centre of gravity: the main location of the majority of economic activity, usually considered on a world scale, and at this scale the centre has shifted towards East Asia.

Economic restructuring: the gradual change from a manufacturing to service sector economy following deindustrialisation.

Economic sanctions: commercial and financial support for a country is withdrawn or blocked by a country or group of countries because of infringement of international laws or agreements; this may include trade barriers, tariffs, and restrictions on financial transactions. Individuals may also be specifically targeted.

Economic sovereignty: the authority to make laws governing economic operations in a country.

Economic water scarcity: occurs when water resources are available but there is insufficient human, institutional and financial capital to access the water in order to meet demand.

Ecosystem functioning: refers to the biological, chemical and physical processes that take place within an ecosystem.

Ecosystem productivity: the use, or fixing, of solar energy by plants to increase their biomass (primary productivity) through photosynthesis, which supports the growth of herbivores (secondary productivity) and carnivores (tertiary productivity).

Ecosystem resilience: refers to the capacity of an ecosystem to recover from disturbance or withstand ongoing pressures.

Emerging power: a country that has rapidly increased its influence and economic position in the world.

Empire: a group of nations and people ruled over by a more powerful foreign government or leadership, creating a territory that is often worldwide in extent.

Energy mix: the range of energy sources used by a country or region, from non-renewable ones such as fossil fuels to renewables such as wind energy.

Energy pathway: the route by which an energy type is transferred from the production area to the consumption area, such as by pipeline or shipping route.

Energy security: a situation where there is a secure and affordable supply of energy to meet the needs of consumers (people and businesses).

ENSO cycle: a naturally occurring phenomenon that involves the movement of a mass of very warm water in the equatorial Pacific due to changes in the surface trade winds, atmospheric circulation and ocean currents. There are two phases, known as El Niño (warm water to the east) and La Niña (warm water to the extreme west).

Environmental quality: the state of a natural environment as measured by various criteria such as biodiversity, or the state of the environment in which people live which would include water, air and land pollution levels.

Evaporation: the process by which liquid water is transformed into water vapour (a gas) when heated.

Evapotranspiration: the total amount of moisture transferred from the Earth to the atmosphere by evaporation and transpiration.

Failed state: a country in which government has lost control of one or more national functions, such as law and order, economy, or democratic recognition.

Foreign policy: a government strategy for dealing with the government or leaders of another country. This will be decided on the basis of what is best for the home country and human rights and development in the foreign country.

Fossil (non-renewable) water: water that has been contained and undisturbed for an extremely long period of time (millennia), usually groundwater in an aquifer. There is little to no significant recharge, perhaps due to a change in climate due to tectonic movement such as in the Sahara, therefore it is a non-renewable resource.

Fragile environment: a natural ecosystem that is sensitive to change due to the delicate balance between all the elements in the ecosystem and systems affecting it.

Free trade: trade that is not controlled by taxes, tariffs or regulations.

Frontal precipitation: occurs frequently in mid-latitudes when a warm tropical air mass meets a cooler polar air mass. The warmer air is less dense and rises over the colder air, which causes the warm air to cool, leading to condensation of water vapour, clouds of different types and precipitation.

G20: the 20 richest countries in the world, which meet semi-regularly to discuss key global economic issues.

Gender equality: a situation where men and women are treated equally in all aspects of life.

Geopolitical intervention: the use of economic, political or military power by countries (superpowers) or groups of countries (African Union) in different regions of the world to bring about change.

Geopolitical relations: the study of the effects of geography on international politics and international relations.

Global governance: the actions of IGOs that collectively provide a framework of rules and practices that attempt to govern the behaviour of players on a global scale.

Global hydrological cycle: the continuous transfer of water between land, atmosphere and oceans and seas on a planetary scale within a closed system, driven by solar radiation.

Global water budget: the amount of water transferred and stored in the Earth's hydrological cycle each year, including fluxes (water flows) and the volume of the water stores (oceans, atmosphere, biosphere, cryosphere, groundwater and surface water).

Globalisation: the growing economic interdependence of countries worldwide through increasing volume and variety of cross-border transactions in goods and services, freer international capital flows, and more rapid and widespread diffusion of technology.

Groundwater flow: water moving sideways through a permeable or porous rock under the influence of gravity.

Hard engineering: the use of man-made, artificial structures to manage flooding or water supply.

Hard power: the ability of a country or group of countries to use military force or direct economic influence to make another country accept a situation or idea.

Hegemony: the ability to establish leadership over a group of people through political or cultural processes.

Human capital: the manual and intellectual skills, knowledge, understanding and experience of people creates a resource that can be used.

Human rights: inalienable fundamental rights to which a person is inherently entitled; all people, regardless of nationality, location, language, religion, ethnic origin or status should have these rights.

Hydrological drought: occurs when there is insufficient soil moisture to meet the needs of vegetation at a particular time.

IGO: Intergovernmental organisation such as the United Nations and its affiliates which involve the participation of many countries.

Imperial era: a period of time when many countries were governed and very strongly influenced by colonial rule, usually by a European country.

Indigenous population: people who have lived in an area for hundreds or thousands of years, with their own language, cultural traditions and their own territory to which they have strong ties.

Indirect control: a subtle way of exercising power such as through trade or cultural influences.

Inequalities: economic and social differences between people that have and those that lack, creating situations regarded as unfair.

Infant mortality: the number of deaths of children under one year of age, compared with the total number of live births in one year in an area. Usually expressed out of 1,000, but sometimes given as a percentage.

Infiltration: the movement of water vertically downwards through the spaces (pores) in the soil.

Integrated drainage basin management: aims to establish a framework for coordination whereby all administrations and stakeholders involved in river basin planning and management can come together to develop an agreed set of policies and strategies, such that a balanced and acceptable approach to land, water, and natural resource management can be achieved (World Bank definition).

Interception: the process by which raindrops are prevented from falling directly onto the ground by the leaves, branches and twigs of vegetation.

Interdependence: the strong links between two or more countries which create a situation where they become dependent on each other.

International agreement: a written or oral form of a treaty but dealing with a narrower range of matters than a full treaty, particularly used for technical or administration matters.

International law: the legal rules that have been established by treaty and recognised by countries as binding in their international behaviour and relations.

Intervention: the use of economic or military power to intervene in a sovereign country due to concerns about human rights issues.

Labour: the workforce in any economic sector.

Life expectancy: the average number of years that a newborn baby would be expected to live if the same conditions in the area of birth remained unchanged.

Low-tax regime: a state that sets lower rates of taxation than the norm, often to attract investment.

Maternal mortality ratio: the number of deaths of females per 100,000 live births in a year while pregnant or within 42 days of pregnancy finishing.

Meteorological drought: occurs when long-term precipitation trend is below the long term average.

Migrant: a person who moves their 'permanent' residence from one place to another for at least one year, this may be long distance or short distance, and voluntary or forced.

Military aid: this is given to countries, usually allies, to help fight terrorist or defend against insurgency groups or combat international crime (such as drugs); sometimes it may be used for geopolitical reasons to change a regime. Military aid may involve money to buy arms, or military equipment, or intelligence and surveillance, or forces for peacekeeping operations.

Mitigation: ways in which people can reduce human impacts on climate by reducing emissions and creating or enhancing stores of greenhouse gases.

Monsoon: a seasonal change in the direction of prevailing winds of a region, causing wet and dry seasons in many sub-tropical areas.

Multipolar: a world where several countries have a highly influential role in the world or their world region and link together in world economic and political systems.

Nation state: a sovereign state that derives its legitimacy from representing a particular people; a people with a cultural identity, often derived from a common history or dominant historical narrative.

National identity: the identity that results from a common background based on historical and modern culture, symbolism and feelings of belonging.

Nationalism: identifying with and developing a sense of belong to a nation through ties to cultural symbols such as flag, sport, foods and drinks and clothing.

Neocolonial: the use of investment, trade and culture to influence independent countries instead of direct governance.

Neoliberal view: an approach that transfers control of economic factors to the private sector away from government control, with the idea that there is an open market for trade with limited protectionism and subsidies and the economy is free of restrictive barriers and regulations.

NGO: non-governmental organisation, often a charity or voluntary citizens' group, operating at a local, national or international level.

Open system: a sequence of linked processes with inputs and outputs, including transfers of energy and matter to and from other systems, for example a drainage basin.

Orographic precipitation: rainfall or other type of precipitation occurring when an air mass is forced to rise over high land (mountains or large hills). As moist air rises, it cools and causes condensation, cloud formation and precipitation, mostly on the windward-facing slopes and highest altitudes.

Out-gassing: the release of a gas that was dissolved or stored due to changes in heat or pressure; for example carbon is released by metamorphic activity at plate boundaries or hot spots.

Patent: the exclusive right given to an inventor or company to use or sell an invention for a period of time, this is important for protecting the property rights of individuals and encouraging entrepreneurship.

Percolation: water moving vertically downwards through and into a permeable or porous rock.

Physical water scarcity: occurs when there is a physical lack of available freshwater resources to meet demand due to over-abstraction by agriculture, industry and domestic purposes.

Phytoplankton: minute plants, such as cyanobacteria, found in upper layers of oceans, which fix large amounts of carbon through photosynthesis and form the base of aquatic food webs.

Political tensions: unrest and distrust arising from unpopular decisions made by the ruling groups of a country.

Primary energy: the main original sources of energy before conversion in alternative forms, such as coal and crude oil.

Primary producers: living organisms that produce their own food, using sunlight, carbon dioxide, water and other chemicals in the process of photosynthesis; they are also sometimes referred to as autotrophs.

Privatisation: the transfer of business or industrial ownership from government to private enterprises.

Protocol: a type of agreement that is less formal than a treaty or convention but helps to interpret parts of a treaty or add clauses, or regulate technical matters; protocols may be added to expand a treaty but that may not be agreeable to all who signed the treaty.

Rare earths: a group of seventeen elements that occur together in geologic deposits and are very difficult to separate from each other; they are very important to the manufacture of electronic devices.

Refugee: a person who has been forced to leave their country to escape likely death from war, persecution or a natural disaster.

Regional power: a country or state that has a power or influence in a part of a continent or world region.

River regime: the pattern of river discharge over a year; usually there are seasonal variations.

Saltwater encroachment: the movement of saltwater into freshwater aquifers due to sea level rise, storm surges and/or human abstraction of groundwater which lowers the water table.

Saturated overland flow: occurs when water accumulates in the soil until the water table reaches the surface, forcing further rainwater to runoff the surface. May also occur when the amount of precipitation exceeds the infiltration capacity of the soil.

Secondary energy: a convenient and more usable energy source, such as electricity, that has been created from a primary energy source.

Sequestration: processes by which carbon is removed from the atmosphere and stored for a long period of time, for example by plants and soil in nature, or through carbon capture and storage (CCS) from power stations.

Smart irrigation: a water conservation scheme that provides crops with a sub-optimal water supply causing mild stress during the crop growth stages when the plants less sensitive to moisture deficiency, and therefore there is no significant reduction in yield.

Soft power: the ability of a country or group of countries to persuade other countries to agree to a situation or idea by making it attractive.

Sovereignty: the rights of a country to have its own government and run its own affairs within an internationally recognised territory.

Storm hydrograph: shows change in a river's discharge at a given point on a river over a short period of time (usually before, during and after a storm).

Superpower: a very powerful country with worldwide influence due to a dominant economy, culture, political persuasion or military strength.

Sustainable water management: schemes that aim to balance economic, social and environmental needs by working with communities to develop soft-engineering schemes, which work with natural processes to restore or extend water security and often involving water conservation.

Tax haven: a place that offers financial services to wealthy individuals and companies, often characterised by secrecy that may enable avoidance, evasion and circumvention of criminal law; also disclosure rules (transparency), financial regulation and inheritance rules of other countries and their legal jurisdiction can be avoided.

Technological fix: a human innovation using technology to solve a problem such as water supply issues.

Thermohaline circulation: slow, large-scale seawater movement between all of the oceans, caused by differences in temperature and density.

Throughflow: water moving sideways through the soil, downslope under the influence of gravity.

Tipping point: the point at which even a small change is significant enough to cause larger, more significant changes which then cannot be prevented or reversed.

Totalitarian regime: a form of government where the ruling political group exercise tight control over all aspects of their citizens' lives to the point where freedom of expression is suppressed (extreme authoritarian).

Trade embargo: a method of banning trade with a country or group of countries, this is usually in response to infringements of international law such as human rights, or breaching trade agreements.

Trading bloc: a type of intergovernmental agreement, often in a world region and part of an intergovernmental organisation (e.g. NAFTA, ASEAN), where regional barriers to trade such as tariffs are reduced or eliminated between the participating states.

Transboundary water: where a river, lake, or aquifer crosses one or more major political borders (such as a state or province border within a nation or an international border).

Transpiration: the biological process by which water is drawn upwards from the soil by plants and evaporated through the pores (called stomata) in the leaves.

Treaty: an agreement signed between states, recognised under international law.

UN Security Council: an important division of the United Nations responsible for maintaining international peace and security and approving changes to the UN Charter; it has five permanent members and ten rotating members.

Uni-polar: a world where one country dominates global politics and economics, and perhaps culture.

War on terror: a global military, political, legal and conceptual struggle against officially recognised terrorist groups, organisations, and the countries supporting them.

Water budget: the annual balance between inputs (precipitation) and outputs (evapotranspiration and channel flow) at a place.

Water conservation: strategies to reduce water usage and demand.

Water insecurity: occurs when the economic, social and environmental requirements for water supplies are not met.

Water recycling: the treatment and purification of waste water using advanced membrane technologies and ultra-violet disinfection so that it is clean and safe to be re-used for industrial or domestic purposes.

Water scarcity: occurs when renewable water resources are only between 500 and 1,000 m^3 per capita per year.

Water security: the capacity of a population to safeguard sustainable access to adequate quantities of acceptable quality water for sustaining livelihoods, human wellbeing, and socio-economic development, for ensuring protection against water-borne pollution and water-related disasters, and for preserving ecosystems in a climate of peace and political stability (UN definition).

Water sharing treaty: international agreements on international water sources such as a river flowing through several countries, such as the Nile, Ganges or Mekong.

Water stress: when renewable water resources are only between 1,000 and 1,700 m^3 per capita per year.

Water transfer: hard engineering projects, such as pipelines or aqueducts, that divert water from drainage basins with surplus water to those with shortages.

Welfare state: a government establishes a system to look after its citizens, including health care, pensions, education, and benefits for those at a disadvantage.

Westernisation: the gradual adoption or conversion to Western economics (capitalism) and culture; sometimes considered synonymous with Americanisation.

Index

Acknowledgements

The publisher would like to thank the following for their kind permission to reproduce material in this product.

Picture credits

(Key: b-bottom; c-centre; l-left; r-right; t-top)

Alamy Images: A.P.S. (UK) 105, Songquan Deng 282, Design Pics Inc 94br, 109, epa european pressphoto agency b.v. 94bl, Brendan Hoffman 185, Jake Lyell 71, John Kellerman 302, robert harding 290, WorldFoto 10–11; **Christian Aid:** Prospery Raymond 226; **Fotolia.com:** 22–23, 134–135, 256, 269, Vasily Merkushev 56; **Generation Indigenous (Gen I):** Photo Courtesy of Sun Rosa / www.wearesubrosa.com / © Seed Communications LLC 218; **Getty Images:** 52, 144, 166l, 166r, 254, Ted Aljibe 208, Barcroft Media 250–251, Bloomberg 230r, Don Cook 292, Yuri Cortez 263, Kevin Frayer 303, Dave Hogan 295, Istock 293, Sia Kambou 243, Sebastian Meyer 233, SERGEI SUPINSKY / AFP 272, Anthony Wallace 237; **ITER Organization:** MatthieuColin.com 110; **Lee Howard Photography:** 176; **Lindsay Frost:** 16; **Médecins Sans Frontières:** Sam Taylor 239; **NASA:** Goddard Space Flight Center 123; **Press Association Images:** Sunday Alamba / AP 230l; **Shutterstock.com:** 107, Christopher Kolaczan 102; **UNOG Library, League of Nations Archives:** JC McIlwaine 186–187; **USGS / Craig D. Allen:** Craig D Allen 50; **Wikimedia Commons:** John Constable 297, Walter Crane 273, DOD Defense Visual Information Center 285bl, Southwesterner 158, U.S. Army – Mountain Ridge Security 285br, U.S. Army photo by Staff Sgt. Kyle Davis. 285tr, U.S. Navy photo by Chief Photographer's Mate Eric J. d 285tl

Cover images: *Front:* **Getty Images:** Lumi Images / Romulic-Stojcic

All other images © Pearson Education

Figures

Figure 1.2 from Estimates of the Global Water Budget and Its Annual Cycle Using Observational and Model Data, *Journal of Hydrometeorology*, Vol. 8 (4), pp.758–769 (Trenberth, K.E. et al. 2007); Figure 1.3 from *Wetlands Overview* EPA 843-F-04-011a, EPA (Office of Water 2004) p.1; Figure 1.5 from Mean Annual Precipitation 1950 – 2000, http://atlas.gwsp.org/atlas/img/map/a1_precip_WSAG1_0_wl.png, By kind permission of C. Vörösmarty, Water Systems Analysis Group (University of NH) (Currently Advanced Science Research Center, City University of New York).; Figure 1.7 adapted from Water cycle of the Earth's surface, showing the individual components of transpiration and evaporation that make up evapotranspiration, https://en.wikipedia.org/wiki/Evapotranspiration#/media/File:Surface_water_cycle.svg, CC BY 3.0; Figure 1.8 from Hydrological cycle with land use impacts, http://soer.justice.tas.gov.au/2009/image/126/index.php; Figure 1.10 from 39031 – Lambourn at Welford, http://nrfa.ceh.ac.uk/data/station/meanflow/39031; Figure 1.11 from 73005 – Kent at Sedgwick, http://nrfa.ceh.ac.uk/data/station/meanflow/73005; Figure 1.14 from *Natural disaster hotspots: A global risk analysis*, World Bank Hazard Management Unit (Dilley, Maxx; Chen, Robert S.; Deichmann, Uwe; Lerner-Lam, Arthur L.; Arnold, Margaret. 2005) Figure 4.1b, p.37, © World Bank. https://openknowledge.worldbank.org/handle/10986/7376 License: CC BY 3.0 IGO; Figure 1.17 adapted from El Niño and Rainfall, *IRI Data Library*, http://iridl.ldeo.columbia.edu/maproom/IFRC/FIC/elninorain.html; Figure 1.18 adapted from La Niña and Rainfall, *IRI Data Library*, http://iridl.ldeo.columbia.edu/maproom/IFRC/FIC/laninarain.html; Figure 1.21 from *Graph: Sahel precipitation anomalies 1900 – 2013, doi:10.6069/H5MW2F2Q* Joint Institute for the Study of the Atmosphere and Ocean, University of Washington http://research.jisao.washington.edu/data_sets/sahel/; Figure 1.22 from Figure 1. The Millennium Drought (1997–2009), *Factsheet 2: The Millennium Drought and 2010/11 Floods*, p.1 (SEACI), © CSIRO, 2016; Figure 1.23 adapted from Wetland Functions and Wetland Mitigation: What's the Connection?, *The Wire*, Fall (Janni, K. 2013); Figure 1.25 from *Natural disaster hotspots: A global risk analysis*, World Bank Hazard Management Unit (Dilley, Maxx; Chen, Robert S.; Deichmann, Uwe; Lerner-Lam, Arthur L.; Arnold, Margaret. 2005) Figure 4.1c, p.38, © World Bank. https://openknowledge.worldbank.org/handle/10986/7376 License: CC BY 3.0 IGO; Figure 1.28 from

Flooding in Cumbria December 2015: Weather Data, http://www.metoffice.gov.uk/climate/uk/interesting/december2015, © Crown copyright. Contains public sector information licensed under the Open Government Licence (OGL) v3.0. http://www.nationalarchives.gov.uk/doc/open-government-licence/version/3/; Figure 1.30 from *Water climate change impacts report card*, Living With Environmental Change (LWEC) Partnership (Watts G and Anderson M (eds.) 2016) p.5, copyright © Living With Environmental Change, adapted from an original by Vasily Merkushev; Figure 1.31 from Drought under global warming: a review, *Wiley Interdisciplinary Reviews: Climate Change*, Vol.2 (1), pp.45–65 (Dai, A. 2011), Reproduced with permission of Blackwell Scientific in the format Republish in a book via Copyright Clearance Center.; Figure 1.32 from Global flood risk under climate change, *Nature Climate Change*, Vol. 3(9), pp.816–821 (Hirabayashi, Y., et.al. 2013), doi:10.1038/nclimate1911; Figure 1.33 from The World Bank: World Development Indicators: Renewable internal freshwater resources per capita: Food and Agriculture Organzation, AQUASTAT data; Figure 1.34 from *OECD Environmental Outlook to 2050: The Consequences of Inaction*, OECD publishing (OECD 2012) p.217, © OECD Publishing, Reproduced with permission of the OECD. Permission conveyed through Copyright Clearance Center, Inc.; Figure 1.35 from *Ranking the World's Most Water-Stressed Countries in 2040*, World Resources Institute (Maddocks, A., Young, R.S. and Reig, P. 2015) Licensed under Creative Commons CC-BY 4.0 https://creativecommons.org/licenses/by/4.0/; Figure 1.36 from *Water for Food, Water for Life: A Comprehensive Assessment of Water Management in Agriculture*, Earthscan/International Water Management Institute (Molden,D. (ed) 2007) Map 2.1, p. 63, 978-1-84407-396-2 © IWMI, Reproduced under licence; Figure 1.38 from China: a blast from the past, *Financial Times*, 14/12/2009 (Anderlini, J.), http://www.ft.com/cms/s/0/d5b9172e-e8ee-11de-a756-00144feab49a.html#axzz4EQIEm3GR; Figure 1.39 from *Green Infrastructure Guide for Water Management: Ecosystem-based management approaches for water-related infrastructure projects*, UNEP (2014) Table1, p.6; Figure 2.4 adapted from *Oceanography: An Invitation to Marine Science*, 7th ed., Cengage Learning (Garrison, T.S. 2010) Figure 5.10, p.139, © 2010 Brooks/Cole, a part of Cengage Learning, Inc. Reproduced by permission. www.cengage.com/permissions; Figure 2.6 adapted from Figure 1.8 The worldwide ocean currents of the thermohaline circulation system, http://worldoceanreview.com/en/wor-1/climate-system/great-ocean-currents/, © Walther-Maria Scheid Berlin, Germany, for maribus gGmbH, World Ocean Review; Figure 2.11 from www.data.worldbank.org/indicator/EG.USE.PCAP.KG.OE/countries?display=map, © OECD/IEA 2015, www.iea.org/statistics, Licence: www.iea.org/t&c; as modified by Pearson Education Ltd; Figure 2.12 from www.data.worldbank.org/indicator/EG.GDP.PUSE.KO.PP/countries?display=map, © OECD/IEA 2015, www.iea.org/statistics, Licence: www.iea.org/t&c; as modified by Pearson Education Ltd; Figure 2.16 from *Map: Major Trade Movements* – Trade flows worldwide (billion cubic metres), *BP Statistical Review of World Energy*, June 2015, p.29 http://www.bp.com/content/dam/bp/en/corporate/pdf/bp-statistical-review-of-world-energy-2015-full-report.pdf; Figure 2.17 adapted from U.S. Gulf Offshore Oil Production: Moving into Deeper Water Horizons, http://www.wri.org/resource/us-gulf-offshore-oil-production-moving-deeper-water-horizons, World Resources Institute, licensed under CC-BY-3.0 https://creativecommons.org/licenses/by/3.0/. Data sources: Oil lease, platforms and pipelines from US DOI Minerals Management Service; National Park Service, NOAA, National Gap Analysis Program, ESRI Data & Maps; Figure 2.20 adapted from Petrobras faces challenges to reap rewards of pre-salt reserves, Financial Times, 09/07/2013 (Leahy, J.), © The Financial Times Limited. All Rights Reserved. Used under licence.; Figure 2.22 from UK Primary energy use by source, http://www.carbonbrief.org/five-charts-show-the-historic-shifts-in-uk-energy-last-year, By kind permission of Carbon Brief; Figure 2.26 from Peak Farmland and the Prospect for Land Sparing, *Population and Development Review*, Vol. 38, Figure 5 p.228 (Ausubel, J. H., Wernick, I. K. and Waggoner, P. E. 2013); Figure 2.28 from *Climate Change 2013: The Physical Science Basis*, IPCC (IPCC Working Group I 2014) Figure 11.4(d) p.987; Figure 2.29 from *Climate Change 2013: The Physical Science Basis*, IPCC (IPCC Working Group I 2014) Figure 11.4(c) p.987; Figure 2.30 from *Climate Change 2014: Impacts, Adaptation, and Vulnerability*, IPCC (IPCC Working Group II 2014) Figure 4-8 p.309; Figure 2.33 from *Climate Change 2013: The Physical Science Basis*, IPCC (IPCC Working Group I 2014) Figure 11.14 p.987; Figure

2.34 from *Climate Change 2014: Impacts, Adaptation, and Vulnerability*, IPCC (IPCC Working Group II 2014) Box 4-4 p.316; Figure 2.36 from *The Climate Change Performance Index Results 2016*, Germanwatch and Climate Action Network Europe (Burck, J Marten, F and Bals C. 2016) pp.10–11, 31; Figure 2.38 from *The Energy Efficiency Strategy: The Energy Efficiency Opportunity in the UK*, Department of Energy and Climate Change (2012) p.14, © Crown copyright 2012, © Crown copyright. Contains public sector information licensed under the Open Government Licence (OGL) v3.0. http://www.nationalarchives.gov.uk/doc/open-government-licence/version/3/; Figure 3.2 from Bowei's travels: Internet firms, http://worldstartupreport.strikingly.com/; Figure 3.3 adapted from The Geographical Pivot of history, *The Geographical Journal*, Vol.23(4), pp.421–437 (Mackinder, H.J. 1904); Figure 3.4 from The British Empire, Trading Routes and Construction, http://www.the-map-as-history.com/demos/tome05/Map-of-the-British-Empire-european-colonization.php; Figure 3.5 from Science research, http://www.worldmapper.org/images/largepng/205.png; Figure 3.6 from Sino-African Trade – An infographic depicting the percentage breakdowns in trade between the People's Republic of China and the African continent. Data from Renaissance Capital, http://afrographique.tumblr.com/post/5387542552/an-infographic-depicting-the-percentage-breakdowns, Reproduced by permission of Ivan Čolić, with kind thanks to Max Kaizen and Dave Duarte of the ODMA; Figures 3.9, 3.11 from *Geography B Evolving Planet*, Pearson Education Ltd (Frost, L. 2013); Figure 3.12a adapted from World GDP per Capita, 2007 Dollar, http://www.singularity2050.com/2007/07/economic-growth.html; Figure 3.12b adapted from Five hundred years of world trade to GDP ratios, retrieved from:, https://ourworldindata.org/data/global-interconnections/international-trade/, Mohamed Nagdy and Max Roser (2016) – 'International Trade'. Published online at OurWorldInData.org. [Online Resource]; Figure 3.16 from *6 Graphs Explain the World's Top 10 Emitters*, World Resources Institute (Ge, M., Friedrich, J., and Damassa, T. 2014) Licensed under Creative Commons CC-BY 4.0 https://creativecommons.org/licenses/by/4.0/; Figure 3.18 from 6GE03 PreRelease from EdExcel A2 Paper, June 2015; Figure 3.19 from Daily chart: Oil at $50, *The Economist* (Data Team), http://www.economist.com/blogs/graphicdetail/2015/01/daily-chart-1; Figure 3.22 from Beijing defiant ahead of court ruling on its claims in South China Sea, *The Guardian*, 12/07/2016, map (Phillips, T.), https://www.theguardian.com/world/2016/jul/12/china-defiant-ahead-of-court-ruling-on-its-claims-in-south-china-sea, Copyright Guardian News & Media Ltd 2016; Figure 3.24 from *Urban world: Cities and the rise of the consuming class*, McKinsey Global Institute (Dobbs, R. et al 2012) Exhibit 3, p.17; Figure 3.25 from *The Middle East and the West: WWI and Beyond*, NPR (Shuster, M. 2004) http://www.npr.org/templates/story/story.php?storyId=3860950; Figure 3.26 from The Arab Spring, five years on: A season that began in hope, but ended in desolation, *The Independent*, 08/01/2016, The New Map of the Arab World (Cockburn, P.), http://www.independent.co.uk/news/world/middle-east/the-arab-spring-five-years-on-a-season-that-began-in-hope-but-ended-in-desolation-a6803161.html; Figure 3.27 from A Guide to Who Is Fighting Whom in Syria, Slate (Keating, J. & Kirk, C.), http://www.slate.com/blogs/the_slatest/2015/10/06/syrian_conflict_relationships_explained.html; Figure 3.28 from For richer, for poorer, *The Economist*; Figure 3.32 adapted from *Eclipse: Living in the Shadow of China's Economic Dominance*, Peterson Institute for International Economics (Subramanian, A. 2011) Figure 2.4, p.45; Figure 4.2 from *Measuring Global Development – an update, Geoactive Online, Vol.22 (456)*, Nelson Thornes (Frost, L. 2011) p.1, reprinted by permission of the publishers, Oxford University Press.; Figure 4.3 adapted from Mapping Paths to Asylum in the EU, *Freedom in the World 2016* (2016), https://freedomhouse.org/report/freedom-world/freedom-world-2016 © 2016 Freedom House; Figure 4.4 from Gapminder World Poster 2013, https://www.gapminder.org/GapminderMedia/wp-uploads/gapminder_world_2013_v8.pdf, Creative Commons Attribution License 3.0 Based on a free chart from www.gapminder.org; Figure 4.5 from Gapminder at https://www.gapminder.org/world/#$majorMode=chart$is;shi=t;ly=2003;lb=f;il=t;fs=11;al=30;stl=t;st=t;nsl=t;se=t$wst;tts=C$ts;sp=5.59290322580644;ti=2013$zpv;v=

0$inc_x;mmid=XCOORDS;iid=phAwcNAVuyj1jiMAkmq1iMg;by=ind$inc_y;mmid=YCOORDS;iid=tKOphM3UPRd94T6C6pmsuXw;by=ind$inc_s;uniValue=8.21;iid=phAwcNAVuyj0XOoBL_n5tAQ;by=ind$inc_c;uniValue=255;gid=CATID0;by=grp$map_x;scale=log;dataMin=194;dataMax=96846$map_y;scale=lin;dataMin=1.1;dataMax=9.4$map_s;sma=49;smi=2.65$cd;bd=0$inds=, Visualization from Gapminder World, powered by Trendalyzer from www.gapminder.org; Figure 4.6 from UNESCO eAtlas of Gender Inequality in Education, http://www.uis.unesco.org/Education/Pages/gender-atlas-en.aspx; Figure 4.9 from *Global Tuberculosis Report* World Health Organization (2015) Figure 2.6, p.18, © World Health Organization 2015; Figure 4.11 adapted from The future of life expectancy and life expectancy inequalities in England and Wales: Bayesian spatiotemporal forecasting, *The Lancet*, Vol. 386 (9989), pp.163–170 (Bennet, J.E. et al 2015), Licenced under Creative Commons, CC BY 4.0; Figure 4.17 from Corruption Perceptions Index Map 2015, http://www.transparency.org/cpi2015, by Transparency International. Licensed under CC-BY-ND 4.0; Figure 4.21 adapted from Sectors, https://euaidexplorer.ec.europa.eu/AidOverview.do, © European Union, 2016; Figure 4.22 from The Plan to Retake Sanaa, https://www.stratfor.com/sample/analysis/plan-retake-sanaa, Reproduced by kind permission of Stratfor.com; Figure 5.2 from "The Global Flow of People" (www.global.migration.info) published as: Quantifying global international migration flows, Science, 343, 1520–1522 (Sander, N., Abel, G. & Bauer, R. 2014); Figure 5.9 from Perils of Perception: Immigrants, https://www.ipsos-mori.com/_assets/sri/perils/interactive/, © Copyright Ipsos MORI 2015; Figure 5.11 from Syria and Iraq main ethnic and religious groups, http://www.geopoliticalatlas.org/syria-and-iraq-main-ethnic-and-religious-groups/, Reproduced by kind permission of geopoliticalatlas.org; Figure 5.16 from Peace Operations in Africa, *CFR Backgrounders* (Renwick, D. 2015), http://www.cfr.org/peacekeeping/peace-operations-africa/p9333, Copyright (2015) by the Council on Foreign Relations. Reprinted with permission.; Figure 5.22 from http://climatekids.nasa.gov/polar-temperatures/satellite-image-of-antarctica-lrg.png; Figure 5.25 from Map: Transnational land acquisitions, http://www.eea.europa.eu/data-and-maps/figures/transnational-land-deals-1/gmt8_fig1_landgrabbing.png/GMT8_Fig1_Synthesis_21878.png.75dpi.png/download; Figure 5.28 from Fragile States Index 2015, http://www.fsi.fundforpeace.org/rankings-2015, Copyright © 2015, The Fund for Peace; Figure on page 75 adapted from Why fresh water shortages will cause the next great global crisis (Graphic), *The Guardian*, 08/03/2015 (McKie, R.), Copyright Guardian News & Media Ltd 2016

Tables

Table 1.1 from Estimates of the Global Water Budget and Its Annual Cycle Using Observational and Model Data, *Journal of Hydrometeorology*, Vol. 8 (4), pp.758–769 (Trenberth, K.E. et al. 2007);Table 1.4 from National River Flow Archive; Table 2.1 from Open University; Table 2.3 IPPC 2014; Tables 2.5, 2.8 and 2.9 from BP Statistical Review 2015; Tables 2.11 and 2.12 from the World Bank; Table 2.13 from NOAA; Table 3.2 from Portland Communications and UN Development Indicators; Table 4.5 from IBGE; Talbe 4.6 from the World Bank; Table 4.7 from World Bank, CIA, Economist Intelligence Unit, Transparency International; Table 4.8 from US Department of State and Freedom House; Table 4.9 from European Commission; 4.14 from the World Bank; Table 5.1 from UNHCR; Table 5.3 from Office for National Statistics; Table 5.4 from UNEP; Table 5.5 from the UK Automotive Council

Text

Extract on pages 76–77 adapted from Why global water shortages pose threat of terror and war, *The Guardian*, 09/02/2014 (Goldenburg, S.), Copyright Guardian News & Media Ltd 2016; Extract on pages 132–133 adapted from The twilight of the resource curse?, *The Economist*, © The Economist Newspaper Limited, London 2015; Extract on pages 184–185 adapted from Showing Haiti on Its Own Terms, *National Geographic*, pp.99–118 (Fuller, A. 2015), Copyright Alexandra Fuller/National Geographic Creative 2015